Introduction to Soil Chemistry

CHEMICAL ANALYSIS

A SERIES OF MONOGRAPHS ON ANALYTICAL CHEMISTRY AND ITS APPLICATIONS

Series Editor
MARK F. VITHA

Volume 178

A complete list of the titles in this series appears at the end of this volume.

Introduction to Soil Chemistry

Analysis and Instrumentation

Second Edition

Alfred R. Conklin, Jr.

Published by John Wiley & Sons, Inc., Hoboken, New Jersey
Published simultaneously in Canada

For general information on our other products and services or for technical support, please contact our Customer Care Department within the United States at (800) 762-2974, outside the United States at (317) 572-3993 or fax (317) 572-4002.

Wiley also publishes its books in a variety of electronic formats. Some content that appears in print may not be available in electronic formats. For more information about Wiley products, visit our web site at www.wiley.com.

Library of Congress Cataloging-in-Publication Data:

Conklin, Alfred R. (Alfred Russel), Jr., 1941-
    Introduction to soil chemistry: analysis and instrumentation / Alfred R. Conklin, Jr. — Second edition.
        pages cm
    Includes bibliographical references and index.
    ISBN 978-1-118-13514-3 (cloth)
    1. Soil chemistry. 2. Soils–Analysis. I. Title.
    S592.5.C655 2013
    631.4′1–dc23

                          2014018809

Printed in the United States of America

10 9 8 7 6 5 4 3 2 1

CONTENTS

PREFACE

The author is both a soil scientist and a chemist. He has taught courses in all areas of chemistry and soil science, analyzed soil, for organic and inorganic compounds, in both soil solids and extracts, using various methods and instruments, for 44 years. *Introduction to Soil Chemistry, Analysis and Instrumentation, 2nd Edition*, is the result of these 44 years of experience in two distinct climatic zones in the Philippines, four countries in Africa, and one in Central and one in South America. In the United States, this experience includes analysis of soils from all sections of the country.

This book is intended as a reference for chemists and environmentalists who find that they need to analyze soil, interpret soil analysis, or develop analytical or instrumental analyses for soil. Soil scientists will also find it valuable when confronted by soil analyses that are not correct or appear to be incorrect or when an analysis does not work.

There are two themes in this work: (1) that all soil is complex and (2) that all soil contains water. The complexity of soil cannot be overemphasized. It contains inorganic and organic atoms, ions, and molecules in the solid, liquid, and gaseous phases. All these phases are both in quasi equilibrium with each other and are constantly changing. This means that the analysis of soil is subject to complex interferences that are not commonly encountered in standard analytical problems. The overlap of emission or absorption bands in spectroscopic analysis is but one example of the types of interferences likely to be encountered.

Soil is the most complicated of materials and is essential to life. It may be thought of as the loose material covering the dry surface of the earth, but it is much more than that. To become soil, this material must be acted upon by the soil-forming factors: time, biota, topography, climate, and parent material. These factors produce a series of horizons in the soil that make it distinct from simply ground-up rock. Simply observing a dark-colored surface layer overlaying a reddish layer shows that changes in the original parent material have taken place. The many organisms growing in and on soil including large, small, and microscopic plants, animals, and microorganisms also make soil different from ground-up rock.

There are physical changes constantly taking place in soil. Soil temperature changes dramatically from day to night, week to week, and season to season. Even in climates where the air temperature is relatively constant, soil temperatures can vary by 20° or more from day to night. Moisture levels can change from saturation to air dry. These changes have dramatic effects on the chemical reactions in the soil. Changes in soil water content change the concentration of soil constituents and, thus, also their solubility and reaction rate.

Not only are soil's physical and observable characteristics different from ground-up rock, so also are its chemical characteristics. Soil is a mixture of inorganic and organic solids, liquids, and gases. In these phases, inorganic and organic molecules, cations, and anions can be found. Inorganic and organic components can be present as simple or complex ions. Crystalline materials occur having different combinations of components, for example, 1:1 and 2:1 clay minerals, leading to different structures with different physical and chemical characteristics with different surface functionalities and chemical reactivities.

Organic components range from simple gaseous compounds, such as methane, to very complex materials such as humus. Included in this mix are gases, liquids, and solids, and hydrophobic and hydrophilic molecules and ions. All organic functional groups are included in soil organic matter, and it is common to find polyfunctional organic molecules as well as simple and complex biochemicals. Humus is an example of a complex molecule that contains many different functional groups. Both polyfunctional organic molecules and biochemicals coordinate and chelate with inorganic materials in soils, particularly metals.

The fact that soil always contains water, or more precisely an aqueous solution, is extremely important to keep in mind when carrying out an analytical procedure because water can adversely affect analytical procedures and instrumentation. This can result in an over- or under-determination of the concentrations of components of interest. Deactivation of chromatographic adsorbents and columns and the destruction of sampling tools such as salt windows used in infrared spectroscopy are examples of the potential deleterious effects of water. This can also result in absorbance or overlap of essential analytical bands in various regions of the spectrum.

This *Second Edition* continues the basic approach of the first with the addition of four chapters. Chapter 1 is an outline of the development of soil chemistry with specific reference to the development of instruments that have been essential to the present understanding of soil chemistry. Chapter 7 is a new chapter dealing with soil sampling, both in the field and in the laboratory, soil water sampling, sample transport, and storage. Chapter 8 discusses direct, modified, and indirect methods of soil analysis. Chapter 15 covers the recent development of hyphenated instrumental methods and their application to soil analysis.

Chapters 11 and 12 are the result of separating Chapter 7 from the *First Edition* into two chapters. Chapter 11 deals specifically with the extraction of inorganic analytes and Chapter 12 deals with organic analyte extraction.

All physical and chemical characteristics of soil have a pronounced effect on its analysis. The intention here is to first investigate some of the most important characteristics of soil and its extracts that impact its analysis, as well

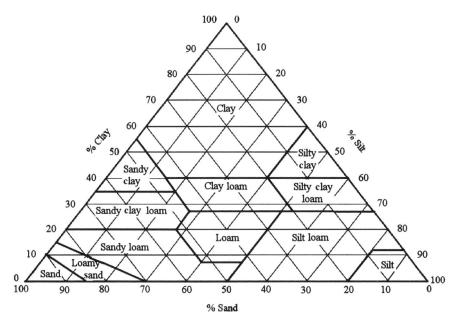

Textural triangle

as the instrumentation applied to its analysis, and to elucidate those interferences that may be most troubling.

Chapters conclude with a list of references followed by a bibliography. The bibliography lists general sources for the material covered in the chapter, while the references give some specific examples illustrating the application to soil. These provide the reader with additional resources and examples of how the material in the chapter is actually used in soil analysis and research. These also provide a source of standard methods and procedures of soil analysis and provide the reader with pitfalls and interferences that may be encountered in the particular analysis being discussed.

The Internet references given have been checked and were found accurate at the time of writing. However, Internet addresses are subject to change. If unable to find an address, try accessing the parent organization and looking for the desired information through its home page. For instance, if the Internet address for a USDA (United States Department of Agriculture) site is not found, one can access the USDA home page and find the needed information from there.

The author wishes to thank D. Meinholtz, J. Bigham, N. Smeck, H. Skipper, B. Ramos, T. Villamayer, J. Brooks, M. Goldcamp, M. Anderson, T. Stilwell, M. Yee, N. Gray, L. Baker, J. Shumaker, Audrey McGowin, and E. Agran for their help in reviewing this manuscript. I would also like to thank M. Vitha for his help in preparing the manuscript.

ALFRED R. CONKLIN, JR.

INSTRUMENTAL METHOD ACRONYMS

AA	Atomic Absorption
DTA	Differential Thermal Analysis
FID	Flame Ionization Detector
FT-IR	Fourier Transform Infrared (spectroscopy)
GC	Gas Chromatography
HG	Hydride Generator
HPLC	High-Precision Liquid Chromatography or High-Pressure Liquid Chromatography
ICP	Inductively Coupled Plasma
LC	Liquid Chromatography
MS	Mass Spectrometry
NMR	Nuclear Magnetic Resonance (spectroscopy)
TA	Thermal Analysis
TCD	Thermal Conductivity Detector
TLC	Thin-Layer Chromatography
UHPLC	Ultra-High-Pressure Liquid Chromatography
UV-Vis	Ultraviolet-Visible (spectroscopy)
XAS	X-ray Spectroscopy
XANS	X-ray Near-Edge Spectroscopy
XRD	X-ray Diffraction
XRF	X-ray Fluorescence

Common Hyphenated Instrumental Method Abbreviations

Hyphenated Method	Separation Method	Modification	Identification Method
GC-MS	Gas chromatography	None	Mass spectrometry
HPLC-MS	High-precision liquid chromatography	None	Mass spectrometry
LC-ICP	Liquid chromatography	None	Inductively coupled plasma
LC-AAS	Liquid chromatography	Hydride derivatization	Atomic absorption spectroscopy
TA-MS	Thermal analysis	None	Mass spectrometry
DTA-MS	Differential thermal analysis	None	Mass spectrometry
LC-ICP-MS	Liquid chromatography	Inductively coupled plasma	Mass spectrometry
GC-IR-MS	Gas chromatography	Infrared spectroscopy	Mass spectrometry

Abbreviated Periodic Table of the Elements

1	2	3	4	5	6	7	8	9	10	11	12	13	14	15	16	17	18
1 H 1.0																	2 He 4.0
3 Li 6.9	4 Be 9.0											5 B 10.8	6 C 12.0	7 N 14.0	8 O 15.9	9 F 18.9	10 Ne 20.1
11 Na 22.9	12 Mg 24.3											13 Al 26.9	14 Si 28.0	15 P 30.9	16 S 32.0	17 Cl 35.5	18 Ar 39.9
19 K 39.1	20 Ca 40.0	21 Sc 44.9	22 Ti 47.8	23 V 50.9	24 Cr 51.9	25 Mn 54.9	26 Fe 55.8	27 Co 58.9	28 Ni 58.7	29 Cu 63.5	30 Zn 65.3	31 Ga 69.7	32 Ge 72.6	33 As 74.9	34 Se 78.9	35 Br 79.9	36 Kr 83.8
37 Rb 85.4	38 Sr 87.6	39 Y 88.9	40 Zr 91.2	41 Nb 92.9	42 Mo 95.9	43 Tc 98	44 Ru 101.0	45 Rh 102.9	46 Pd 106.4	47 Ag 107.9	48 Cd 112.4	49 In 114.8	50 Sn 118.7	51 Sb 121.8	52 Te 127.6	53 I 126.9	54 Xe 131.3
55 Cs 132.9	56 Ba 137.3	71 Lu 175.0	72 Hf 178.5	73 Ta 180.9	74 W 183.8	75 Re 186.2	76 Os 190.2	77 Ir 192.2	78 Pt 195.1	79 Au 196.9	80 Hg 200.6	81 Tl 204.4	82 Pb 207.2	83 Bi 208.9	84 Po 209	85 At 210	86 Rn 222
87 Fr 223	88 Ra 226.0	103 Lr 257	104 Rf 261	105 Db 262	106 Sg 263	107 Bh 262											

Lanthanide and Actinide series not shown

CHAPTER

1

SUMMARY OF THE HISTORY OF SOIL CHEMISTRY

Soil is essential to life. All life supporting ingredients derive, either directly or indirectly, from soil. Plants growing in soil are directly used for food or are fed to animals, which are then used for food. These same plants take in carbon dioxide produced by animals and give off oxygen. Soil and the plants it supports moderate the amount of liquid and gaseous water in the environment by serving as a reservoir controlling its movement. Elements essential to life, even life in water, are released from soil solids and are recycled by soil chemical and biologically mediated reactions. Thus, an understanding of soil characteristics, the chemistry occurring in soil, and the chemical and instrumental methods used to study soil is important.

As the field of chemistry developed, so did the interest in the chemistry of soil. This was natural because the early chemists extracted elements from geological sources and, in the broadest sense, from soil itself. In fact, the development of the periodic table required the extraction, isolation, and identification of all of the elements, many of which are found abundantly in soil.

The total elemental composition of different soils was studied for some time. This involved a great deal of work on the part of chemists because methods for separating and identifying the elements were long and complicated. As knowledge accumulated, the relationship between the elements found in plants and those found in soil became of greater interest. This was sparked by an interest in increasing agricultural productivity.

At the end of the 19th and beginning of the 20th century, much of the theoretical work and discoveries that would be necessary for the further development of soil chemistry were in place. This included the fundamental scientific basis for various types of instrumentation that would be necessary to elucidate more fully the basic characteristics and chemistry of soil.

Introduction to Soil Chemistry: Analysis and Instrumentation, Second Edition.
Alfred R. Conklin, Jr.
© 2014 John Wiley & Sons, Inc. Published 2014 by John Wiley & Sons, Inc.

1

Toward the end of the 20th and into the 21st century, basic knowledge of soil chemistry was well developed, although much was and still is not understood. Instrumentation had matured and had been married to computers providing even more powerful tools for the investigation of chemistry in general and soil chemistry in particular. Instruments were being combined sequentially to allow for both separation and identification of components of samples at the same time. Instrumentation that could be used in the field was developed and applied.

A time line for discoveries, development of ideas, and instrumentation essential for our present-day understanding of soil chemistry is given in Table 1.1. It is interesting to note that, in some cases, it took several years to develop ideas and instrumentation for studying specific components of soil, such as

TABLE 1.1. Time Line for the Development of Ideas and Instrumentation Essential to the Understanding of Soil Chemistry

19th Century	
1800	Discovery of infrared light—Herschel
1835	Spectrum of volatilized metal—Wheatstone
1840	Chemistry and its application to agriculture—Liebig
1855	Principle of agricultural chemistry with special reference to the late researches made in England—Liebig
1852	On the power of soils to absorb manure—Way
1860	Spectroscope—Kirchoff and Bunsen
1863	The natural laws of husbandry—Liebig
End 19th Beginning 20th Century	
1895	X-rays—Röntgen
1897	Existence of electron—Thomson
1907	Lectures describing ions—Arrhenius
1909	pH scale—Sörenson
1913	Mass spectrometry—Thompson
1933	Electron lens—Ruska
1934	pH meter—Beckman
20th Century	
1940	Chromatography (described earlier but lay dormant until this time)—Tswett
1941	Column chromatography—Martin and Synge
1945	Spin of electron (leads to NMR spectroscopy)—Pauli
1959	Hyphenated instrumentation GC-MS

Sources:

Coetzee JF. A brief history of atomic emission spectrochemical analysis, 1666–1950. *J. Chem. Edu.* 2000; **77**: 573–576.

http://www.nndb.com. Accessed June 3, 2013.

http://www.nobelprize.org/nobel_prizes/physics/laureates/1945/pauli-bio.html. Accessed May 28, 2013.

Gohlke RS. Time-of-flight mass spectrometry and gas-liquid partition chromatography. *Anal. Chem.* 1959; **31**: 535–541.

ions and pH, and to apply them to soil chemistry. In other cases, such as visible and ultraviolet spectroscopy, application was almost immediate. Although tremendous strides have been made in the development of some instrumentation, such as nuclear magnetic resonance (NMR), it is still in its infancy with regard to application to soil chemistry.

1.1 THE 19TH CENTURY

The 19th century is considered the century of the beginnings of the application of chemistry to the study of soil. However, foundations for these advances had been laid with the discoveries of the previous century. Antoine-Laurent de Lavoisier, Joseph Priestley, and John Dalton are well-known scientists whose discoveries paved the way for the developments in agricultural chemistry in the 19th century [1,2].

At the end of the 18th century and the beginning of the 19th, Joseph Fraunhofer invented spectroscopy. At that time, spectroscopy was largely used to investigate the spectra of stars [3]. William Herschel discovered infrared radiation that would later be used in infrared spectroscopy to investigate soil organic matter. Also in the early part of the 19th century, Sir Charles Wheatstone was actively investigating electricity. His most prominent work involved the development of the telegraph. But he also invented the Wheatstone bridge, which would become an important detector for chromatography. A lesser known observation was of the spectrum of electrical sparks, which he attributed to vaporized metal from the wires across which the spark jumped. These were important steps in the eventual development of spectrographic methods of studying metals, especially metals in soil [4].

The result of 19th century chemical analysis of soil was twofold. The soil was found to be largely made up of a few elements among which were silicon, aluminum, iron, oxygen, nitrogen, and hydrogen. The second result was that different soils largely had the same elemental composition. Along with this were the investigations of the elemental content of plants and the relationship between those elements found in soil and those found in plants [5]. As these investigations advanced, it became evident that the inorganic components in soil were essential to plant growth and that crop production could be increased by increasing certain mineral components in soil. It did not take too long to determine that ammonia, phosphorous, and potassium are three essentials that, when added to soil, increase plant productivity. At this early point, chemists were largely interested in studying changes in and the activities of nitrogen, phosphorus, and potassium in soil. Two things about these components were discovered. One was that they needed to be soluble to be used by plants, and the second was that not all forms were available to plants.

Although observations about agriculture in general and soils specifically had been made for centuries, it was the chemist Justus von Liebig who is generally credited with the beginnings of the application of chemistry to the systematic study of soils. That beginning is usually dated as 1840, when Liebig published his book titled *Chemistry and Its Application to Agriculture*. This was followed by *Principles of Agricultural Chemistry with Special Reference to the Late Researches Made in England*, published in 1855, and *The Natural Laws of Husbandry*, published in 1863.

Three ideas either developed by Liebig or popularized by him are the use of inorganic fertilizers, the law of the minimum, and the cycling of nutrients, which foreshowed the present-day concern for sustainability. One interesting aspect of this is the fact that Liebig is generally cited as being an organic chemist, while his work on soil chemistry, if not wholly inorganic, at least is largely based on or revolves around the characteristics and use of inorganic chemicals. Perhaps Liebig's involvement could be attributed to the fact that during this time, it was widely thought that organic matter was the most important constituent needed for plant growth, that is, that plants got their nutrients directly from the organic matter or humus in soil.

In the middle of the 19th century, Liebig espoused the idea that fields could be fertilized with inorganic compounds and salts, particularly those of phosphate [6]. In addition, other chemicals needed by plants and frequently mentioned by Liebig are sulfuric acid, phosphoric acid, silicic acid, potash, soda, lime, magnesia, iron, chloride of sodium, carbonic acid, and ammonia [7].

During this time, both organic and inorganic materials added to soil were called manure and exactly what was being added is sometimes confusing. Both organic manure from any source and inorganic compounds and salts added to the soil to increase yields were referred to as manures.

Today, manure refers to excretory products of animals and finds its most common usage in reference to farm animals. This organic material was and is used as fertilizer to provide necessary elements for plants. In the past, it was practically the only material readily available for increasing plant or crop production. The general idea, however, is to add something to soil that will improve plant production. Thus, it remains common in popular agriculture literature to find that material added to soil to improve crop production is called manure even if the material is not organic [8,9].

Organic materials were seen as a potential source of plant nutrients and of interest to agricultural chemist and the world at large. Sewage sludge, compost, and indeed any organic material became a potential source of nutrients for plants. There was little or no understanding of microorganism involvement in organic material in general or in manure in particular, and so there was no understanding of the possibility of spreading diseases by using untreated or uncomposted organic matter [10].

One excretory organic product of particular interest and importance, discovered on islands off the coast of Peru by Alexander von Humboldt in 1802, was guano. He studied this product, which became a widely exploited fertilizer material that was transported and sold around the world [11].

One of the important components of guano is ammonia and because of the observed beneficial effect of ammonia on plant growth, there was early interest in the ammonia content of the organic matter in general and its availability to plants. This led to an interest in understanding the composition of soil organic matter. Unfortunately, full understanding is yet to be had. Organic matter in soil can be extracted and classified in various ways on the basis of the extraction method used, and various components can be isolated. The extracted organic matter can be broken down and the individual parts analyzed. All of this leads to a lot of information but has not led to a great deal of understanding of the molecular arrangement and geometry of soil organic matter, particularly humus.

Humus is what remains after all the organic matter added to soil is decomposed. This seemingly contradictory statement means that the organic matter remaining in soil after decomposition of added organic matter has been synthesized during the decomposition process. Thus, components released during decomposition react to form a new material generally called humus. The most common extraction of organic matter from soil is with base, and this leads to the isolation of humus.

Humus is important in soil because soil must contain humus along with many other diverse components. Humus plays an important role in any soil analysis because it can contain and release both inorganic and organic components. It has groups that can coordinate with transition metals, thus taking them out of solution. It can also absorb and release a wide variety of organic molecules that can affect analytical results. Humus also has a relatively high cation exchange capacity which is important in understanding the chemistry of soil.

Humus has been and is often referred to as a polymer; however, it is not a polymer. It is not a collection of mers such as in polybutene, where butene is the mer and there are many butene mers bonded together. It is rather a complex mixture of different components bonded together. Up to 12 different types or fractions of soil organic matter are recognized. Humus itself can be subdivided into eight fractions depending on the extraction procedure used. Thus, the existence of organic matter, humus, and its various constituents in soil must be recognized and taken into account in any analysis. Organic matter may augment or obfuscate analytical results.

The importance of organic matter and humus was recognized by Liebig as illustrated in his book, *Chemistry and Its Application to Agriculture and Physiology*, which presents the results of the analysis of plants, ash, humus, and soil. This analysis of plants led to the conclusion that plants take up certain elements from the soil ([12], pp. 155–156). Other elements were also found, but the importance, if any, of these was unknown [12].

Included in Liebig's work are 29 pages ([12], pp. 217–246) of the results of the analysis of 47 soils, in some cases including subsoils, which are generally indicated as being 45 cm deep. The analytical results are given as a percentage of the total, which is a logical approach but is somewhat lacking when relating these results to the environment or soils in general. Also, the analyses do not give a picture or results that could be easily related to the productivity of soil.

This, in part, was probably due to the fact that the major constituent of the analysis was silica, which is the major constituent of all soils. The other constituents of soil are important to plants, but their relationship to plant nutrition is not understood.

These soils were described on the basis of their place in the landscape, texture in a general sense, general productivity, and specific crops grown. The components listed in the analytical report were, most often, silica, alumina, iron oxides, manganese, lime, magnesia, potash, soda, phosphoric acid, sulfuric acid, and chlorine. Also included were humus, carbonic acid, and organic nitrogen. Some of these constituents are presented in different ways in different analyses but generally are present in all analyses. Most of the analyses were for soils of England, but analyses of soils from Germany, the United States, specifically Ohio, Puerto Rico, and Java are also included.

Liebig's attempts to use chemical analysis to identify low productivity soils and to differentiate them from highly productive soils were not very successful. Many of Liebig's analytical results were hard to understand in relationship to plant growth and soil chemistry, because it was not until the development of the concept of pH by Sörenson and its application to soil chemistry using the pH meter developed by Beckman that much of what was occurring could be understood [13,14]. However, these studies did lead to the development of the law of the minimum, which Liebig is often credited with inventing and which has been stated in many different ways but is commonly given as

> A manure containing several ingredients acts in this wise: The effect of all of them in the soil accommodates itself to that one among them which, in comparison to the wants of the plant, is present in the smallest quantity.
>
> Justus von Liebig, 1863 [15]

Apparently, Carl Sprengel also stated the same or a similar concept as Liebig perhaps at an earlier date. However, from the literature, it appears that Liebig and Sprengel were not on the best of terms and it is unclear as to who originated the idea of the law of the minimum first. On the other hand, Liebig did do a great deal to popularize the law of the minimum, which is why it is mostly associated with him [16].

The law of the minimum is easy to understand on a cellular, molecular, and atomic level and is applicable to all organisms. Organisms cannot grow by producing partial cells and to produce a whole cell they must have all the constituents in the correct proportion. It is not possible to have a functioning cell that does not have a complete cell membrane, complete genetic material, functional mitochondria, and all the other parts of a functioning cell.

A specific example of the law of the minimum on a molecular level is the essential amino acids. The need for essential amino acids in human nutrition shows that one amino acid cannot substitute for another. Substitution of one amino acid for another can lead to disease. This is shown, for example, in sickle cell anemia, where a substitution in two amino acids leads to a different hemo-globin and a different cell structure [17].

If a sulfur atom is needed for the functioning of an enzyme, none of the elements surrounding sulfur in the periodic table can substitute for sulfur. One of the reasons some elements can be toxic is that they take the place of the correct, needed element, thus making the molecule they are in nonfunctional.

Liebig's original arguments for this law are well laid out in his book, *The Natural Laws of Husbandry*, published in 1863 [8]. Although specific parts of the book are given as the source of some specific ideas, the ideas are, in reality, developed over a number of pages. As with the law of the minimum, the ideas of cycling of nutrients and conservation of soil fertility are covered in the same work over a number of pages. His basic argument is that if soil is to remain productive, farmers must replenish the nutrients they remove from it when they harvest crops.

Along with Liebig, John Way and John Bennet Lawes (a soil chemist) were important agriculturalists of the time, although they are not as well known in chemistry circles. Way and Laws established the Rothamsted Experimental Station in England and carried out both chemical and field experiments designed to explain the chemistry occurring in soil. Sometimes, they supported and, at other times, were at odds with Liebig. In either case, they were tremendously important in the development of modern agricultural chemistry.

The work of Liebig, Way, and Lawes provided the basic understanding of many of the constituents in soil including both inorganic and organic acids and bases, but not of pH (which was not known at the time). The importance of lime, phosphate, and sulfur was understood, if only incompletely. Sulfur was reported as being applied as sulfuric acid. It is, however, unclear as to exactly how it was applied. John Laws clearly described the changes in ammonia compounds when they were applied to soil. Unfortunately, he was unable to fully explain what was going on in this process [18]. Regardless, two things about these components were discovered. One was that they needed to be soluble to be used by plants, and the second was that not all chemical forms or species were available to plants.

He also observed and described base exchange in soil, which would later be understood to be cation exchange. However, this understanding had to wait until further development of chemistry and the concepts of ions and exchange reactions.

Along with this were investigations of the elemental content of plants and the relationship between those elements found in soil and those found in plants. As these investigations advanced, it became evident that the inorganic components in soil were essential to plant growth and that crop production could be increased by increasing certain mineral components in soil. It did not take too long to determine that ammonia, phosphorous, and potassium are the three essential components that increase plant productivity. At this early point, chemists were largely studying changes in and the activities of these components in soil.

During this time, all the prominent scientists and chemists were directly involved in either research and discoveries related to soil chemistry or discoveries that formed the basis for a much deeper understanding of processes occurring in agricultural systems including soils.

However, toward the end of the 19th century, there were still some things about soils and chemistry that inhibited an understanding of much of soil chemistry. The concepts of pH and ions had not yet been developed. Although clay was known and had been known for centuries, the varieties of clays in soil were not known and thus their effect on soil chemistry was unknown. The basic concepts of ion exchange and buffering were also not yet understood either in chemistry or in soils.

1.2   THE END OF THE 19TH AND THE BEGINNING OF THE 20TH CENTURY

In 1904, King published a rather extensive collection of work regarding the composition of soils and manure [19]. At this time, a number of analyses involved extracting soil with water, often over an extended period of time, analyzing the components extracted, and reporting the results on the basis of 1,000,000 L of water used in the extraction. This represented a useful way of presenting data, which is essentially reporting it as parts per million (ppm). An acre of soil (6 in. or 15 cm deep) is taken as containing either 2,000,000 lb or a hectare 2,000,000 kg. Thus, simply multiplying the amount in 1,000,000 by two gives the amount of a constituent in an acre or hectare.

The results of this type of experiment are illustrated by water extractions conducted by Schultze ([20], p. 328) involving the sequential extraction of soil. Part of the data from this work is graphed in Figure 1.1 and shows the results for six extractions. Only the results for total solids and the inorganic fraction are graphed (other data are given in the table from which these data are taken). The pattern shown is commonly seen in soil extractions even when a contaminant is being extracted. It is to be expected that, with multiple extractions, the level of analyte[1] would constantly decrease and eventually become zero. That does not often happen, but rather, the level of analyte drops to some low level and seesaws around this low level.

Subsequently, E. W. Hilgard took up the cause of soil chemistry. He carried out research using data from King and others in order to find a chemical characterization of soil that would differentiate between productive and unproductive soils. Additionally, soils were extracted with acids of various strengths as indicated by their specific gravity. Results of this type of extraction were expected to indicate the long-term productivity of a particular soil. Also during this time, the concept of a minimum level of a particular component,

---

[1] The specific element, compound, or species to be analyzed for.

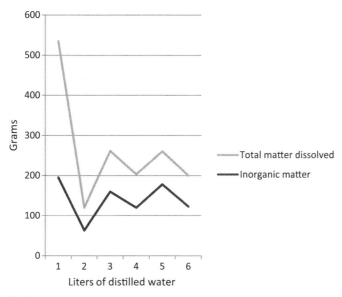

Figure 1.1. Total and inorganic matter extracted from soil by water. Adapted from Schultze ([20], p. 328).

for instance, phosphate (reported as phosphoric acid), needed for a soil to be proactive was developed [20].

The transition between the two centuries also saw the discovery and development of two concepts essential to the further development of the understanding of soil chemistry. One was the discovery by J. J. Thomson of the electron, a subatomic particle. This work occurred around 1897 and culminated in the determination of the electron charge-to-mass ratio, which made it possible to develop the idea of ions [21]. This was basic to the concept of ions discussed and developed by Svante Arrhenius in a series of lectures given at the University of California at Berkeley in 1907 [22]. In this series of lectures, he clearly describes ions of hydrogen and chlorine. The basic idea of a hydrogen ion and its application to enzyme chemistry would be further developed by S. Sörenson [13].

Interestingly, previous to Arrhenius's development of the theory of ions, John Way described the exchange process after a series of experiments in which he added potassium chloride to soil and then passed water through it. Potassium was retained by the soil and an equivalent amount of calcium and magnesium was found in the water exiting the soil. The elements retained and eluted from soil were the bases, and so he called this base exchange. Today, we call it cation exchange because we know that the elements are in the form of cations [23]. However, the understanding of this process had to wait until the development of knowledge of ions.

An ion is any species that has lost or gained an electron. Of particular interest and importance are the inorganic ions of nitrogen (i.e., ammonium and nitrate), phosphate, hydrogen, calcium, and potassium as shown in Figure 1.2. Many organic molecules can and do exist as ions and are important in soil chemistry; however, they have not been studied as intensively as have been the inorganic ions. The development of the concept of ions is essential to the understanding of pH and to understanding ion exchange, particularly cation (positive ion) exchange in soil.

In 1909, Sörenson described the development of the pH scale based on the work of Arrhenius and the characteristics of water. Experiments and the resulting calculations show that water dissociates into hydrogen ions (H^+) and hydroxide ions (HO^-) and that the product of their concentrations equals close to 10^{-14} ions in aqueous solution. From this, a pH scale from 0 to 14 was developed and the scale describing this relationship using the abbreviation pH was developed (in the older literature [13], one may encounter both the "p" and the "h" capitalized, i.e., as PH). Today, it is universally designated as pH [13,22].

Chemists were limited to determining the pH using litmus paper. Blue litmus turns red at pH 4.5 and red litmus turns blue at pH 8.3. Thus, the region between 4.5 and 8.3 (see Figure 1.3) was unavailable. Many important chemical reactions including conditions affecting the availability of plant nutrients occur in this region of the pH scale. Understanding these important reactions thus

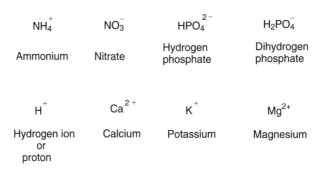

Figure 1.2. Common important inorganic anions and cations ions in soil.

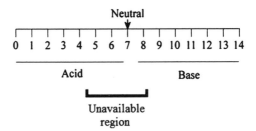

Figure 1.3. The pH scale showing the region unavailable to chemists before the development of the pH scale and the pH meter.

had to wait for the full development of the pH scale and the instrumentation necessary to investigate these pHs.

It would not be until the development of the pH meter and pH electrode that soil scientists had a good way to measure soil pH. In 1934, A. O. Beckman introduced the first pH meter and started a company to build and sell the meter. This sparked an intense study of soil pH and its relationship to plant nutrient availability [14].

Two fundamental discoveries about the structure of the atom and electromagnetic radiation also occurred during this period and provided a foundation for instrumentation that would be fundamental in furthering our understanding of soil chemistry. One was the discovery of X-rays, also sometimes called Röntgen rays, discovered in 1895, by W. Röntgen [24]. The second was made by J. J. Thomson in 1912. He observed positive rays and described how these could be used to identify compounds and elements. Subsequently, he presented a clear description of the process in 1913. This led to the development of mass spectrometry [25].

These discoveries allowed for important increases in the understanding of soil chemistry. The concept of ions and the fact that some elements could exist as ions were an essential step forward. This led to an understanding of the phenomenon John Way clearly described in his work of what he called base exchange, as cited earlier. It led not only to an understanding of ion exchange but also of soil buffering. The discovery of X-rays would eventually lead to the ability to describe and identify soil clays that are the source of much of the cation exchange in soils. The idea of soil pH as opposed to soil being simply acidic or basic based on litmus paper was essential to understanding soil fertility and contamination.

1.3   THE 20TH CENTURY

At the very beginning of the 20th century, two very important discoveries or inventions were made. M. S. Tswett discovered and developed chromatography and Fritz Haber demonstrated the chemical production of ammonia. Both of these would dramatically affect soil chemistry. Chromatography provided a method of separating the myriad organic and inorganic compounds and ions found in soil. Development of the Haber process led to the widespread use of ammonia and nitrate fertilizers and the intense study of the chemical changes that nitrogen undergoes in soil.

The development of chromatography was first described by M. S. Tswett and is generally credited to him [26]. He initially separated chlorophylls using a column of calcium carbonate and various solvents. His basic setup for chromatography was, and still is, a stationary phase and a mobile phase. As the mobile phase carries components of a mixture across the stationary phase, they are separated from each other and come out of the setup at different times [27]. The term chromatography came about because the compounds initially separated by Tswett, the chlorophylls, are colored. The term continues to be used, although most mixtures separated today are colorless. His general method, however, lay dormant until the 1940s.

It was in this time period that A. J. P. Martin and R. L. M. Synge again brought column chromatography to the forefront. Beyond this, they opened the flood gates for the development of various chromatographic techniques with further development of paper chromatography and the development of thin-layer chromatography [28]. All the basic variants of chromatography, elution, gas (also divided into gas-liquid and gas-solid chromatography), ion exchange, thin layer, high-performance, supercritical, capillary electrochromatography, and size exclusion were developed and put to practical use.

Even though it is sometimes not thought of as a chromatographic technique, we should also include electrophoresis. In this instance, paper or a gel is the stationary phase and electricity is the mobile phase. Although all types of chromatography are in extensive use in all kinds of investigations, electrophoresis has a particular prominence today because of DNA analysis [29].

One particularly important application of chromatography has been to the analysis of pesticides, their degradation and movement. Small amounts of pesticides can be determined and their interaction with soil can be modeled using chromatographic methods [30]. It is unlikely that all types of chromatographic separation have been developed or even conceived. New variants such as ultrahigh-pressure liquid and hydrolitic interaction liquid chromatography are but two examples.

It was well established that nitrogen in a "fixed" form—that is, combined with another atom other than nitrogen—was essential for plants. Before the development of the Haber process for chemically "fixing" nitrogen, the sources of nitrogen fertilizer were nitrogen-fixing plants and bird droppings, particularly guano, mined mostly off the coast of Peru. Nitrogen-fixing plants, the legumes, were used in rotations or as green manure crops to provide nitrogen for subsequent crops. Guano was widely mined and exported as fertilizer all around the world.

In addition to nitrate and other nitrogen compounds, guano contains phosphate and potassium, thus making it a good fertilizer. Deposits of guano were discovered and studied by Alexander von Humboldt in 1802 [11]. This source of fertilizer had thus been known for some time before the development of the Haber process for producing ammonia. However, guano was and is limited and unsustainable as a source of fertilizer for crops, particularly on a worldwide basis.

Although the atmosphere is 78% nitrogen gas (N_2), it is not available (i.e., able to be used) to plants or animals except after it has been "fixed." Thus, the development of the process for making ammonia from hydrogen and atmospheric nitrogen by Haber was extremely important. The first reaction (1) in Figure 1.4 shows the reaction carried out in the Haber process. This reaction is reversible so ammonia is compressed and cooled, and liquid ammonia is removed from the reaction mixture to drive the reaction to the right.

$$3H_2 + N_2 \quad \underset{450°C, \ 300 \ bar}{\overset{Catalyst}{\rightleftharpoons}} \quad 2H_3N \qquad\qquad 1$$

$$NH_3 + H_2O \quad \longrightarrow \quad NH_4OH \qquad\qquad 2$$

$$2NH_3 + 4O_2 \quad \longrightarrow \quad 2HNO_3 + 2H_2O \qquad\qquad 3$$

Figure 1.4. Reaction and conditions used by Fritz Haber to produce ammonia from hydrogen and nitrogen. The reaction of ammonia with water to form ammonium and the oxidation of ammonia to nitric acid, a common reaction in soil, are also given.

Figure 1.4 also shows two other reactions. In reaction 2, ammonia reacts with water to form ammonium hydroxide. Reaction 3 shows that ammonia can also be oxidized to form nitric acid from which all forms of nitrates can be produced. All three forms of nitrogen (ammonia, ammonium hydroxide, and nitrates in various forms) are commonly found in soil and can be added to soil to supply nitrogen to plants (see also Figure 6.5). This process thus opened up an inexpensive method of producing nitrogen compounds that would be used as fertilizers.

It was during this time, the 1940s, that the spin of electrons and protons was observed by Wolfgang Pauli [31,32]. This discovery would eventually lead to the development of NMR spectroscopy, better known simply as NMR. It is also the basis of magnetic resonance imaging (MRI) [32]. Also during this time, Ernst Ruska experimented with and developed a "lens" that could be used to focus a beam of electrons. This led him to develop an electron microscope with a 400× magnification [33]. Thus, all the basic knowledge necessary for the further development of our understanding of soil chemistry including the instrumentation needed to explore it was in place at the beginning of the century.

Soil pH differences between pH 4.5 and 8.3, the pHs where litmus changes, could also now be studied and their effect on plants refined. With this knowledge also came the concept of soil buffering and its importance in understanding soil chemistry. A better understanding of the cation exchange of soil was developed. Various extraction procedures were developed to extract specific analytes from soil. One of the questions, which is ongoing, was and is the specificity of extractants, particularly as to the biological activity or availability of the analyte extracted. It was also during this time that experimentation with various phosphate extractants was occurring. The objective was to find an extractant that would extract that portion of phosphate available to plants during the growing season. These ideas, combined with the availability of nitrogen, allowed for the study of nitrogen changes in soil and eventually the understanding of the nitrogen cycle. A much deeper understanding of soil chemistry developed from all of this basic knowledge.

The middle of the century saw the development of instrumentation based on these discoveries and the evolution of functioning instruments that were available from a number of manufacturers. Extracts could be analyzed using these instruments. They were and are constantly being improved in terms of detection, particularly with relationship to sensitivity. Thus, they provide the tools necessary for an even deeper understanding of soil chemistry.

1.4   THE END OF THE 20TH AND THE BEGINNING OF THE 21ST CENTURY

During the past few centuries, interest shifted from simply determining if something was present in soil to the form or "species" it was in. This was driven by the fact that the form, often, if not always, determines its biological availability, danger, or toxicity. This has been described as "speciation" and is often thought of as referring to the ionic state of the analyte in question. However, it should also be applied to combinations of inorganic and organic compounds and ions and their environments.

A number of advances in instrumentation occurred at the end of the 20th and the beginning of the 21st centuries. Computers were being used not only to acquire signals generated by instrumentation but also to display the data and to manipulate it. As this process continued, they were also used to control instruments and to allow for automatic sample changers to be added to instruments. Autosamplers allow samples to be continuously analyzed even in the absence of an instrument operator. This significantly increases the number of samples analyzed, which greatly increases the amount of chemical information available about a wide variety of soils.

Additionally, with the inclusion of computers as part of an instrument, mathematical manipulation of data was possible. Not only could retention times be recorded automatically in chromatograms but areas under curves could also be calculated and data deconvoluted. In addition, computers made the development of Fourier transform instrumentation, of all kinds, practical. This type of instrument acquires data in one pass of the sample beam. The data are in what is termed the time domain, and application of the Fourier transform mathematical operation converts this data into the frequency domain, producing a frequency spectrum. The value of this methodology is that because it is rapid, multiple scans can be added together to reduce noise and interference, and the data are in a form that can easily be added to reports.

A type of radiation that was not available earlier came into existence and eventually became available to soil scientists. This is the radiation given off by synchrotrons that emit what is called synchrotron radiation (originally considered a waste product of acceleration electrons close to the speed of light). It is described as similar to bright X-rays. This electromagnetic radiation has been used to successfully elucidate the structure and oxidation states of metals in soil and thus their likelihood of becoming environmental pollutants [34].

Once into the 21st century, hyphenated instrumentation (i.e., those that couple two instruments together) became prevalent in laboratories. This is the combination of two or more, often different, instruments. In simple terms, the purpose is to first separate the analyte of interest and then to identify it. This takes place using a sample injected into the combined instruments. The most common of the hyphenated instruments is the gas chromatograph, the output of which is fed into a mass spectrometer to produce a gas chromatography–mass spectrometry (GC-MS) [35].

Other combinations are available. For example, liquid chromatographs connected to mass spectrometers (known as liquid chromatography–mass spectrometry [LC-MS]) are fairly common. Almost any combination of two instruments that can be thought of has been built. In addition, two of the same instruments can be connected so that the output from one is fed directly into the other for further separation and analysis. Examples include two mass spectrometers in an MS-MS arrangement and two different gas chromatography columns connected in a series, known as GC-GC. To keep up with these advances, one needs to have a working knowledge of the fundamental principles involved in the techniques and of the abbreviations used for the various instrumentation methods.

Although field portable instrumentation, which is battery operated and self-contained, has been developed in the past, there has been a resurgence of interest in field portable instruments, some of which are small enough to be handheld.

1.5   CONCLUSION

Our understanding of the chemistry of soil in terms of both plant and biological availability of inorganic and organic constituents and in terms of environmental interactions and reactions has increased greatly over the past three decades. The inorganic constituents of soil are well known and easily analyzed. A great many of the organic constituents of soil are known; however, there is still much to learn about their occurrence and reactivity. This is particularly true of humus. New instrumentation and combinations of instruments has led to greater insight into the chemistry of metals, nonmetals, and organic matter in soil. This includes their solubility, movement, and reactions. Modern instrumentation is used to analyze soils and soil chemical reactions not only in the laboratory but also in the field. As instrumentation develops and new instrumentation is discovered and brought to bear on the mysteries of soil, an even greater understanding of soil chemistry will be gained.

PROBLEMS

1.1.   Describe the observations made by early soil chemist that led to the understanding that soil possessed cation exchange capacity.

1.2. What discovery led to a significant increase in the understanding of soil clays?

1.3. Give some examples where the development of chromatography led to an increased understanding of soil chemistry. What is the power of chromatographic techniques?

1.4. What is a hyphenated instrumental technique? Give the most common example of such a technique.

1.5. Explain why the development of the pH scale was so important in understanding plant nutrition. What did this development "open up" to the soil chemist?

1.6. Look up and list the various types of spectroscopy that have been developed. Describe the various parts of the electromagnetic spectrum available to each type of spectroscopy.

1.7. Describe the basic characteristics that a field portable instrument must have to be useful in the field.

1.8. Explain why Figure 1.1 is important to anyone extracting soil.

1.9. Is the importance of organic matter to crop production a new concept? Explain.

1.10. Is the concept of sustainability a new concept? Explain.

REFERENCES

1. http://www.chemheritage.org/chemistry-in-history/themes. Accessed May 27, 2012.

2. Moore FJ. *A History of Chemistry*. Charleston, SC: Nabu Press; 2010, pp. 33–82.

3. Brand JCD. *Lines of Light the Sources of Dispersive Spectroscopy 1800–1930*. Australia: Gordon and Breach Publishers; 1995, pp. 37–56.

4. Bowers B. *Sir Charles Wheatstone FRS, 1802–1875*. London, England: Science Museum; 1975, p. 204.

5. Daubeny GGB. Memoir on the rotation of crops, and on the quantity of inorganic matters abstracted from the soil by various plants under different circumstances. *Roy. Soc. (London) Phil. Trans.* 1845; **135**: 179–253.

6. von Liebig J. *Chemistry and Its Application to Agriculture*, Playfair L (ed.). Cambridge, England: John Owen; 1842.[2]

7. von Liebig J. *Principles of Agricultural Chemistry and with Special Reference to the Late Researches Made in England*. London, England: Walton and Maberly; 1855.

8. von Liebig J. *Natural Laws of Husbandry*, Blyth J (ed.). London, England: Walton and Maberly; 1863, p. 2.

9. Brock WH. *Justus von Liebig: The Chemical Gatekeeper*. Cambridge, England: Cambridge University Press; 1997, p. 145.

10. Barker TH. The chemistry of manures. *Trans. Yorkshire Agric. Soc.* 1842; **5**: 31–52.

---

[2] There are many versions of this book; however, this is the one that has the cited data.

11. Egerton FN. A history of the ecological sciences part 32: Humboldt Nature's Geographer. *Bull. Ecol. Soc. Am.* 2009; **90**: 253–282. Available at: http://dx.doi.org/10.1890/0012-9623-90.3.253. Accessed June 3, 2013.

12. Liebig J. *Organic Chemistry and Its Application to Vegetable Physiology and Agriculture.* Philadelphia: J. Campbell; 1845.

13. Sörenson SPL. Enzyme studies II. The measurement and meaning of hydrogen ion concentrations in enzymatic processes. *Biochem. Z.* 1909; **21**: 131–200.

14. http://www.chemheritage.org/. Accessed June 3, 2013.

15. *Justus von Liebig, 1863.* http://www.todayinsci.com. Accessed June 3, 2013.

16. Ploeg RR, van der Böhm W, Kirkham MB. On the origin of the theory of mineral nutrition of plants and the law of the minimum. *Soil Sci. Soc. Am. J.* 1999; **63**: 1055–1062.

17. Pauling L. Molecular structure and disease. October 9, 1958. Lowell Lecture, Massachusetts General Hospital, Boston, Massachusetts. Available at: http://osulibrary.oregonstate.edu/specialcollections/coll/pauling/blood/notes/1958s2.8-ts-02.html. Accessed May 28, 2013.

18. Way JT. On the power of soils to absorb manure. Consulting chemist to the society. (Second paper). *J. R. Agric. Soc. Engl.* 1852; **13**: 123–143.

19. King FH. *Investigations in Soil Management Being Three of Six Papers Influence of Soil Management upon the Water-Soluble Salts in Soil and the Yield of Crops.* Madison, WI: Author; 1904.

20. Hilgard EW. *Soils, Their Formation, Properties, Composition, and Relations to Climate and Plant Growth in the Humid and Arid Regions.* London: Macmillan & Co., Ltd., 1906, pp. 313–371.

21. Dahl PF. *A History of JJ Thomson's Electron Flash of the Cathode Rays.* London, England: IOP Publishing; 1997.

22. Arrhenius S. *Theories of Chemistry Being Lectures Delivered at the University of California, in Berkley,* Slater T (ed.). London, England: Longmans, Green and Co; 1907, p. 48.

23. Way JT. On the power of soils to absorb manure. *J. Roy. Agric. Soc.* 1850; **ix**(1): 128–143.

24. Glasser O. *Wilhelm Conrad Röntgen and the Early History of the Roentgen Rays.* San Francisco, CA: Norman Publishing; 1993, p. 1.

25. Thomson JJ. Rays of positive electricity. *Proc. R. Soc. A* 1913; **89**: 1–20.

26. Issaq HJ, Berezkin VG. Mikhail Semenovich Tswett: the father of modern chromatography. In Issaq HJ (ed.), *A Century of Separation Science.* New York: Marcel Dekker, Inc.; 2002, pp. 19–26.

27. Ettre LS. Chromatography: the separation technique of the twentieth century. In Issaq HJ (ed.), *A Century of Separation Science.* New York: Marcel Dekker, Inc.; 2002, pp. 1–17.

28. Martin AJP, Synge RLM. A new form of chromatogram employing two liquid phases A theory of chromatography. 2. Application to the micro-determination of the higher monoamino-acids in proteins. *Biochem. J.* 1941; **35**: 1358–1368.

29. Miller DN, Bryant JE, Madsen EL, Ghiorse WC. Evaluation and optimization of DNA extraction and purification procedures for soil and sediment samples. *Appl. Environ. Microbiol.* 1999; **65**: 4715–4724.

30. Helling CS, Turner BC. Pesticide mobility: determination by soil thin-layer chromatography. *Science* 1968; **162**: 562–563.

31. "Wolfgang Pauli—Biography." Nobelprize.org. 18 Jan 2013. Available at: www.nobelprize.org. Accessed June 3, 2013.

32. Pauli W. *Collected Scientific Papers*, Vols. 1 and 2, Kroning R, Weisskopf VF (eds.). Hoboken, NJ: Wiley Interscience; 1964.

33. www.nobelprize.org. Accessed June 3, 2013.

34. Fisher M. Advanced spectroscopy in soil biogeochemical research. *CSA News* 2011; **56**: 4–10.

35. Gohlke RS. Time-of-flight mass spectrometry and gas liquid partition chromatography. *Anal. Chem.* 1959; **31**: 535–541.

BIBLIOGRAPHY

Brand JCD. *Lines of Light: The Sources of Dispersive Spectroscopy, 1800–1930*. Australia: Gordon and Breach Publishers; 1995.

Brock WH. *Justus von Liebig: The Chemical Gatekeeper*. New York: Cambridge University Press; 2002.

Hilgard EW. *Soils, Their Formation, Properties, Composition, and Relations to Climate and Plant Growth in the Humid and Arid Regions*. London, Macmillan & Co., Ltd.; 1906.

Issaq H, ed. *A Century of Separation Science*. New York: Marcel Dekker, Inc.; 2002.

Stoltzenburg D. *Fritz Haber: Chemist, Nobel Laureate, German, Jew*. Philadelphia: Chemical Heritage Foundation; 2005.

CHAPTER

2

SOIL BASICS PART I
LARGE FEATURES

Soil is vastly more complex than simply ground-up rock. It contains solid inorganic and organic components in various stages of decomposition and disintegration, an aqueous solution of elements, inorganic and organic ions and molecules, and a gaseous phase containing nitrogen, oxygen, carbon dioxide, water, argon, methane, and other gases. In addition, it contains a large and varied population of macro-, meso-, and microscale animals, plants, and micro-organisms. If any of these components is missing, it is not soil!

The solid portion of soil is composed of inorganic sand, silt, clay, and organic matter (OM), which interact to produce the large soil features[1] (i.e., peds, profiles, pedons, and landscapes). These features, not considering rock, are discussed in this chapter. In Chapter 3, components smaller than rock, which soil scientists define as those inorganic particles smaller than 2.00 mm in

---

[1] Many soils contain gravel, stones, and rock. However, these components, because of their low reactivity, will not be considered in this book.

Introduction to Soil Chemistry: Analysis and Instrumentation, Second Edition.
Alfred R. Conklin, Jr.
© 2014 John Wiley & Sons, Inc. Published 2014 by John Wiley & Sons, Inc.

diameter, are discussed. Geologic features and gravel, stones and rock, and other substances are not discussed.

Large soil components consisting of sand, silt, clay, and OM are peds, profiles, pedons, and landscapes. Peds are formed by the aggregation of sand, silt, and clay particles to form larger (secondary) soil structures that result from the action of the soil forming factors (see Figure 2.2, Figure 2.3, Figure 2.4, Figure 2.5, Figure 2.6, Figure 2.7, and Figure 2.8). Profiles develop in the loose material on the earth's surface, the regolith, and are composed of horizons of varying texture, structure, color, bulk density, and other properties. The relationship between the soil and the regolith is illustrated in Figure 2.1. Typically, horizons are, as the name implies, horizontal and are of varying thickness. A pedon is the smallest unit that can be considered "a soil" and consists of all horizons from the soil surface to the underlying geologic strata. An area consisting of similar pedons is called *a polypedon*.

Soil features, texture, peds, profiles, and other properties and materials go together in different ways to form different soils. Soil scientists in the United States classify soils into 12 orders, some of which are illustrated in Figure 2.2, Figure 2.3, Figure 2.4, Figure 2.5, Figure 2.6, Figure 2.7, and Figure 2.8. The orders are differentiated by their characteristics, which are a result of the

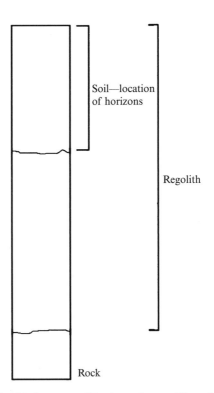

Figure 2.1. The relationship between soil horizons, the regolith, and the underlying rock.

Alfisol

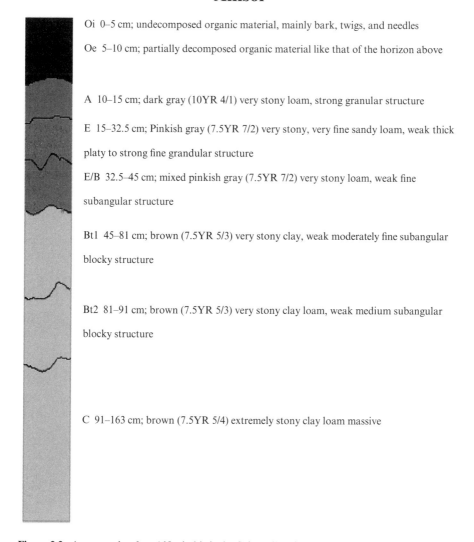

Oi 0–5 cm; undecomposed organic material, mainly bark, twigs, and needles

Oe 5–10 cm; partially decomposed organic material like that of the horizon above

A 10–15 cm; dark gray (10YR 4/1) very stony loam, strong granular structure

E 15–32.5 cm; Pinkish gray (7.5YR 7/2) very stony, very fine sandy loam, weak thick platy to strong fine grandular structure

E/B 32.5–45 cm; mixed pinkish gray (7.5YR 7/2) very stony loam, weak fine subangular structure

Bt1 45–81 cm; brown (7.5YR 5/3) very stony clay, weak moderately fine subangular blocky structure

Bt2 81–91 cm; brown (7.5YR 5/3) very stony clay loam, weak medium subangular blocky structure

C 91–163 cm; brown (7.5YR 5/4) extremely stony clay loam massive

Figure 2.2. An example of an Alfisol; this is the Seitz soil series, the state soil of Colorado [2].

Mollisol

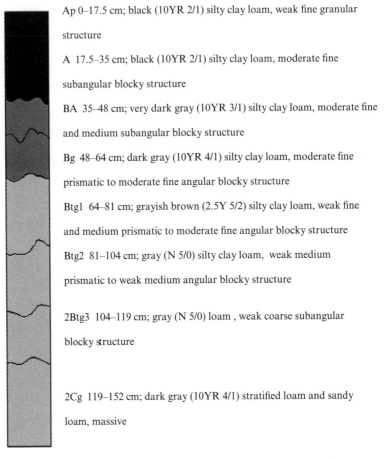

Ap 0–17.5 cm; black (10YR 2/1) silty clay loam, weak fine granular structure

A 17.5–35 cm; black (10YR 2/1) silty clay loam, moderate fine subangular blocky structure

BA 35–48 cm; very dark gray (10YR 3/1) silty clay loam, moderate fine and medium subangular blocky structure

Bg 48–64 cm; dark gray (10YR 4/1) silty clay loam, moderate fine prismatic to moderate fine angular blocky structure

Btg1 64–81 cm; grayish brown (2.5Y 5/2) silty clay loam, weak fine and medium prismatic to moderate fine angular blocky structure

Btg2 81–104 cm; gray (N 5/0) silty clay loam, weak medium prismatic to weak medium angular blocky structure

2Btg3 104–119 cm; gray (N 5/0) loam , weak coarse subangular blocky structure

2Cg 119–152 cm; dark gray (10YR 4/1) stratified loam and sandy loam, massive

Figure 2.3. An example of a Mollisol; this is the Drummer soil series, which is the state soil of Illinois [2].

Ultisol

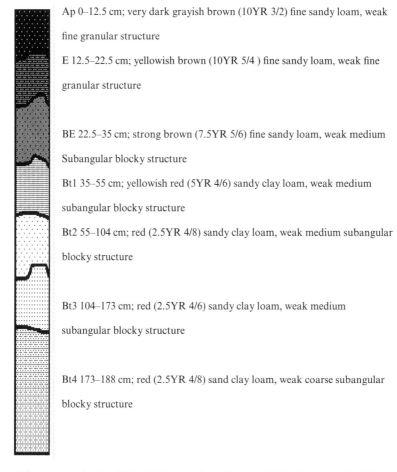

Ap 0–12.5 cm; very dark grayish brown (10YR 3/2) fine sandy loam, weak

fine granular structure

E 12.5–22.5 cm; yellowish brown (10YR 5/4) fine sandy loam, weak fine

granular structure

BE 22.5–35 cm; strong brown (7.5YR 5/6) fine sandy loam, weak medium

Subangular blocky structure

Bt1 35–55 cm; yellowish red (5YR 4/6) sandy clay loam, weak medium

subangular blocky structure

Bt2 55–104 cm; red (2.5YR 4/8) sandy clay loam, weak medium subangular

blocky structure

Bt3 104–173 cm; red (2.5YR 4/6) sandy clay loam, weak medium

subangular blocky structure

Bt4 173–188 cm; red (2.5YR 4/8) sand clay loam, weak coarse subangular

blocky structure

Figure 2.4. An example of an Ultisol; this is the Bama Series, which is the state soil of Alabama [2].

Spodosol

Oi 0–2 cm; partially decomposed forest litter, strongly acid

A   2–5 cm; black (7.5YR 2.5/1) sand, fine granular structure

E   5–13 cm; brown (7.5YR 5/2) sand, weak fine granular structure

Bhs 13–18 cm; dark reddish brown (5YR 3/3) sand, weak fine granular structure

Bs1 18–56 cm; dark brown (7.5YR 3/4) sand, weak fine granular structure

Bs2 56–91 cm; strong brown (7.5YR 4/6) sand, weak fine granular structure

BC 91–130 cm; yellow brown (10YR 5/6) sand, weak fine granular structure

C 130–203 cm; light yellowish brown (10YR 6/4) sand, single grain

Figure 2.5. An example of a Spodosol; the Kalkaska soil series, the state soil of Michigan (single-grain structure in horizon C means that all the sand particles act independently of each other) [2].

Aridisol

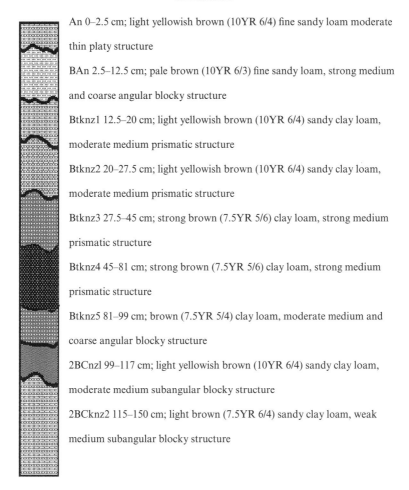

An 0–2.5 cm; light yellowish brown (10YR 6/4) fine sandy loam moderate thin platy structure

BAn 2.5–12.5 cm; pale brown (10YR 6/3) fine sandy loam, strong medium and coarse angular blocky structure

Btknz1 12.5–20 cm; light yellowish brown (10YR 6/4) sandy clay loam, moderate medium prismatic structure

Btknz2 20–27.5 cm; light yellowish brown (10YR 6/4) sandy clay loam, moderate medium prismatic structure

Btknz3 27.5–45 cm; strong brown (7.5YR 5/6) clay loam, strong medium prismatic structure

Btknz4 45–81 cm; strong brown (7.5YR 5/6) clay loam, strong medium prismatic structure

Btknz5 81–99 cm; brown (7.5YR 5/4) clay loam, moderate medium and coarse angular blocky structure

2BCnzl 99–117 cm; light yellowish brown (10YR 6/4) sandy clay loam, moderate medium subangular blocky structure

2BCknz2 115–150 cm; light brown (7.5YR 6/4) sandy clay loam, weak medium subangular blocky structure

Figure 2.6. An example of an Aridisol; the Casa Grande soil series, the state soil of Arizona.

Andisol

Oi 0–0.8 cm; fresh coniferous needles and twigs.
Oe 0.8–2.8 cm; partially decayed needles and twigs
Oa 2.8–3.3 cm; well decayed OM/Mt. St. Helens volcanic ash
A 3.3–12.7 cm; pale brown (10YR 6/3) gravelly ashy silt loam, very fine granularstructure

Bw 12.7–53.3 cm; pale brown (10YR 6/3) gravely ashy silt loam, weak very fine grandular structure

2BC 53.3–73.7 cm; very pale brown (10YR 7/3) gravely sandy loam, massive

3C1 73.7–99.1 cm; very pale brown (10YR 7/3) very gravely loamy sand, massive

Figure 2.7. An example of a Andisol; this is the Bonner soil series, found in northern Idaho, eastern Washington, and western Montana [3].

Vertisol

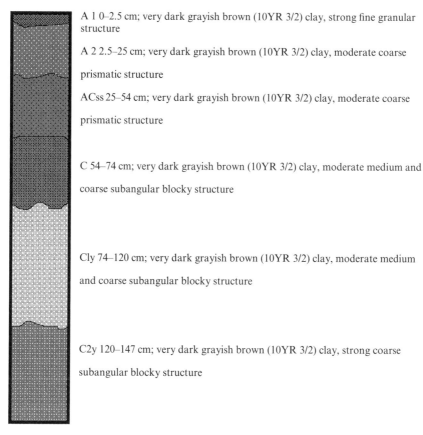

A 1 0–2.5 cm; very dark grayish brown (10YR 3/2) clay, strong fine granular structure

A 2 2.5–25 cm; very dark grayish brown (10YR 3/2) clay, moderate coarse prismatic structure

ACss 25–54 cm; very dark grayish brown (10YR 3/2) clay, moderate coarse prismatic structure

C 54–74 cm; very dark grayish brown (10YR 3/2) clay, moderate medium and coarse subangular blocky structure

Cly 74–120 cm; very dark grayish brown (10YR 3/2) clay, moderate medium and coarse subangular blocky structure

C2y 120–147 cm; very dark grayish brown (10YR 3/2) clay, strong coarse subangular blocky structure

Figure 2.8. An example of a Vertisol; this is the Lualualei series, which is found in Hawaii [2].

action of the soil forming factors: climate, parent material, topography, biota, and time, which all interact during soil formation. Climate, moisture, and temperature determine the biota, as well as a soil's age, and whether it can develop in a certain locality. The soil parent material will be acted upon by other factors but will in turn provide the minerals needed for plant growth and other biological activity.

Consideration of the larger components of soil might seem like a strange place to start a discussion of soil chemistry, analysis, and instrumental methods. However, these larger structures can and do affect the chemistry of a soil. For instance, in many cases, the larger features control the movement of air and water in soil. Sandy textured soils will have higher infiltration and percolation rates than clayey soils. The same can be said for soils with good, strong structure, versus soils with poor or weak structure. As water moves into soil, it displaces air and, as it moves out, air replaces it. Thus, soil with poor or slow infiltration and percolation will have generally lower oxygen contents. This directly affects soil aeration and thus its oxidation–reduction status.

Infiltration and percolation rates also determine which salts have been leached out of the soil. For instance, high infiltration and percolation rates leach calcium and magnesium out of soil and they become acidic. Where calcium and magnesium are not leached out, the soils are neutral or basic. Thus, the type and amount of salts present will affect a soil's pH, which will in turn affect the solubility and availability of both natural and contaminating inorganic and organic compounds.

The species of components present will also be affected by oxidation–reduction, and pH. For example, iron is primarily in the Fe^{3+} (oxidized) or the Fe^{2+} (reduced) state depending on the oxidation–reduction potential of the soil. Speciation, which depends, in part, on the oxygen status of soil, is of environmental concern because some species are more soluble, such as Fe^{2+}, and are thus more biologically available than others. The occurrence of a specific species is related to the chemistry occurring in a soil, which is related to its features. Thus, large features must be taken into consideration when studying soil chemistry and when developing analytical and instrumental methods.

2.1 HORIZONATION

The most striking large feature of soil is the occurrence of distinct horizons. For the analytical chemist, three soil horizonation conditions exist: (1) high-rainfall areas that typically have tree or tall grass vegetation and extensive horizon development, (2) low-rainfall and desert areas with sparse and desert vegetation and little horizon development, and (3) areas with rainfall between these extremes will have variable vegetation and horizonation. It is not possible to draw sharp boundaries between these areas because local conditions such as rainfall frequency, time of year, and intensity of rainfall dramatically affect the climate in transition areas. Each of these situations presents unique challenges to the analytical chemist.

2.1.1 Horizon Development under Moist Conditions

Parent material salts are dissolved when water falls on soil. These salts are leached into and eventually out of the soil. Plants grow, produce biomass, exude OM (including acids), die, and thus add OM to both the soil surface and subsurface. Silica and alumina, which are relatively immobile, slowly dissolve and are eluted into the lower areas of a developing soil. OM is decomposed, mixed by organisms, and leached into the soil. As this process continues, horizons develop in the soil parent material and, eventually, a recognizable soil profile is produced (see Figure 2.1, Figure 2.2, Figure 2.3, Figure 2.4, Figure 2.5, Figure 2.6, Figure 2.7, Figure 2.8, and Figure 2.9). The depth of the soil and

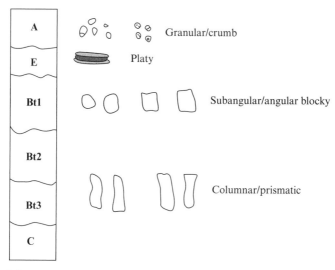

Figure 2.9. Soil structure and its most common location in a soil profile. Platy structure can be found in any horizon. Subangular and angular blocky structures can be found both higher and lower in the profile than indicated.

the number and type of horizons that develop depends on the soil forming factors, the most active of which are climate and biota, although parent material, as mentioned earlier, is also important [1,2].

Soil parent material is not always derived from the underlying rock. In some cases, rock is covered by other geologic materials deposited by ice (glacial till), water (alluvium), wind (loess), gravity (colluvium), or a combination of these transporting agents. A more complete list of transporting agents and the geomorphic features they form is given in Table 2.1. Once deposited, these materials become the parent material from which soil develops.

While it may appear to be logical that the soil surface horizons will be made up of material derived from the parent material and plants growing on the soil, in most cases, this is not true. Many soils have continuous additions of both inorganic and organic material deposited from both water and wind. In Ohio (United States), many soils develop from an underlying glacial till and an overlying silt loess, wind-transported silt, cap. In other areas, occasional or regular flooding may deposit material on the soil surface, which then becomes part of the soil surface horizons. Even in areas where there would appear to be little wind- or water-transported material, small amounts of both inorganic and organic compounds will be added to soil from the atmosphere.

The first horizon develops on the surface of the soil and is called the **A** horizon. Because it has OM deposited on and in it, and is the first to have salts dissolved and eluviated, it is higher in OM and lower in salts than the lower horizons. Clay is eluviated out of this horizon and is deposited lower in the

TABLE 2.1. Soil Parent Material Transporting Agents, The Name of the Material, and the Geomorphic Features They Form[a]

Agent	Name Applied to Material	Geomorphic Features
Gravity	Colluvium	Toe slope of hills
Air	Loess	Loess and loess cap
	Dune sand	Sand dunes
	Volcanic cinders	Volcanic ash layers
Water	Alluvium	Flood plains
		Deltas and alluvial fans
	Lacustrine	Lake sediments
	Outwash	Terraces, outwash plains, kames, and esker
Ice	Glacial till	Till plains
Ocean water	Marine sediments	Coastal plains

[a]This is not an exhaustive list.

profile. The A horizon also has finer granular and crumb structure. This is the horizon that is plowed in preparation for planting crops[2] and is the one most commonly sampled for analysis.

The **A** horizons are slightly different from other horizons because they are plowed and planted. In many cases, a soil may have an **Ap** horizon, the "p" indicating that it is or has been plowed. The plowing need not have been done recently for the horizon to be described as **Ap**. Even a soil so newly developing that it has no horizons will have an **Ap** horizon designated if it is plowed.

With the passage of time, a full set of horizons will develop in a regolith or parent material. Major (master) horizon designations and some of their distinguishing characteristics are given in Table 2.2. Each horizon may be subdivided on the basis of other distinguishing characteristics. The O, or organic, horizons can be subdivided into those where the OM is just beginning, Oi, is well along, Oe, or highly decomposed, Oa. Examples of these horizons can be seen in Figure 2.2 and Figure 2.5. Likewise, the B horizons can be characterized by increased clay content, **Bt**, or reducing conditions, **Bg** (the meaning of the upper- and lowercase letters are described in Table 2.2). When two horizons are similar but still distinguishably different, a number following the major horizon designation indicates this. For example, a soil may have several Bt horizons designated by Bt1, Bt2, and Bt3 as in Figure 2.4. A number in front of a horizon designation indicates that it is formed or is forming from a different parent material. Examples of these different horizons can be seen in Figure 2.3, Figure 2.6, and Figure 2.7.

Figure 2.2 also has two other distinctive horizons. The E horizon (which also occurs in Figure 2.4), which stands for *eluviated*, is characterized by a depletion

[2] Soils may be worked in a number of different ways other than plowing. In addition, they may not be worked at all as when no till planting is used. In this case, there is minimal soil disturbance.

TABLE 2.2. Major and Minor Horizon Designations in Soil Profiles[a]

Major Designations	Characteristic
O	Organic horizon may be partially or decomposed OM
A	Topmost inorganic horizon characterized by relatively high OM content
E	Interior horizon characterized by eluviation
B	Subsoil horizon that shows evidence of weathering or illuviation
C	Lowest horizon of the geologic material underlying the soil that shows little evidence of soil formation.

Subdesignations	
w	Minimal soil horizon development
p	Plowed—only applied to topmost or **A** horizon
t	Accumulation of illuvial clay
a	Highly decomposed OM
h	Illuvial accumulation of OM
s	Illuvial accumulation of oxides of iron and aluminum
ss	Slickensides are smoothed smeared sides on peds
i	Undecomposed OM
e	Intermediately decomposed OM
g	Gleying or mottling
k	Accumulation of carbonates (lime)
n	Accumulation of sodium ions
y	Gypsum deposit in soil
z	Accumulation of soluble salts

[a]This is only a partial list relevant to the soils discussed in this chapter.

of clay and is lighter in color and coarser in texture than the over- or underlying horizons. The second distinctive horizon is the E/B. This designation indicates a transition horizon, which contains characteristics of both the overlying E and underlying B horizons. There are other designations of transition horizons such as AB. Whichever letter comes first indicates that that horizon characteristic dominates the transition horizon.

The small (lowercase) letter subordinate horizon designators are used wherever appropriate to designate significant differences or distinctions in a horizon (see Table 2.2). A common designation is the small letter **t**, which stands for the German word *ton*, which means clay. This is a B horizon, where there is a significant increase in clay when compared to the overlying horizons. Another designation shown in Figure 2.3 is the small letter **g (Bg)**, which stands for *gleying*, also called *mottling*. This is a situation where varying colors develop under reducing conditions, which result when the soil is saturated with water for a significant period of time. In many cases, the colors are varying shades of red or rust colored spots. However, sometimes grays and even blues can occur depending on the soil.

In Figure 2.5, the profile description has an **E** horizon and a **B** horizon with small letter **h** and **hs** designations and a **B** horizon below this with a small letter **s**. The E horizon is light in color and, if light enough, might also be called an *albic horizon*. The small letter **h** indicates an accumulation of highly decomposed illuvial OM and the small letter **s** refers to the accumulation of oxides of illuvial iron and aluminum.

Many other designators are used for various horizons and horizon conditions. A complete list of these will not be given here but can be found in any introductory soils text or book on soil classification.

In naming soils, diagnostic horizons are used. These are different from those used in a profile description such as those given in Figure 2.2, Figure 2.3, Figure 2.4, Figure 2.5, Figure 2.6, Figure 2.7, and Figure 2.8. However, these diagnostic horizons are indicated in the name given to the soil and are not necessarily designations that the researcher doing soil chemistry, developing analytical or instrumental procedures for soil, is likely to have occasion to see or use. If the need arises, they are easy to find in the literature.

2.1.2   Horizon Development under Low-Rainfall and Desert Conditions

Under low-rainfall and desert conditions, the lack of rainfall results in very different soil conditions. Figure 2.6 shows the profile description of an Aridisol (an arid, desert, region soil) (see Section 2.4 for the explanation of the naming system). All lower horizons contain a high content of sodium ions and most also contain other soluble salts of various kinds.

In low-rainfall regions, salts are not readily washed out of soils and may even build up on the soil surface. This happens when rain dissolves salts in the soil surface and leaches them a little way into the soil. Subsequent evaporation of water from the soil surface causes water and dissolved salts to move to and be deposited on the soil surface. No salts or only the most soluble salts are leached out of the soil, in low-rainfall regions, and the pH is typically basic.

As would be expected, only limited horizon development will occur and, frequently, the horizons will be thin. However, it is possible to find soils with significant horizon development in desert regions. In some cases, these soils develop when the particular area receives more rainfall.

Salts, in addition to causing the soil to be basic, can have deleterious effects on analytical procedures. For example, significant error can occur if a potassium-selective electrode is used to determine potassium in a high-sodium soil (see Chapter 9). As discussed in Chapter 14, other salts could cause inaccurate results when atomic absorption analysis of a soil extract is carried out.

2.1.3   Horizon Development in Areas between High- and Low-Rainfall Conditions

In areas between high and low rainfalls, horizonation may be well or poorly developed, soil pH may be either acidic or basic, and there may or may not be salt buildup in the soil. As the amount of rainfall an area receives decreases, the depth to a layer of salt deposition decreases. In these areas, the analyst must be aware of this potential variation and must have the reagents necessary for all of these eventualities.

2.1.4 Erosion

It should be mentioned that, in addition to deposition of material on a forming soil, erosion is also occurring. Even under the best protection, soil is always undergoing what is called *geologic erosion*. Because of this, soil development will always reach a quasi equilibrium between deposition, development, and erosion leading to a soil where the horizons change very little over time.

2.2 PEDS

Profile descriptions also detail the structure found in a horizon and indicate its strength. Figure 2.9 shows a soil profile with an indication of the location of the various structure types. Looking at Figure 2.2, Figure 2.3, Figure 2.4, Figure 2.5, Figure 2.6, Figure 2.7, and Figure 2.8, an example of each major structure type is indicated. In most cases, granular and crumb structure can be found to occur only in the top 25 cm of soil. Platy structure can be found in any horizon, although it is frequently found in E horizons or in the transition zones between the A and lower horizons. Traffic, farming, or other activity will promote the formation of platy structure in the A horizon and often results in an angular, blocky structure, at the base of the A horizon. In C horizons, platy structure is a remnant of the original character of the parent material from which the soil is forming. Subangular and angular blocky structure is typically found in the upper part of the B horizons and the prismatic structure in the lower part. However, blocky and prismatic structure can be found in any part of the B horizons.

Peds are formed by natural aggregation of sand, silt, and clay particles. Although a soil's texture is defined by the relative proportions of sand, silt, and clay, these components almost never act independently of each other. The binding agents that hold them together are clay, OM (particularly microbial gums), and various inorganic ions, notably calcium, magnesium, iron, and aluminum. Planes of weakness occur between peds and are extremely important in determining a soil's characteristics because they are areas where air, water, and roots penetrate the soil easily. When observing soil peds, roots are seen growing in the voids between structures (roots can be seen on the second from the bottom ped in Figure 2.10). Because of their effect on air, water, and roots, peds help to determine a soil's chemistry.

Figure 2.10 shows actual peds isolated from a soil profile. Generally speaking, structure is considered to be a positive component in soil, except for platy peds. Platy structure can retard the movement of air and water down through

Figure 2.10. Peds isolated form soil. From top to bottom spheroidal, platelike, blocklike, and prismlike.

the soil profile and so is considered a negative component. Because it restricts air and water movement, it may cause areas of water saturation and thus foster reducing conditions.

Because large voids between peds allow ready movement of water down through the soil profile, ped surfaces often become coated with material carried in the water. The most common coating is clay and is call "clay skins." If there is rubbing between the peds, the skins will have a smeared surface and will often be shiny. Under such conditions, the surfaces are said to be "slickensides." Less common are surface coatings of silt and OM carried by leaching into the lower horizons. Silt coatings are common in horizons from which clay particles have been translocated to lower horizons. OM coatings are common in acid sandy soils especially those called *Spodosols* (Figure 2.5).

Coating of primary and secondary soil structures is an extremely important phenomenon for the analysis of soil. Without coating, the chemical activity of soil primary and secondary particles could be determined by knowledge of their structure and makeup. With coating, the chemistry becomes vastly more complex. This complexity is discussed in detail in later chapters.

For the soil separates (sand, silt, clay) to become aggregated, they must come close together. The forces that cause particles to come together are the

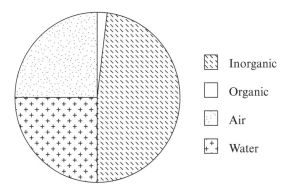

Figure 2.11. Idealized general composition of a soil sample.

weight of the overlying soil, plant roots moving through the soil, heating and cooling, freezing and thawing, and wetting and drying, It might also be argued that cultivation forces particles together and thus tends to improve structure. However, it is equally true that cultivation breaks down structure. Because of these two competing effects, the net effect of cultivation on soil structure can be either positive or negative depending on the condition of the soil, particularly wetness, when cultivated.

Ideally, a well-aerated soil is considered to be half-solid and half-void space. The void space is half-filled with air and half with water. This idealized condition is illustrated in Figure 2.11. Such a soil is under general oxidizing condition, and oxidation of all components, particularly OM is expected. In real soil samples, the total amount of void space depends on soil texture and structure, and the amount of air space is inversely related to the water content. When the void space becomes filled with water, the soil becomes reducing. This takes a little time because dissolved and trapped oxygen must be used up before full reducing conditions are reached. Reducing conditions are accompanied by the production of methane, hydrogen, reduced forms of metals, and sulfur, if present, producing sulfides.

At this point, oxidation and reduction in soil seems simple. When the soil is not saturated with water, it is oxidizing, and when saturated, it is reducing. However, even under oxidizing conditions, reduced compounds and species are produced. How can methane be produced even in a soil under oxidizing conditions? The answer is that, in addition to simple pores, which are large in diameter and open at both ends, there are pores that have restricted openings, have only one opening, or have only a restricted opening as illustrated in Figure 2.12.

Where the pores have only one opening or restricted openings, they do not drain and the interiors are reducing even when the soil in well aerated. The reducing conditions lead to the production of species such as methane and Fe^{2+}. Although methane and iron are two of the most easily found and identified, other reduced species, both organic and inorganic, are commonly present.

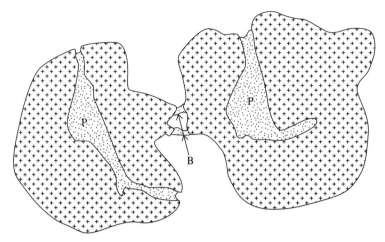

Figure 2.12. Two soil peds showing a pore with restricted openings on the left and a pore with only one opening on the right. In between are bridges holding the two peds together. P, pores; B, bridges.

Two conditions exist where the soil separates are not aggregated to form secondary particles or peds. One is where the individual separates act independently from each other. This condition is called *single-grained*, like beach sand, and occurs in course sandy soil. The second condition is where there are no lines of weakness between units and the soil is said to be *massive*; that is, it acts as one massive block. Soil becomes massive if it is worked when it is too wet, forming hard clods that, when dry, form smaller irregular clods when broken. Neither of these conditions is naturally common in soil, although the massive condition occurs in the soil parent material or regolith before soil formation begins.

2.3   SOIL COLOR

In Figure 2.2, Figure 2.3, Figure 2.4, Figure 2.5, Figure 2.6, Figure 2.7, and Figure 2.8, the horizon descriptions give a color in both words and in Munsell color book designations. The Munsell system describes color in terms of hue (i.e., the basic color), value (lightness), and chroma, or the purity of the color. The hue is the first number followed by a color designation. In Figure 2.2, the A horizon is 10YR (where YR indicates a yellow-red color); in Figure 2.3, the Btg1 horizon has a color designated as 2.5Y (where Y indicates yellow). In the Munsell system, the value for white is given the number 10 and blacks the number 0. The chroma increases from left (1) to right (8). However, not all colors in the Munsell system are used to describe soil colors.

The typical Munsell color chart for soils covers only those colors described as being yellow or red or some combination of the two. This is not to say that

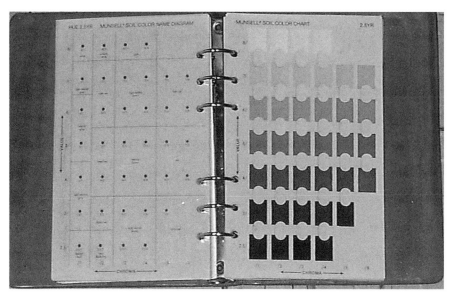

Figure 2.13. The 2.5YR (yellow-red) page of a Munsell color book. The value is from white at the top to black at the bottom. Chroma becomes higher from left to right.

other colors do not occur in soil; they do. It is common to find gray soils in which no hue occurs and, in this case, only the value and chroma are indicated as shown in Figure 2.3, horizon Btg2. Most books used by soil scientists include a gley page for such soils. Under highly reducing conditions, blues will sometimes be found. A picture of a page in a Munsell color book commonly used by soil scientist is shown in Figure 2.13 [3].

Soil color deserves special attention. When rock is ground, it typically produces a gray powder. Gray occurs in E and gleyed horizons under acid and reducing conditions, and thus this color provides important information when analyzing such as soil. The normal soil colors red and black are derived from iron and OM, respectively. Iron is in the form of oxides and hydroxy oxides (Fe_2O_3, FeOOH, etc.). Black and brown colors are normally derived from highly decomposed OM and humus.

As a soil develops, OM decomposes to produce humus, which is black. Additionally, release of iron from minerals by weathering yields various reds and yellows. Both mechanisms yield soil coloring agents. Under oxidizing conditions, where soil is not saturated with water, the iron will be oxidized and thus in the ferric state [Fe(III)]. When the iron and OM are deposited on the surfaces of sand, silt, clay, and peds, they develop a coat that gives them a surface color. However, soil color is not only a surface characteristic but extends through the soil matrix. Under oxidizing conditions, soil has a reddish color. The chroma of this color depends to some extent on the amount of and the particular iron oxide present.

Under most conditions, little OM is eluviated down through the soil. However, many soils have dark or black thick upper horizons, which are high in OM. These horizons are found in soils developing under high water conditions and grass vegetation where the grass and associated roots die each year and contribute OM to the profile to a depth of ≥ 0.5 m depending on the type of grass and other environmental conditions.

When soil is saturated with water, the soil environment becomes reducing and, under these conditions, iron is reduced to the ferrous [Fe(II)] state. The soil color becomes lighter and more yellow. Under reducing conditions, soil also develops variations in color called *mottling* or *gleying*. Thus, any soil horizon description that includes a small "g" designation indicates that the soil is under reducing conditions for a significant period of time during the year. It might be expected that mottling or gleying would occur only in the lower horizons; however, it can occur anywhere in a soil profile, even quite near the surface.

Iron in the ferrous state is more soluble than iron in the ferric state. Indeed, in the ferric state, it is insoluble under most soil conditions. Under reducing conditions, ferrous iron may be leached out of soil, leaving it gray in color. This is the origin of the term gleying.

Some types of vegetation and environmental conditions result in acid-producing litter on the soil surface, including leaves, needles, and bark. Under these conditions, OM decomposition products (humic and fulvic acids, etc.) can eluviate and be deposited deeper in the soil. In Figure 2.5, the litter in the Oi horizon produces acid, which allows the illuviation of aluminum, iron and OM decomposition products into the B horizons to form the Bhs horizon. The leaching of aluminum, iron, and OM out of an area of the soil profile results in it becoming a light gray or white, giving rise to the potential development of an albic horizon.

To some extent, the description of a soil profile, including color, gives an indication of some of the chemistry and chemical conditions occurring in that profile. This, in turn, provides the researcher and analyst with information about the types of compounds and species likely to be found and the conditions necessary to isolate them [4].

2.4   SOIL NAMING

Various soil naming systems are used throughout the world. In all of these, the horizons and their subdesignations vary somewhat according to the different classification systems. In the United States, the United States Department of Agriculture (USDA) has developed the USDA Soil Taxonomy system (simply referred to as *Soil Taxonomy*), which recognizes 12 soil orders. The United Nations, through its Food and Agriculture Organization and United Nations Educational, Scientific and Cultural Organization (FAO-UNESCO) and the

International Society of Soil Science, has a system that includes 26 soil groupings. There are many other systems including those developed by Canada, France, the former Soviet Union, China, Australia, and other countries. Each recognizes a different number of major soil groups or soil orders.

Two basic concepts are used in developing the naming of different soils. First is the idea that a soil's characteristics will determine which group it falls into and its name. Another idea is that soils will be related to a reference soil. In both cases, the natural soil horizons are used in the soil description. In Soil Taxonomy, horizons used for naming the soil are called *diagnostic horizons*. In the reference soils, the horizons are called *reference horizons*. The general concepts are similar in all systems, and so are many names for soil characteristics. The names are often descriptive in that they give an idea of the characteristic of the soil.

In the FAO soil system and Soil Taxonomy, several soils have the same or similar names or names that can easily be seen as being related to each other. Examples would be Histosols and Vertisols (see Figure 2.8), which carry the same name in both systems. Andosols and Andisols (see Figure 2.7) only differ by one letter. Ferralsols and Oxisols should be understood as being similar if one knows a little about highly weathered soils. If one is working on an international scale or wishes to work globally, it is important to know that there are different ways of naming soils and familiarity with the characteristics of various soil types is necessary. This is particularly important if one is developing a soil testing method or instrumental method, which might be used on an international scale.

2.5 THE LANDSCAPE

In a landscape, scientists can differentiate one soil from another by the soil's physical and chemical characteristics. The smallest unit that can be considered a soil is called a *pedon* and has an area of 1–10 m^2 and is normally 1.5–2 m deep. If an area contains contiguous pedons of similar characteristics, then they are said to be a *polypedon* (multipedons) region.

In USDA Soil Taxonomy and in field reports of soils, the description given is that of a pedon—considered the "typical" pedon. In the first case, the field description is that of a pedon, and when mapping soils, the soil mapping unit strives to capture the distribution of similar polypedons. A soil mapping unit applied to the field may contain "inclusions" of dissimilar ploypedons. Both Soil Taxonomy and the soil mapping unit nomenclature contain the terminology likely to be associated with soil samples that the soil chemist, environmental analyst, or instrument procedure developer will encounter. Knowing where the soil comes from, its Soil Taxonomy designation, and field description give the soil researcher and analyst a great deal of information about the chemical characteristics of the soil. In addition, reporting the soil name and its analysis together provide the reader with information that is invaluable in applying the analytical results to that particular soil as well as other environmental conditions.

A landscape will contain many different soils, which can be expected to change with position in the landscape, vegetation, slope, climate, and parent material. It is also to be expected that the soil will change only over long periods of time, namely, decades and centuries.

However, groupings of soils can be and are made for all soils on this planet, and all soil types can be found on all continents except Antartica. Ultisols, Mollisols, Alfisols, and Spodosols, and other soil types occur in North and South America, Asia, Europe, and Africa, Australia, and New Zealand. Thus, researchers and analysts all over the world need to know about soils, soil types, and their nomenclature.

2.6   RELATIONSHIP OF LARGE FEATURES TO SOIL CHEMISTRY, SOIL ANALYSIS, AND INSTRUMENTATION

Each topic discussed earlier provides a great deal of information about the soil and its chemistry. This information is invaluable to the soil chemist, the person developing a soil analytical procedure, or a person wishing to make an instrumental analysis of some soil characteristic or component. Knowing which types of soils are present tells much about its chemistry, the likely pH or pH range of the soil, and its salt content. The occurrence of clay, especially in the **Bt** horizon, will affect the availability and solubility of both inorganic and organic components. For instance, plant available phosphorus is decreased by both low (acid, $pH < 7$) and high (basic or alkaline, $pH < 7$) soil pH. Likewise, the type of clay present will dramatically affect the extractability of a soil component or contaminant. Clay types and their chemistry are discussed in Chapter 3.

The discussion of soil types that follows should not be taken as exhaustive but only as providing examples of the types of information available and why knowledge of the large features of soil is important in understanding its chemistry and analysis. The following chapters refer back to these characteristics where appropriate.

2.6.1   Soil Types

What do the names Alfisol and Mollisol tell us about the chemistry of these soils? Alfisols develop under humid climates and so are acidic and have a medium base saturation. All the easily leached salts have been removed. However, basic parent materials often underlie them. Mollisols develop under lower rainfall conditions than do Alfisols and so have higher pH levels, higher concentrations of more easily leached salts, and—although they contain significant levels of calcium—they are still often slightly acidic. Mollisols have higher OM throughout the upper portions of the soil profile than do Alfisols. Both soils have well-developed B and Bt horizons.

Ultisols and Spodosols (see Figure 2.4 and Figure 2.5) develop under different conditions but have some important similarities. Ultisols are the ultimate in soil development in terms of horizon development and leaching of base cations. Salts have been leached out, but they are generally considered to be soils having maximum clay formation. The soils are typically very acidic (here, "very" is used relative to other soils) and there is the occurrence of aluminum as Al^{3+}. This is of note because aluminum in this form is toxic to plants. Although there is a Bt horizon, the clay is the simplest and least active of the clays (see Chapter 3). Spodosols develop in coarse-textured soil under trees, the detritus of which provides acid to water leaching through the soil profile. Its **B** horizon contains illuvial accumulation of highly decomposed OM and oxides of iron and aluminum. In this case, the B horizons as shown in Figure 2.5 do not have an increase in clay.

The pH and clay content of these soils are extremely important in understanding their chemistry. The lack of salts, low pH, and dominance of low activity clays will greatly affect the retention and extraction of components from both Ultisols and Spodosols.

In Aridisols (Figure 2.6), both the occurrence of clay in the lower horizon and the occurrence of high pH and salt contents greatly affect the retention of components. For these soils, analytical procedures must be potentially impervious to high pH and salt content, or steps must be taken to remove salts or to change the pH before analysis.

2.6.2 Soil Color

The color of soil gives an indication of its oxidation–reduction conditions and the amount of OM present. Well-aerated soils will be under oxidizing conditions; iron will be in the Fe^{3+} state, less soluble and thus less available for chemical reaction. Under water-saturated conditions, soil will be under reducing conditions as indicated by increased yellow colorings, gleying, and mottling. Iron will be in the Fe^{2+} state, which is more soluble and thus more available for chemical reaction. Under these conditions, reduced species such as methane (CH_4), hydrogen, (H_2), and sulfides will be found.

Under saturated or very wet conditions, soils tend to have increased amounts of OM. This results in dark colors and dramatically changes the chemical characteristics of a soil. OM increases a soil's sorptive and cation exchange capacities and thus alters the movement and extraction of components present. OM increases ped formation and stability, thus increasing both aeration and percolation, but under saturated conditions, reduction reactions prevail (see Figure 2.12).

While color is often discussed as being closely related to iron and its oxidation state, it is also related to all other soil components. When soil color indicates oxidizing conditions, all multiple oxidative state capable cations are expected to be in their highly oxidized states, and when soil color indicates reducing conditions, low oxidative states will occur [4].

2.6.3 Soil Structure

Increasing soil structure results in increases in both small and large pores, which means improved water and oxygen movement in and out of soil. This increases aeration and oxidative conditions. The use of words such as "strong" in describing soil structure such as in the soil profile descriptions in Figure 2.6 and Figure 2.8 indicates ped formation that is strong enough to resist breakdown under wet conditions, thus leading to pores that drain well even when wet. Strong structure favors oxidation of soil constituents as long as the soil does not have a high water table.

2.7 CONCLUSIONS

Soils develop by the action of the soil forming factors on soil parent materials including material transported by different agents. The result of these soil forming factors is the formation of soil horizons, different colors, and peds. Each of these factors has a pronounced effect on a soil's chemistry. Knowledge of the soil type and profile description can provide the soil chemist, analyst, or researcher with valuable information about the characteristics of soil relevant to the development of extraction, analytical, and instrumental analytical procedures. It also is the place to start when investigating the failure of a procedure.

PROBLEMS

2.1. Describe how horizons form in soil. How would knowing the horizon help a soil chemist understand differences in the extractability of a soil component?

2.2. List three major horizon designations and three subordination designations and explain what they reveal about the horizon.

2.3. What common soil characteristics are indicated by soil color? Explain how a soil's color is described.

2.4. Name two major different types of soil structure. How does soil structure relate to a soil's oxidative or reductive condition?

2.5. What major soil types have high salt and pH levels? Explain how this comes about.

2.6. Describe the effect of small pores on the chemistry of a soil and the types of compounds that are likely to be found.

2.7. What colors do Fe^{2+} and Fe^{3+} impart to soil?

2.8. What do the subordinate soil descriptors **p** and **t** stand for?

2.9. Explain how and where methane (CH_4) is formed in aerobic soils.

2.10. Immediately after a soil becomes saturated with water, some oxygen may be present in the soil. In what forms is this oxygen and where is it?

REFERENCES

1. State soils pictures. Available at: http://soils.usda.gov/gallery/state_soils/. Accessed June 3, 2013.
2. State soils description. Available at: http://soils.usda.gov/technical/classification/. Accessed June 3, 2013.
3. *Munsell Soil Color Charts*. Baltimore, MD: Kollmorgen Corporation; 1975.
4. Bigham JM, Ciolkosz EJ (eds.). *Soil Color*. Madison, WI: Soil Science Society of America; 1993.

BIBLIOGRAPHY

Bigham JM, Ciolkosz EJ, Luxmoore RJ. *Soil Color*. Madison, WI: Soil Science Society of America; 1993.

Bradey NC, Weil RR. *The Nature and Properties of Soils*, 12th ed. Upper Saddle River, NJ: Prentice Hall; 1999.

Dent D, Anthony Y. *Soil Survey and Land Evaluation*. Boston: Allen & Unwin; 1981.

Douglas LA. *Soil Micromorphology and Soil Classification*, Thompson ML (ed.). Madison, WI: Soil Science Society of America; 1985.

Soil Survey Staff. *Soil Taxonomy: A Basic System of Soil Classification for Making and Interpreting Soil Surveys*. Washington, DC: U.S. Government Printing Office; 1975.

CHAPTER

3

SOIL BASICS PART II
MICROSCOPIC TO ATOMIC ORBITAL DESCRIPTION OF
SOIL CHEMICAL CHARACTERISTICS

Many of the remaining chapters discuss the removal of ions and compounds from soil samples so that they can be analyzed and quantified through a variety of techniques. This is generally achieved through an extraction process in which a liquid extractant is mixed with the soil. Analytes of interest that are

Introduction to Soil Chemistry: Analysis and Instrumentation, Second Edition.
Alfred R. Conklin, Jr.
© 2014 John Wiley & Sons, Inc. Published 2014 by John Wiley & Sons, Inc.

in the soil are transferred to the extraction solution and subsequently analyzed. An understanding of the forces that hold ions and organic molecules to the soil or in soil water is necessary in order to perform such extractions. The chemical characteristics of soils and common ions and compounds found in soil are therefore the focus of this chapter because the chemical characteristics influence the type and strength of the interaction forces.

In this chapter, soil components will first be considered as individual, independent, noninteracting entities. Then, the interaction between the various components in soil will be discussed. However, it is essential to know and remember that components in soil do not act independently of each other. In addition, surfaces always have a coating of some type that is not continuous, varies in thickness, and sometimes exposes the underlying surface. Sometimes, this first coating will have another, different coating on top of it.

Before an understanding of the interactions between soil components and surfaces is possible, it is essential to understand the composition of uncoated soil components. Once this is known, it is then possible to discern the interactions and bonding patterns of these components with and without coatings.

The solid portion of soil is made up of sand, silt, clays, and organic matter. Elements, inorganic and organic molecules, and ions are also present. The soil solution is a combination of elements and inorganic and organic ions and molecules. The gaseous portion contains gases commonly found in the atmosphere. However, the concentrations of the gases are very different in soil air than in the atmosphere. All components are subject to being partitioned between these three phases.

The chemistry of soil is contained in the chemistry of these three phases. For the solid phase, the chemistry will depend on the amount and type of surface available for reaction. In the liquid phase, solubility will be the most important characteristic for determining the chemistry occurring. In the gaseous phase, gas solubility and the likelihood that the component can be in the gaseous form (i.e., vapor pressure) will control reactivity.

SOIL COMPONENTS INDEPENDENT

3.1 SOIL SOLIDS

Soil scientists define sand particles 2 mm in diameter as the largest "regular" component of soil. Many soils contain gravel, rocks, and other materials that are larger than 2 mm in diameter, but because of their relatively large size, they have limited surface area per unit mass, and provide little in the way of chemical reactivity in soil. For this reason, these particles will not be considered. Particles that are smaller than 2 mm are divided into groups in numerous ways by different disciplines and researchers. For our purposes, it will suffice to view them as belonging to three groups: sand, silt, and clay. The breakdown of rock to form sand, silt, and clay is illustrated in Figure 3.1.

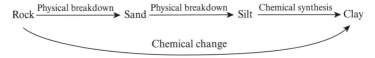

Figure 3.1. Rock decomposition pathway that produces sand, silt, and clay.

3.1.1 Sand

The sand fraction of soil is composed of particles 2–0.02 mm in diameter.[1] It is derived from the physical breakdown of rock, commonly due to wave action in oceans and lakes, tumbling in a river, or grinding in a glacier. Heating and cooling, wetting and drying, and freezing and thawing are also common physical ways that sand-size particles are produced. Plant roots growing through cracks in rocks also break them down into smaller particles that may eventually become sand size. Many soil scientists find it convenient to subdivide sand into fine, medium, and coarse, or even more groups. The surface area-to-mass ratio of each group is different; however, the surfaces and the chemical reactions at them are not.

In terms of chemistry, sand is assumed to be made up of silica, the empirical formula of which is SiO_2 (silicon dioxide), and can be considered the anhydride[2] of silicic acid, H_4SiO_4. However, in the environment in general and the soil specifically, silica is polymerized to form sheets and three-dimensional structures such as sand particles. Here, each silicon atom is tetrahedrally surrounded by four oxygen atoms (see Figure 3.2) and the oxygen atoms also have two lone pairs of electrons. All surfaces of silica will have these electrons that attract any positive or partially positive species near the surface. Thus, water, through its partially positive hydrogen atoms and any positive species it contains, will be attracted to the surfaces of silica. Figure 3.2 shows a water molecule illustrating the partially positive hydrogen atoms and the partially negative oxygen atoms.

At broken edges, it is possible to find oxygen atoms attached to silicon but have no silicon bonding partner. These are often associated with hydrogen forming hydroxyl (–OH) groups. The bonds and orbitals of silicon, oxygen, and hydroxyl groups bonded to silicon determine the chemical reactivity of freshly formed surfaces of this soil fraction.

Quantifying silica interactions with its surroundings is difficult. First, the surfaces are not regular, and thus it is impossible to calculate their area. Surface areas must be measured, and although surface area measurement is not difficult, it is time-consuming and open to inaccuracies. Second, as noted earlier, the surfaces are irregularly covered and it is impossible to know the extent, type, and thickness of materials covering all the surfaces. However,

---

[1] This is the International Soil Science Society designation. The United States Department of Agriculture (USDA) defines sand as ranging from 2 to 0.05 mm in diameter.

[2] An anhydride is a compound that has lost the elements of water.

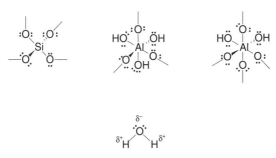

Figure 3.2. A silicon tetrahedron (left), an aluminum octahedron (middle) as a central layer in a 2:1 clay, and an aluminum octahedron (right) as a surface layer in a 1:1 clay (right). The oxygen atoms are bonded to other silicon and aluminum atoms in the clay (bonds are not intended to be shown at the correct angles). Below is a water molecule showing partially positive hydrogen atoms and partially negative oxygen atoms. Also shown are the two lone pairs of electrons on all the oxygen atoms.

silica bonds and electron pairs are important in any chemical analysis, analytical procedure, or instrumental procedure applied to soil.

3.1.2 Silt

The silt fraction is particles 0.20–0.002 mm in diameter. This fraction or separate is produced by the same physical breakdown as described earlier for the formation of sand. Silt is more finely divided silica, but the surfaces are basically the same as those of sand (i.e., silicon and oxygen), and oxygen lone pairs of electrons and hydroxyl groups control its chemistry. Because the particles are smaller, they have more surface area per unit mass. This results in the availability of a greater number of bonds for chemical reactions [1].

3.1.3 Clay

The next smaller separate is actually a group of particles of different types collectively called *clay* and are particles measuring less than 0.002 mm in diameter. They are significantly different from sand and silt separates both physically and chemically. Physically, they are mostly colloidal in size and thus remain suspended in water and have large surface areas per unit mass. There are a large and varied number of clays that differ in terms of both the components they contain and in the arrangement of those components. Here, we will consider the common types of clays found in soil that are models for all soil clays. These can be grouped into two classes: (1) those composed of layers of alumina octahedral sheets and silica tetrahedral sheets and (2) the amorphous clays. All clays are synthesized in soil (see Figure 3.1) from the following three general types of reactions:

1. Clay that is present in rock and is released when the rock breaks down as a result of physical processes. These clay structures are subject to change once released from rock.

2. Clays released by chemical breakdown of rock, which can lead not only to the release but also to the formation of clays.
3. Clays formed in soil by chemical reactions occurring after rock decomposition.

Soil components, silica, and alumina are solubilized, in low concentration, and can react, or crystallize, to form new clays. In addition, clays from any source change over time and become simpler and simpler. Silica is more soluble than alumina and so the silica:alumina ratio decreases over time. Eventually, this leads to deposits of alumina that are used as an aluminum ore for the production of aluminum metal. Although these reactions are considered to be very slow on a human timescale, they do occur.

Clays composed of layers are called *layered silicates*. The most common sheets are tetrahedral silicon and octahedral aluminum (see Figure 3.2, Figure 3.3, and Figure 3.4). Three common representative clays in soil are 1:1 kaolinite, 2:1 fine-grained micas, and 2:1 smectites; that is, kaolinites have one sheet of silicon tetrahedra and one sheet of aluminum octahedra. The fine-grained mica and smectites have two sheets of silicon tetrahedra and one sheet

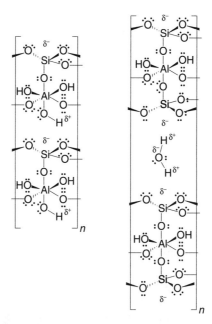

Figure 3.3. The left structure represents kaolinite, a 1:1 clay mineral, and the right structure, a 2:1 clay mineral. These representations are intended to show surface groups, surface pairs of electrons, unsatisfied bonds, and associations between clay particles. Note that clay structures are three-dimensional and these representations are not intended to accurately represent the three-dimensional nature nor the actual bond lengths. Also, the brackets are not intended to represent crystal unit cells.

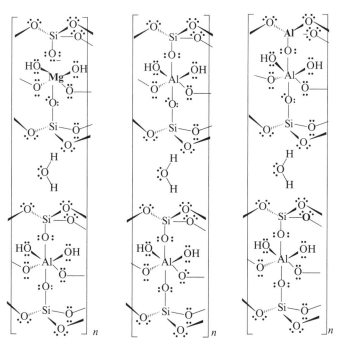

Figure 3.4. Two types of isomorphous substitution. The middle structures are two-dimensional representations of clay without isomorphous substitution. On the left is an isomorphous substitution of Mg for Al in the aluminum octahedral sheet. On the right is isomorphous Al substitution for Si in the silicon tetrahedral sheet. Clays are three-dimensional and –OH on the surface may be protonated or deprotonated depending on the pH of the surrounding soil solution. There will be additional water molecules and ions between many clay structures. Note that clay structures are three-dimensional and these representations are not intended to accurately represent the three-dimensional nature nor the actual bond lengths; also, the brackets are not intended to represent crystal unit cells.

of aluminum octahedra. These latter clays have a sandwich-like arrangement with the aluminum octahedral sandwiched between two silica tetrahedral layers.

In terms of soil development and the development of soil horizons, the smectites and fine-grained micas are found in younger, less weathered soils. Kaolinite and amorphous clays are found in highly weathered soils. Considering a time sequence, at the beginning of formation, soil will contain more complex clays that weather to simpler forms over time. However, it is convenient to start with a description of the simpler layer silicate clays and then describe the more complex clays.

An additional important characteristic of clays is their surfaces, which are distinguished as being either external or internal. Internal surfaces occur, for example, in nonexpanding 2:1 clays, such as the fine-grained micas, and are generally not available for adsorption, chemical, or exchange reactions.

3.1.3.1   1:1 Clay—Kaolinite

The 1:1 kaolinite clay is depicted in Figure 3.3. One surface is composed of oxygens with lone pairs of electrons. These electrons are in **p** orbitals and thus extend away from the oxygens into space. The oxygens are partially negative, as indicated by the symbol δ^-, because they are more electronegative than the surrounding atoms, namely, Si. The other surface is composed of –OH groups where the hydrogens are partially positive as indicated by the symbol δ^+. The electrons in the sigma bond between the oxygen and hydrogen are drawn closer to the oxygen because it is more electronegative, thus leaving the hydrogen slightly positive.

The consequence of these partial charges is that one surface of kaolinite is compatible with and attractive to the other surface. This results in increased stability of kaolinite and the formation of relatively stable structures. Some kaolinite particles can be larger than the 0.002 mm upper limit for clay! Both surfaces also attract and hold water through these partial charges. The absorptive activity of kaolinite is associated with its surface electrons and partially positive hydrogens, and thus the two faces of kaolinite can attract anions, cations, water, and electrophilic and nucleophilic organic compounds.

When a particle of kaolinite is broken or at the edges of particles, bonds will be unsatisfied. This leaves negative charges, which attract cations. Thus, these charges are satisfied by metal cations, particularly Ca^{2+}, Mg^{2+}, K^+, and Na^+. Other cations can be attracted to these sites including positively charged organic molecules such as quaternary amines and ammonium. Because the crystals are relatively large and the charges develop only at edges, the attraction between cations and kaolinite is small, on a unit mass basis, compared to that of other clays.

The phenomena of cations being attracted to surfaces is commonly referred to as cation exchange because cations that are more attracted to the surface can displace those that are less attracted. For example, a +2 cation can displace a +1 cation from a surface because the +2 cation has a greater chemical drive to interact with a negatively charged surface than does the +1 ion. Thus, the +1 ion is exchanged for the +2 ion.

3.1.3.2   2:1 Clays—Fine-Grained Micas and Smectites

The 2:1 clays are smaller and much more complex than the 1:1 clays. Because they are smaller, they have larger surface area-to-mass ratios and more edges. This results in both increased cation exchange and adsorption. The adsorption in 2:1 clays will be different from that occurring in kaolinite because both surfaces of the particles are the same. In 2:1 clays, cation exchange is not limited by edge effects because of a phenomenon called *isomorphous substitution*, which results in increased cation exchange, changes in shape of the particles, and changes in the way they interact with water and cations.

A first approximation to the makeup of 2:1 clays can be seen in Figure 3.3 and Figure 3.4. The particles consist of a sandwich of octahedral alumina between two sheets of tetrahedral silica. One result of this arrangement is that both surfaces are made up of only oxygen. This means that both surfaces are slightly negative, and thus two particles repel each other if there is no positive or partially positive species, such as the hydrogens on water molecules, between them. Figure 3.4 shows water molecules between clay particles. This allows the clay particles to come closer together and, in reality, water molecules are always present between the clay layers. As with all of the soil solutions, this water contains cations (not shown) further enabling the particles to come together.

Another characteristic of 2:1 clays is isomorphous substitution, where *iso* means "same" and *morphous* means "shape." During the formation of clay, cations other than aluminum and silicon become incorporated into the structure. In order for this to work and still ensure a stable clay, the cation must be about the same size as either aluminum or silicon, hence the term *isomorphous*. There are a limited number of cations that satisfy this requirement. For silicon, aluminum as Al^{3+} and iron as Fe^{3+} will fit without causing too much distortion of the clay structure. For aluminum, iron as Fe^{3+}, magnesium as Mg^{2+}, zinc as Zn^{2+}, and iron as Fe^{2+} will fit without causing too much structural distortion (see Figure 3.4).

The two 2:1 clay types are distinguished by sheets in which substitution occurs. Nonexpanding fine-grained mica-type clay has isomorphous substitution of Al^{3+} in the silicon tetrahedal sheet. Thus, a tetrahedral sheet of fine-grained mica may have an aluminum atom substituted for a silicon atom. The expanding clays might have substitution of Mg^{2+} for aluminum in the octahedral sheet. These two substitutions were chosen to illustrate that, with substitution, some bonds in the clay structure will go unsatisfied. This means that some bonding electrons will not be shared between two atoms, resulting in the clay having a negative charge that is satisfied by the same external cations discussed for kaolinite (Section 3.1.3.1). This results in cation exchange capacity (CEC) greater than that ascribed to edge effects alone.

The Fine-Grained Micas. The fine micas are distinguished by having 2:1 structure and are nonexpanding when the water content of their surroundings changes. Isomorphous substitution is in the silica tetrahedral sheets and causes a change in the shape of the crystal. Thus, this portion of the surface has a defect in it, which is the correct size to allow either or both ammonia, as NH_4^+, or potassium, as K^+, to fit between the layers.[3] Also, the unsatisfied bonds resulting from isomorphous substitution are close to the surface and thus the charge is closer to that of the cations found on the surface, resulting in a relatively strong attraction for both cations and water. The two phenomena result in trapping of both water and cations between crystals.

---

[3] Although it seems strange, NH_4^+ and K^+ are approximately the same size.

Water and cations trapped between layers are held so strongly that they do not exchange with water and cations in the surrounding environment and thus are not biologically available. For instance, ammonium trapped between the clay layers is not a source of nitrogen for plants. Also, it will not be oxidized by bacteria to nitrite and nitrate, and thus will not be a potential source of nitrate or nitrite pollution.

Soil containing large amounts of fine-grained mica clay can result in surprising analytical results. An analytical procedure that breaks apart the fine-grained mica, such as strong-acid digestion, will show that the soil contains a large amount of ammonium or potassium that can cause concern about their effect on the environment. This can happen when the analyst is not aware that the NH_4^+ and K^+ are biologically unavailable and stable under normal environmental conditions.

The Smectite Clays. The smectite-type clays are distinctive in that they expand and cause significant destruction to synthetic (human-made) structures. In this type of 2:1 clay, isomorphous substitution occurs in the aluminum sheet. If there is substitution of lower-oxidation-state metal such as magnesium, there will be an unsatisfied pair of bonding electrons in the interior of the crystal and there will be no noticeable change in the surface. Because the charge is in the interior of the crystal, its attraction for cations is diminished by distance. Thus, smectite crystals are not held together strongly by cations and are able to incorporate more water and ions between sheets when the environment is wet and less when it is dry.

The amount of swelling depends on the cations and the number of waters of hydration (i.e., water molecules) they have. Cations with large numbers of hydration waters cause a higher degree of swelling. Lithium thus causes more swelling than does sodium, and so on. The greater the swelling, the easier it is for water and ions to move in and out between the clay sheets, making them both environmentally and biologically available.

The amount of swelling and contraction in smectites is quite dramatic. Typically, soils containing large amounts of this type of clay will develop cracks that are 30 cm wide at the surface and greater than 100 cm deep, and these cracks will allow surface material to fall into them during dry periods. This characteristic is so unique that these types of soils are given their own name. They are called *Vertisols*; the name is taken from the concept that material from the surface falls to the bottom of the cracks, resulting in inversion of the soil.

When a soil containing 2:1 clays becomes wet, the clays swell shut and water movement through it is extremely slow. In a soil profile, wetting and swelling of this type of clay will prevent downward movement of water and associated contaminants. For this reason, swelling clays are used to seal both landfills and ponds to prevent leaching or leaking.

The crystals of 2:1 swelling clays are typically smaller than either kaolinite or fine-grained mica and thus have higher adsorption capacity and cation exchange capacities. However, surface adsorption will be of the same chemistry as that in the fine-grained micas.

3.1.3.3 Amorphous Clays

In addition to the crystalline clays described earlier, there are some materials that act like clays but do not have crystalline structure. Amorphous clays do not have a definite X-ray diffraction pattern and are differentiated from the crystalline clays on this basis. They are composed of mixtures of alumina, silica, and other oxides and generally have high sorptive and cation exchange capacities. Few soils contain large amounts of amorphous clays [2].

3.1.4 Soil Texture

Sand, silt, and clay are the three components of soil texture. Various relative compositions, expressed on a percentage basis, are used to give soils a textural name such as sandy loam, or loamy clay. For soils containing significant amounts of silt, the term *loam* is used, although with high levels (>88% silt, 20% sand, and 12% clay), the soil would be called *silt*. Thus, a soil containing high amounts of sand would be called sandy loam, loamy sand, or sandy clay. Clay soils are designated as clay, clay loam, and sandy clay. Silty soils are designated as silt loam, silty clay loam, and so on. The textural name of a soil is established by determining its relative percentage of sand, silt, and clay and then finding the percentage on a textural triangle, included in all standard texts on soil [1, 3] (see also this book's Preface), to find the name. The textural name of a soil often accompanies the name, (e.g., Milton silt loam) or can be obtained from a soil scientist familiar with an area's soils.

Sandy soils are easiest to extract and analyze, while clay soils are the hardest. Drying, crushing, and sieving will aid in extraction and analysis, although it may not be necessary to crush very sandy soils. Clayey soils may retain small amounts of contamination even after extensive extraction (see Figure 1.1). In all cases, extraction and analysis procedures must be robust enough to handle all textures containing all clay types [4].

SOIL COMPONENTS INTERACTING

3.2 BONDING CONSIDERATIONS

As noted in Chapter 2, sand, silt, clay, and organic matter do not act independently of each other in soil. Thus, one or several types of chemical bonds or interactions—ionic, polar covalent, covalent, hydrogen, polar–polar interactions, and van der Waals interactions—will be important in holding soil components together. The whole area of chemical bonding is extremely complex, and thus, in addition to specific bonding considerations, there are also more general ways of investigating the interaction between components in soil. These involve, for instance, graphing the adsorption of organic compounds to soil constituents at increasing levels of organic compound. The shape of the graph is used as an indication of the type(s) of interaction(s) between constituents at various concentrations.

3.2.1   Orbital Overlap

Interaction between silicon or aluminum surfaces or edges and surrounding components will entail an overlap of surface orbitals and available orbitals in the approaching species forming covalent or polar covalent bonds. The strength of the interaction will depend on the strength of the overlap of the available orbitals. Bonding energies of orbitals are well known, and so the strength of the bonds can be either directly known or estimated. Orbital overlap is not the only factor affecting the interaction or strength of bonding. In addition to energy, reaction path, steric, and rate factors play a role in any attraction.

Each reaction species must have molecular orbitals available and with the correct symmetry to allow bonding. These will be called "frontier orbitals" composed of the highest occupied molecular orbital (HOMO) and the lowest unoccupied molecular orbital (LUMO). In addition to their involvement in bonding between species, these orbitals are of considerable interest in that they are largely responsible for many of the chemical and spectroscopic characteristics of molecules and species and are thus important in analytical procedures and spectroscopic methods of analysis [5–7].

3.2.2   Ionic Bonding

Ionic bonding is generally thought of as being the predominant type of bonding between the ions that make up salts or the compounds formed between metals and nonmetals. The basic concept is always illustrated as a compound such as sodium chloride and is explained by saying that sodium donates its outermost electron to chlorine such that both have a noble gas electron configuration. The two oppositely charged species then attract each other, forming a compound. Although the seminal characteristic of compounds held together by ionic bonds is that they dissolve in water, giving a solution containing ions, it is essential to keep in mind that many ionic compounds are insoluble. The solubility of these compounds depends on the relative strength of solvation and bonding energy.

Some ionic compounds contain a combination of bonds. For instance, in polyatomic ions such as ammonium (NH_4^+), the hydrogen atoms are bonded to the nitrogen atom by polar covalent bonds. The ionic bond is thus between this covalently bonded moiety and another oppositely charged ion such as chloride (Cl^-).

Ionic bonds are typical of inorganic compounds, and thus, the mineral or inorganic components of soil often contain ionic bonds and are soluble in water. This means two things: (1) the soil solution should always be expected to contain salts and their corresponding ions; and (2) the inorganic components of soil should be expected to dissolve in the soil solution, some at a very slow rate, resulting in their ions being present in low concentration. A third aspect of this is that analysis of inorganic or ionic compounds must take into account not only their solubility in the soil solution but also the possibility that they may be present as exchangeable ions (see also Chapter 11) [8].

3.2.3 Ligands

Ligands are species, neutral or ionic, that bond with metals. Terms such as organic acid, chelate, siderophores, and phytosiderophores are all terms that are often used in describing this general type of chemistry. The bonding occurs through either lone pairs of electrons or π bonds of the ligand and the **d** and **f** orbitals of the metal. Ligands can involve relatively simple inorganic molecules and ions such as carbon dioxide, nitrous oxide, hydroxide, or water. It is also common for the ligand to be a larger organic ion or molecule, such as citrate, acetylacetonate, and bipyridine.

Organic acids, particularly in the ionized state, can act as ligands and can have a significant effect on the solubility of metals found in soil. Acids that are good ligands have several carboxylic functional groups (COO^- groups), including citrate, fumarate, tartrate, and succinate, to name only a few (see Figure 3.5). These types of ligand are of particular interest because they are common in the root exudates of plants and thus potentially play a role in plant cationic micronutrition [9].

Organic molecules and ions not containing carboxylic functionality but containing lone pairs of electron, such as those on hydroxyl (–OH) and carbonyl functionalities (–C=O), and π bonds, such as compounds containing

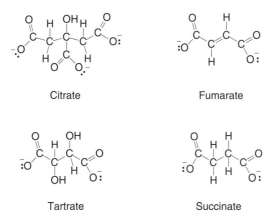

Citrate

Fumarate

Tartrate

Succinate

Figure 3.5. Typical organic acids that form ligands with metals in soil. Shown is the ionized form that is commonly found in root exudates.

benzene rings either singly or condensed, are common constituents of soil organic matter. For this reason, they can be associated with metals in soil as ligands. Analysis of soil for either metals or compounds that can act as ligands requires that the metal and its associated ligand be separated. This is accomplished by using an extractant that can replace the ligand on the metal [10].

3.2.4   Ion Exchange Interactions

The soil colloids, both inorganic (i.e., clay) and organic (i.e., humus), contain charges that are balanced by cations and anions associated with these charged sites. Most soil clays and humus contain a predominance of negative charges and thus act as cation exchangers. Some clays also have significant numbers of positive charges and thus will act as anion exchangers. Thus, soil can have both *cation exchange capacity* (CEC) and *anion exchange capacity* (AEC). See Chapter 11 for a further description and illustration of cation exchange reactions.

When considering CEC or AEC, the pH of the soil solution is extremely important. There will be competition for binding sites between H_3O^+ and other cations in the soil solution. Therefore, the observed CEC will be lower at high proton concentrations, that is, at low or acid pH levels, and higher at basic pH levels. For analytical measurements, the CEC of soil at the pH being used for extraction is the important value, not a CEC determined at a higher or lower pH.

Exchangeable cations must be removed from exchange sites to be detected and quantified. To accomplish this, the soil sample is extracted with a solution containing a cation having multiple positive charges or present at high concentration. At the same concentration, cations having more positive charges will replace those on exchange sites having less charge. This condition can be overcome if the less charged cation is at high concentration when even a cation with one positive charge can replace a cation with multiple charges.

Soils with AEC can be expected to exchange anions in the same way. However, in many soils, anions are present as oxyanions, which often react with soil components to form permanent covalent bonds and thus do not act as exchangeable anions. Phosphate anions are excellent examples of this type of interaction [11].

3.2.5   Hydrogen Bonding

Hydrogen bonding is typified by the attraction of a partially positive hydrogen atom, attached to a partially negative oxygen, which is then attracted to a partially negative oxygen on another molecule (see Figure 3.6). A common example of hydrogen bonding occurs in water, where the hydrogen of one water molecule is attracted to the oxygen in another water molecule. Whenever a hydrogen atom is bonded to a significantly more electronegative

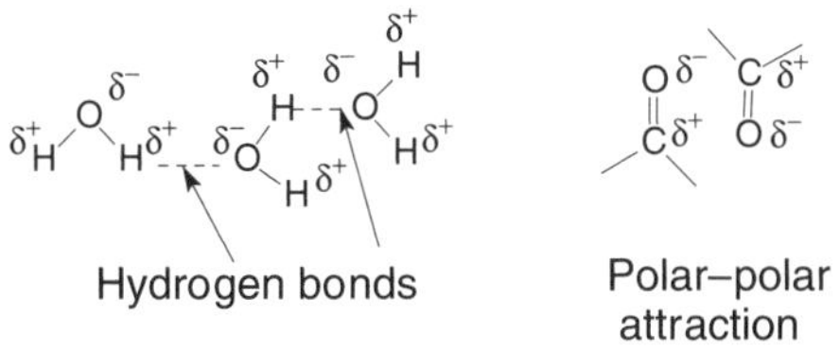

Figure 3.6. Examples of hydrogen bonding and polar–polar attraction.

element, it will have a partial positive charge. It can then be attracted to lone pairs of electrons on other elements in other molecules and thus produce an interaction force between the two molecules. Nitrogen, phosphorus, and silicon atoms would all fall into this category of atoms having lone pairs of electrons.

Although hydrogen bonding is considered to be a much weaker interaction than covalent or ionic bonding, it is nevertheless a relatively strong interaction. When molecules have multiple sites for hydrogen bonding, there can be significant strength in the association; for instance, the molecules that are in a sheet of paper are held to one another, in part, by hydrogen bonding [12, 13].

3.2.6 Polar–Polar Interactions

Whereas hydrogen bonding can be considered as a type of polar–polar interaction, I define polar–polar interactions as the intermolecular attraction of polar groups, which does not involve hydrogen (see Figure 3.6). An example would be the attraction between two propanone (acetone) molecules, where a partially positive carbon atom of the carbonyl group in one molecule is attracted to the partially negative oxygen atom of another molecule. This will be a weaker interaction than hydrogen bonding but a stronger interaction than van der Waals.

3.2.7 van der Waals Interactions

van der Waals attractions are described as the development of instantaneous polar regions in one molecule that induce the development of polar regions in another molecule and result in an attraction between the molecules. These are considered to be instantaneous or short-lived polar regions that are in a constant state of flux. The most common example is the attraction between hydrocarbon molecules where this is the only discernible attracting force between molecules.

Because different components are held to soil by different types of bonding and attractions, the interaction can be relatively strong or weak. Thus, extraction procedures must be capable of extracting the desired component when it is held by different forms of bonding to different soils.

3.2.8   Other Ways of Investigating Bonding

Bonding and other interactions between the components in soil (i.e., the clays) and organic components can be investigated by conducting adsorption experiments. An organic molecule is added to a suspension of clay and the amount adsorbed after a fixed amount of time determined. The amount adsorbed is plotted against the total amount added to the suspension to produce adsorption isotherms. The shape of the graph is then used to indicate the type of interaction between the molecule and the clay. With this type of investigation, various types of adsorption phenomena can be distinguished.

The most common way of handling such data is to try to fit the data to either a Langmuir or a Freundlich type of equation [14], or alternately to simply determine which of these two equations best describes the data obtained. Although some useful information can be obtained about the interactions between the components being studied, neither provides specific information about the type of bonding in terms of orbitals, or interactions such as those discussed in the previous sections. Spectroscopy, as discussed in Chapter 14, is typically the method used to determine bonding details [3, 14].

SOIL COMPONENTS IN COMBINATION

3.3   SURFACE FEATURES

Both sand and silt surfaces are dominated by oxygen and its lone pairs of electrons in **p** orbitals. In some instances, broken surfaces may also have silicon-hybridized sp^3 orbitals[4] available for bonding. Comparison of sand, silt, and clay reveals the surface area of sand and silt to be low and the interaction between surface bonding orbitals and components in the surrounding medium relatively weak.

As a first approximation, the surfaces of the clays can be grouped into three types: (1) surfaces consisting exclusively of oxygen atoms with their lone pairs of electrons, in **p** orbitals, extending at an angle away from the surface into the surrounding medium; (2) surfaces containing –OH groups with the partially positive hydrogen atoms extending into the surrounding medium—because of the bonding angle between oxygen and hydrogen, the partially negative oxygen and its lone pair of electrons will also be exposed to the medium; and (3) surfaces with broken edges, which can present a number of different orbitals, depending on where the break occurs.

It can be imagined that the bonds can be broken at any location, that is, with an oxygen, hydroxy, silicon, or aluminum exposed. In this case, it could further be imagined that s-, p-, and sp^3-hybridized orbitals would be on the surface. This

---

[4] This would be hybridization of one 3s and three 3p orbitals as opposed to 2s,2p hybridization in carbon.

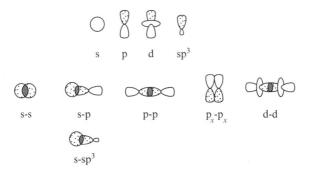

Figure 3.7. Common representations of the s, p, d and atomic orbitals, sp³ hybridized orbitals, and some representations of how they overlap to form bonds between atoms.

will lead to complex bonding and reactivity, resulting in bonds of varying strengths and interactions and of varying types. The issue is then how these surfaces interact with components commonly found in the soil solution.

Molecular orbital depictions of the orbitals described previously are given in Figure 3.7. These types of diagrams can be confusing. The orbitals and their shapes are calculated and the calculations result in the orbitals having plus (+) and minus (–) signs, and will often be represented in this fashion. It is common to think of negative and positive signs as representing negative and positive charges with negative and positive attracting and two negatives or two positives repelling each other. However, the signs (+, –) in this case do not represent charges. In molecular orbital diagrams, the overlap of two atomic orbitals having the same sign denotes a positive interaction leading to bonding, that is, electrons holding the two atoms together. This can also be indicated by the two orbitals having the same shading as in Figure 3.7.

When the two orbitals have different signs, they do not overlap (nor do they cancel each other) but result in the formation of antibonding orbitals. In this case, the electrons are not shared between two atoms and do not hold the atoms together. For each bonding molecular orbital, there is an antibonding orbital. Antibonding orbitals have higher energy than bonding orbitals.

The overlap of p and d orbitals seen in Figure 3.7 can be of two types. The overlap can be *end on end*, as depicted in the p–p and d–d representation, or *side by side*, as depicted in the p_x–p_x representation. This is possible because the p and d orbitals are directed in space along the x-, y-, and z-axes. End-on-end overlaps are generally stronger and the bonds produced are referred to as sigma (σ) bonds. Side-by-side overlap results in pi (π) bonds.

In some atoms, the p and s orbitals are mixed together to form several equivalent, hybridized orbitals. The most common example is carbon, where there are four orbitals that are formed by mixing one s orbital with three p orbitals to give four equivalent orbitals designated as sp³ orbitals.

Silicon as SiO_2, or its polymeric form $[SiO_4]_n$, has only two types of orbitals available for bonding. In Figure 3.8, sp³ bonding occurs in SiO_4 and $[SiO_4]_n$; however, they are tetrahedral, not planar, as shown. In SiO_2, the Si and O

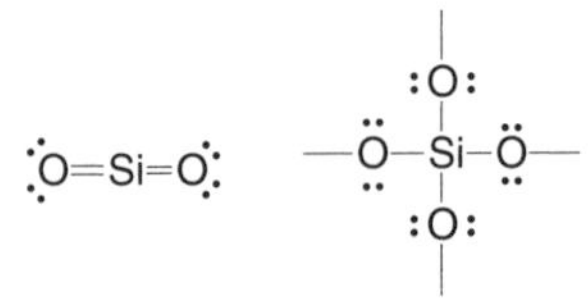

Figure 3.8. Silicon oxide (left) and polymeric silicon (right).

atoms have sp^2-hybridized orbitals[5] forming sigma bonds and p orbitals forming π bonds. The oxygen also has p orbitals containing lone pairs of electrons. Polymeric silicon has sp^3-hybridized orbitals available for bonding, while oxygen still has lone pairs of electrons in p orbitals. These, then, are the orbitals available for bonding to elements or compounds that come into contact with the particle's surface. The type of interaction could be either end on end or π type (side by side). In all cases, steric hindrances may limit the type of interaction occurring.

Clays contain aluminum oxides in addition to silicon as SiO_2 and its polymeric forms. Again, there are the **p** orbitals of oxygen and sp-hybridized orbitals from aluminum, which may result in end-on-end or side-by-side bonding with the same restrictions encountered with silicon.

For removal of compounds bonded to silica or alumina, π and σ bonds must be broken and new bonds formed. This requires energy and the correct orientation of attacking groups along with an effective attacking species. In addition, the rate of species removal will depend on the reaction path and the steric factors involved. All these put together will determine the overall rate of the reaction and the time needed for extraction.

3.4 ENERGY CONSIDERATIONS

The primary determinant in terms of the stability or strength of bonding interaction is energy. A general equation is $\Delta H = H_e - H_i$. This equation simply states that the change in enthalpy (designated as H) is the difference between the energy at the end of the reaction, H_e, and at the beginning, H_i. If a reaction is endothermic, it will be at a higher energy level at the end than at the beginning. The reverse is true for an exothermic reaction. It is expected that exothermic reactions lead to more stable products and are thus more likely to be formed.

In addition to enthalpy, there is also the consideration of entropy, or "randomness." Reactions generally tend to go to a more random state. It would seem that any reaction involving the formation of a highly ordered crystal would be going to a state of decreased randomness. Thus, there would be a

[5] sp^2-hybridized orbitals are formed by mixing one s orbital with two p orbitals to produce sp^2 orbitals.

decrease in entropy, which would not favor the reaction. However, if the overall entropy of the system increases, then the reaction is favored.

The total energy of a system involves both enthalpy and entropy. Thus, whichever causes the greater change in overall energy during the reaction will be the one controlling the reaction and determining whether it is exothermic, endothermic, spontaneous, or not spontaneous [11].

3.5 REACTION PATHS

Reaction paths involve following the energy path during a reaction. Typically, reactants have a starting energy (R + R in Figure 3.9) that increases through either the same transition state (TS) or reaction intermediate (RI) or both, leading to a final product energy (P). In Figure 3.9, the TSs are hypothetical species or structures that are in transition and cannot be isolated and identified. As indicated by the slight energy trough, the reactive intermediates are stable enough to be isolated and studied. Often, reactive intermediates are ions such as carbanions.

In Figure 3.9, the path to the right involves lower energy TSs and reactive intermediates, and so the reaction will initially follow this path, although the product (P_r) is at higher energy than the one obtained by following the left path (P_l). The formation of product P_r will be controlled by kinetics, will be produced first, and will initially be at the highest concentration. After some more time, reactions will take place and the product on the right, P_r, will reeact to form the more thermodynamically stable product on the left, P_l. This means that upon standing, the product on the right will decrease to a low concentration, while the one on the left will dominate the product mix.

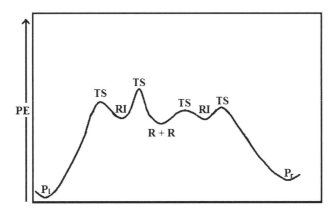

Figure 3.9. A potential energy diagram for a reaction that can occur in two different ways, producing two different products (P), one kinetically and the other thermodynamically controlled. R, reactant; TS, transition state; RI, reactive intermediate; and P (P_l and P_r), product.

It is common to find that a chemical added to soil becomes increasingly difficult to remove with time. The mechanism described earlier is one that would account for this observation. A compound added to soil may initially crystallize rapidly into a structure that is easily dissolved and extracted from soil. With time, its crystalline structure rearranges to a less soluble, lower energy, and thus less easily extractable form [7].

3.6 STERIC FACTORS

Large, bulky groups will slow or prevent nearby reactive sites from being attacked. They may also cause a reactive species to assume a shape that is not accessible to attack. The two common bulky groups, shown in Figure 3.10, are used extensively in organic synthesis.

The position of such groups can affect the reactivity, access to reactive sites, and the conformation of the species to which they are attached. Thus, in considering the extractability of a species, the occurrence of steric hindrances must be considered [7].

3.7 RATE FACTORS

The rate of a reaction can also be an important factor in both the amount and longevity of a component in soil. A very common and interesting example is the microbial oxidation of ammonia first to nitrite and then to nitrate.

$$NH_4 \xrightarrow{\textit{Nitrosomonas} \text{ spp.}} NO_2^- \xrightarrow{\textit{Nitrobacter} \text{ spp.}} NO_3^- \qquad (3.1a)$$

$$NH_4^+ + 1.5\,O_2 \xrightarrow{\textit{Nitrosomonas} \text{ spp.}} NO_2^- + H_2O + 2\,H^+ \quad -275\,kJ \quad (3.1b)$$

$$NO_2^- + 0.5\,O_2 \xrightarrow{\textit{Nitrobacter} \text{ spp.}} NO_3^- \qquad\qquad -74\,kJ \quad (3.1c)$$

Energy is given as ΔG° for each reaction. The microbial oxidation of ammonia to nitrite in soil is found to be slower than the oxidation of nitrite

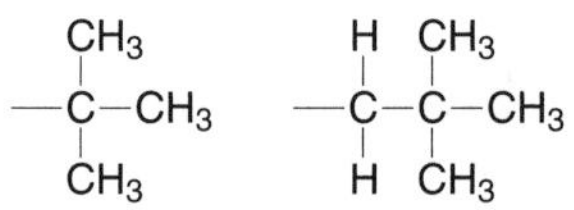

Figure 3.10. Common bulky groups involved in steric hindrance.

to nitrate. When the energy available from each of these reactions is considered, as shown in reactions (3.1b) and (3.1c), it is obvious that this observation is directly related to the amount of energy obtained [15].

In this case, nitrite is not expected to occur or build up to appreciable levels in the environment. *Nitrobacter* species, which obtain energy from the reaction shown in equation (3.1a) must react with approximately 3.7 times as many NO_2^- ions in order to obtain the same amount of energy as *Nitrosomonas* species reacting with NH_4^+. Thus, it can be expected to take up nitrite at a higher rate to compete. This type of energy calculation is simply done by looking at the amount or energy required for bond breaking and the amount released in bond making. Alternately, the energy can be measured directly by calorimetry.

3.8 ALL FACTORED TOGETHER

In a purely chemical approach to how reactions take place, all of these factors come together; specifically, the rate is related to the total energy used and released, the energy of activation required, steric effects, and the types of bonds being broken and formed. For this reason, it is most common to measure the energy and rate quantities directly for the conditions of the reaction. Because of the complexity of soil, it is even more important to measure these directly.

3.9 MICELLES

In Chapter 2, it is observed that sand, silt, and clay do not act independently from each other. In a similar fashion, clay particles do not act independently; rather, they form groups of particles called micelles. The model for a micelle is a group of long-chain fatty acid salts in water. The hydrophobic ends associate with the ionic "salt" end exposed to water. An idealized micelle with some associated water is shown in Figure 3.11. The structure is often described as being ball-like. This ideal is hard to visualize when looking at the typical shape and size of a clay particle. However, it is possible to envision individual clay crystals associated with each other through hydrogen bonding and ion–charge interactions to form small conglomerates of clay particles that act like micelles as shown in Figure 3.11 [16].

3.10 COATED SURFACES

The discussion to this point assumes that all the surfaces of all the components in soil are clean and available for reaction. This never happens. All surfaces

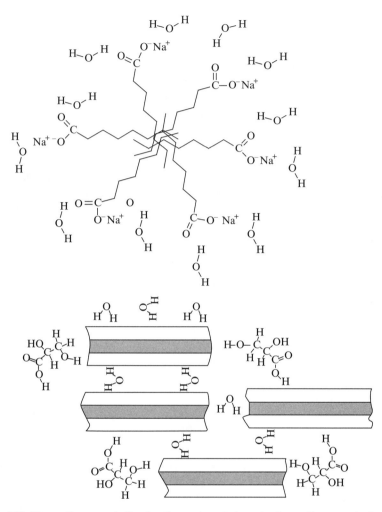

Figure 3.11. Upper diagram: micelle of sodium octanoate in water; lower diagram: micelle of 2:1 clay (the center gray layer is an aluminum octahedral sheet; the upper and lower are silicon tetrahedral sheets).

are covered by coatings of various types; the number, types, and amounts of possible coatings are quite varied. However, four of the most important coatings are water, iron oxides, clays, and organic compounds. All soil surfaces will be "contaminated" with a combination of these compounds plus smaller amounts of others, such as manganese oxides and elemental carbon, if present. Thus, the orbitals, bonding, energy, and other characteristics of these surface coatings will also come into play when considering their reactivity and extraction of components from them [4, 17, 18, 19].

3.11   CONCLUSIONS

Soil inorganic solids are composed of particles of decreasing size from sand to clay. The clay fraction is further divided into principally 1:1 and 1:2 clays. The 1:1 clays are typified by kaolinite, which, compared with other clays, exhibits lower activity. The 2:1 clays are typified by fine-grained micas that are not expanding and smectites that are expanding. Clays have high sorptive capacity and are one source of cation exchange in soils. Bonding of soil components to each other and to surfaces involves all the standard types of bonding, namely, ionic, polar covalent, covalent, hydrogen, polar–polar, and van der Waals. Reactions thus involve s, p, d, and sp orbital overlaps and ionic and partial charges such as those involved in hydrogen bonding. These bonding consider-ations also involve the common surface features such as surface oxygen atoms and hydroxy groups along with exposed aluminum and silicon. Extraction of components from soil therefore involves all types of bonding along with energy, reaction path, steric, and rate factors. Also involved will be micelle formation along with coatings on all soil surfaces.

PROBLEMS

3.1. Indentify the three primary particles in soil and describe the chemical differences between them.

3.2. Identify the three major types of clays in soils and explain how they differ chemically.

3.3. Describe the different types of bonding and their primary occurrences in soil.

3.4. Describe the surface features, with particular reference to orbital avail-ability, involved in surface binding of components in soil.

3.5. In terms of bonding energy, the bonds formed in an exothermic reaction must be lower than those of the reactants. How is this known?

3.6. Explain how both reaction mechanisms and energy considerations con-tribute to the abundance of certain species of ions and compounds found in soils.

3.7. Explain how the rate of a particular reaction may affect the amount of a particular component in soil.

3.8. Diagram the expected structure of an organic and an inorganic (clay) micelle in soil.

3.9. In general terms, describe the condition of the surfaces of all soil components.

3.10. Describe the expected differences between sand and silt surfaces and clay surfaces.

REFERENCES

1. Brady NC, Weil RR. *Elements of the Nature and Properties of Soils*, 13th ed. Upper Saddle River, NJ: Pearson Prentice–Hall; 2002, p. 125.

2. Van Olphen H. *An Introduction to Clay Colloid Chemistry; For Clay Technologist, Geologist and Soil Scientist*. Malabar, FL: Krieger; 1991.

3. Sparks DL. *Environmental Chemistry of Soils*, 2nd ed. New York: Academic Press; 2003, p. 151.

4. Gardiner DT, Miller RW. *Soils in Our Environment*, 10th ed. Upper Saddle River, NJ: Pearson Prentice–Hall; 2004, pp. 27, 105.

5. Sanderson RT. *Polar Covalence*. New York: Academic Press; 1983.

6. Woodward RB, Hofmann R. *The Conservation of Orbital Symmetry*. Weinheim: Verlag Chemie; 1970.

7. Atkins P, de Paula L. *Physical Chemistry*, 7th ed. New York: Freeman; 2001, pp. 432, 960.

8. Bowser JR. *Inorganic Chemistry*. Pacific Grove, CA: Brooks/Cole; 1993.

9. Lipton DS, Blanchar RW, Blevins DG. Citrate, malate, and succinate concentration in exudates from P-deficient and P-stressed *Medicago sativa* L. seedlings. *Plant Physiol.* 1987; **85**: 315–317.

10. Lombnæs P, Chang AC, Singh BR. Organic ligand, competing cation, and pH effects on dissolution of zinc in soils. *Pedosphere* 2008; **18**: 92–101.

11. Bohn HL, McNeal BL, O'Connor GA. *Soil Chemistry*, 3rd ed. New York: Wiley; 2001, pp. 87, 206.

12. Hamilton WC, Ibers JA. *Hydrogen Bonding in Solids: Methods of Molecular Structure Determination*. New York: W.A. Benjamin; 1968.

13. Scheiner S. *Hydrogen Bonding: A Theoretical Perspective*. New York: Oxford University Press; 1997.

14. Harter RD, Smith G. Langmuir equation and alternate methods of studying "adsorption" reactions in soil. In Stelly M (ed.), *Chemistry and the Soil Environment*. Madison, WI: American Society of Agronomy, Soil Science Society of America; 1981, pp. 167–182.

15. Texier AC, Zepeda A, Gómez J, Cuervo-López F. Simultaneous elimination of carbon and nitrogen compounds of petrochemical effluents by nitrification and denitrification. Chapter 6. In Patel V (ed.), *Petrochemicals*. Rijeka, Croatia: Intech; 2012, pp. 101–134. Available at: http://www.intechopen.com. Accessed June 3, 2013.

16. Heil D, Sposito G. Organic-matter role in illitic soil colloids flocculation. II. Surface charge. *J. Soil Sci. Soc. Am.* 1993; **57**: 1246–1253.

17. Bisdom EBA, Dekker LW, Schoute JFT. Water repellency of sieve fractions from sandy soils with organic material and soil structure. *Geoderma* 1993; **56**: 105–118.

18. Courchesne F, Turmel MC, Beauchemin P. Magnesium and potassium release by weathering Spodosols: grain surface coating effects. *J. Soil Sci. Soc. Am.* 1996; **60**: 1188–1196.

19. Jongmans AG, Verburg P, Nieuwenhuyse A, van Oort F. Allophane, imogolite and gibbsite in coatings in a Costa Rican Andisol. *Geoderma* 1995; **64**: 327–342.

BIBLIOGRAPHY

Bowser JR. *Inorganic Chemistry*. Pacific Grove, CA: Brooks/Cole; 1993.

Companion AL. *Chemical Bonding*. New York: McGraw-Hill; 1979.

Earnshaw A, Greenwood NN. *Chemistry of the Elements*, 2nd ed. Portsmouth, NH: Butterworth-Heinemann; 1997.

Hamilton WC, Ibers JA. *Hydrogen Bonding in Solids; Methods of Molecular Structure Determination*. New York: W.A. Benjamin; 1968.

Pauling L. *The Nature of the Chemical Bond and Structure of Molecules and Crystals: An Introduction to Modern Structural Chemistry*, 3rd ed. Ithaca, NY: Cornell University Press; 1960.

Sanderson RT. *Polar Covalence*. New York: Academic Press; 1983.

Scheiner S. *Hydrogen Bonding: A Theoretical Perspective*. New York: Oxford University Press; 1997.

Woodward RB, Hoffmann R. *The Conservation of Orbital Symmetry*. Weinheim/ Bergstr: Verlag Chemie; 1971.

CHAPTER

4

SOIL BASICS PART III
THE BIOLOGICAL AND ORGANIC COMPONENTS IN SOIL

It might seem that soil is basically a dead, inert, nonliving material, but a more useful approach is to view soil as being alive in some way. It has living organisms like plants in and in intimate contact with it. Plants have two parts that interact with soil: the tops and the roots. Both are important, although roots have a more intimate and immediate effect on the biological, physical, and chemical properties of soil. Many other types of organisms (i.e., microorganisms and animals), both large and small, live in soil. Microorganisms come in a wide variety of sizes, shapes, and abilities in terms of the types of environments they live in and the reactions they carry out. It is common knowledge that worms and prairie dogs live in soil, but these serve only as representatives of the variety of animals present.

Animals and plants, especially roots and microorganisms, provide biochemical, bioorganic, and organic compounds to soil. These may be in the form of cellular components, such as cell walls, membranes, enzymes, and complex and simple organic compounds. Decomposition of complex cellular material and

Introduction to Soil Chemistry: Analysis and Instrumentation, Second Edition.
Alfred R. Conklin, Jr.
© 2014 John Wiley & Sons, Inc. Published 2014 by John Wiley & Sons, Inc.

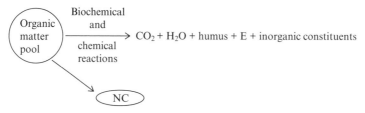

Figure 4.1. Organic matter (OM) breakdown in soil under aerobic conditions. These reactions lead to the formation of humus and are carried out to release energy (E), which is used by microorganisms. Heterotrophic microorganisms use OM to construct new cells (NCs). They also lead to a greater randomness in the system.

biochemicals leads to the formation of simpler intermediate bioorganic and organic compounds. Aerobic decomposition of any organic matter in soil eventually leads to the production of carbon dioxide and water; anaerobic decomposition leads to the same products plus methane; and both result in the synthesis of humus, which is a dark-colored, high-molecular-weight, and highly complex material. The overall reaction of organic matter in soil is illustrated in Figure 4.1.

Organic compounds interact with each other and with the inorganic, macro, micro, and colloidal components in soil. Complex and simple organic molecules form complexes with inorganic cations and anions and with colloidal carbon and clay in the soil solution. Some enzyme–soil complexes may have increased catalytic activity, while others have decreased activity or are completely inactive. Also, some combinations and complexes may render organic matter resistant to decomposition. Eventually, organic matter and complexes may become stabilized such that they have long residence times in soil.

All of these factors are active in all soil at all times and may affect any analysis. Simpler organic and inorganic molecules, compounds, and ions may be associated with more complex organic materials, with humus, or with inorganic carbon and clay as part of the soil matrix. These interactions can mask the occurrence and concentration of both organic and inorganic components, obscuring analysis and confusing analytical results.

BIOTA OF SOIL

4.1 ANIMALS

Animals living in soil range from as large as groundhogs to as small as the smallest insect and arthropod. However, for our purposes, when animals become so small that a microscope is needed to see them, they are classified as microorganisms. It is assumed that the main additions to soil from animals are urine and feces. These are indeed common additions; however, all animals add hair, skin, saliva, and eventually the dead and decaying bodies of the organisms themselves. In addition to biochemicals and bioorganic and organic molecules, animals cause both large and small changes in the physical characteristics of soil that can change its chemistry and the results of chemical and instrumental analysis.

The effects of animals on the physical and chemical characteristics of soil tend to be locally distributed in the sense that the holes are dug and urine and feces and other waste products are deposited in localized areas. Animal holes can lead to the movement of surface soil lower in the soil profile such that **A** horizon material can be found in what would be expected to be a **B** horizon. This occurs most often with the larger burrowing animals. With smaller animals, material from lower levels in a soil may be brought to the surface such as in the case of ants and termites. An ant colony and a grub found about 9 cm deep in soil are shown in Figure 4.2. In other cases, such as with worms, soil may be intimately mixed with organic matter but not moved long distances. There are many different types of worms in soil, such as nematodes, but only a few cause this important type of change.

The movement of animals through soil and their deposition of organic matter can dramatically affect the soil's structure. As explained in Chapter 2, pushing together soil separates results in the formation of peds, increasing air and water movement through soil and changing the oxidation–reduction

Figure 4.2. An ant colony: A is soil brought to surface by ants; B is ant holes in the subsurface; C is a grub found at about 10 cm.

conditions and the inorganic and organic species present. Animals, such as worms, which ingest solid soil as they move through it, cause the degradation and rounding of these particles and thus have a direct effect on inorganic solid components.

Deposition of organic matter will affect areas under and around the deposition. The primary effects are increased microbial activity, increased carbon dioxide, and decreased oxygen content as well as dissolution of molecules. There are also secondary effects caused by the interaction of these molecules with both the inorganic and organic components present. Organic molecules may "dissolve" in the existing soil organic matter. They may also form complexes with inorganic constituents, such as chelate metals present on soil particle surfaces and in the soil solution. Thus, soil samples taken from an area of high animal activity will constitute a matrix that is significantly different from soil taken from an area of low animal activity.

Animals also deposit organic matter on soil even when they do not live in it. Although these deposits are widespread in most cases, they can be concentrated in watering and feeding areas. In all cases, soil near and under an organic matter deposition will be affected biologically, chemically, and physically. Before decomposition begins, rain can move both inorganic and organic constituents into and sometimes through the soil profile. During the decomposition process, additional organic and inorganic compounds and ions will be produced and leached into the soil.

In addition to the physical effects previously described, animals can change the soil's biological and chemical reactions. For instance, animal paths become devoid of plants and compacted, thereby changing water infiltration and percolation and oxidation–reduction reactions, particularly when there is continual use of the paths [1–4].

4.2 PLANTS

Plants have two parts: the tops and the roots. Both have different effects on soil chemistry and analysis. Because the effects are so different, each part will be discussed separately. All plants can be divided into algae, fungi, mosses, liverworts, and vascular plants, while the dominant agriculture plants are commonly divided into grasses and legumes. In addition, these types of plants can be annual, biennial, or perennial in their life cycles. Annual plants are particularly interesting in that both the tops and bottoms die each year and thus add organic matter to soil from both sources.

4.2.1 Tops

Shrubs and trees have woody stems or trunks and moderate to tall growth habit and, along with tall growing woody grasses, such as coconut, are long-lived and typically only add leaves to the soil each year. Evergreens and needle-bearing trees keep their needles all year long; however, needles are continuously lost throughout the year, as are leaves from tropical plants. Organic matter from woody annuals and biennials is added in a similar fashion. Thus, organic matter from roots, stems, and branches is only occasionally added to soil after relatively long periods of time. Addition of organic matter from these types of plants does not generally lead to the development of a thick O or A horizon.

Although it often seems like leaves, particularly those of deciduous trees, do not decompose, it is observed that the layer of leaves on the ground never becomes thick. In the tropics where trees grow all year long, the same thing happens; leaves fall continuously during the year and decompose. Thus, the leaves of all plants decompose over a year's period of time, adding organic matter to the soil.

Grasses and other similar plants, which may be annual, biennial, or perennial in their growth habit, do not have woody components, but also add leaves and stems to the soil each year. These leaves and stems decompose over a 1-year period, adding organic matter to the soil surface. Often these leaves seem to decompose faster than tree leaves; however, in all cases, the rate of decomposition will largely depend on both the characteristics of the plant material and local environmental conditions.

All components in organic matter affect its decomposition, however, the carbon:nitrogen (C/N) ratio is particularly important. Soil organic matter has a C/N ratio in the range of 10:1–12:1. When organic matter with high C/N ratios (e.g., 100:1) is added to soil, microorganisms decomposing it take nitrogen from the soil solution. Analysis of this soil will result in very low values for inorganic nitrogen. As this organic matter is slowly decomposed, nitrogen will be released. Organic matter with low C/N ratios will rapidly release nitrogen to the soil solution. Thus, organic matter will have a dramatic effect on the results of soil analysis. Because of these effects, actively decomposing organic matter will result in changing analytical results over time.

It might be assumed that there will be different organic matter in soil if there are different plants growing on it. This is true when the fresh organic matter and its breakdown products are being investigated. It is particularly evident with Spodisols and Mollisols. Spodisols have a subsurface spodic horizon that results from decomposing acid detritus. This leads to leaching of aluminum and highly decomposed organic matter and, often but not necessarily, iron oxides, to form this horizon. In Mollisols, the deposition of both grass tops and roots each year leads to the development of a thick, dark surface that is referred to as a mollic horizon.

Despite these dramatic effects on soil, the organic matter remaining after the breakdown of plant residues (i.e., humus) is generally similar the world over. The interaction of humus with chemicals (e.g., adsorption, cation exchange), including those used in analytical procedures, is generally the same. Thus, often the type of organic matter being added to soil, except as noted earlier, is not as important as the amount of decomposed organic matter already present. However, the components present in humus, specifically, humic and fulvic acids, vary considerably and can change some of its characteristics. Soil humus will be discussed further in more detail [5].

4.2.2 Roots

It is reasonable to assume the effect of plant roots and tops on the soil would be the same. However, this is not the case. Plant roots profoundly affect the chemical characteristics of soil. Because they are in intimate contact with the soil, roots are constantly extracting nutrients and water from the soil and exuding materials into it. The intimate relationship, which includes physical, microbiological, biochemical, bioorganic, and chemical interactions between roots and soil, is illustrated in Figure 4.3.

Any one of these interactions can be used to distinguish an area around the roots, called the *rhizosphere*, from bulk soil. In addition to these general characteristics, the rhizosphere is the area around plant roots where there is high

Figure 4.3. Plant roots with adhering soil illustrating the interaction between plant roots and soil.

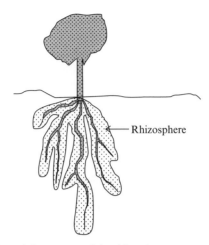

Figure 4.4. Illustration of the concept of the rhizosphere as an area around plant roots.

microbial activity, increased carbon dioxide, decreased oxygen, decreased water and nutrient content, and decreased pH (see Figure 4.4). These conditions develop because of the metabolic activity of roots, exudates, and cells sloughed off by the roots. Root exudates are specific for specific plants and contain a wide range of organic and inorganic compounds. Both cells and exudates provide "food" for increased microbial activity. The rhizosphere is also an area where contaminants that are mobile in soil but that are not taken up or slowly taken up by plants and roots will accumulate.

There are many types of roots, including thick fibrous, deep tap, shallow, and tubers, all in one plant community. Some roots explore the soil to significant depth (i.e., as much as 250 cm deep), while others are shallow (i.e., only 25 cm deep). Different rooting depths are found in all plant types: grasses, legumes, shrubs, and trees. Each root type will contribute its own unique exudates and characteristics to its unique volume of soil and the associated soil solution.

Plant roots *respire*, taking in oxygen and giving off carbon dioxide. This is a simple but essential process, and most land plants die if their (root) oxygen source is interrupted for even a short period of time. There are, however, some plants, including crops, that grow with their roots submerged in water, that is, under anaerobic and reducing conditions. Cyprus is an example of such a tree, while rice is an example of a crop. These plants have developed a vascular system that conducts oxygen to the roots, allowing them to function. However, it is important to note that the environment of these roots is very different from that encountered by roots in unsaturated soil. For instance, anaerobic conditions result in increased solubility of iron and other metals along with reduced forms of carbon and sulfur.

Reduced forms are generally more soluble and thus are more easily extracted than are oxidized forms. High levels of reduced forms, including some plant nutrients, for instance, iron, may result in levels toxic to plants or other soil organisms. Thus, not only the species but also the extractability and, thus, the apparent level of a soil constituent in the rooting zone will be affected by whether the soil is under oxidizing or reducing conditions at the time of sampling and analysis. Storing soil samples from an aerobic soil under reducing conditions will drastically alter analytical results, as will storing a soil sample from an anaerobic soil under aerobic conditions [6–10].

One well-studied (aerobic) root type is that of legumes, which are infected by a microorganism called *rhizobia*, resulting in the formation of nodules on the roots. A symbiotic relationship exists between the rhizobia and the legume wherein rhizobia supply fixed nitrogen to plants and plants provide carbohydrates, an energy source, to the rhizobia. Fixed nitrogen is any nitrogen atom bonded to another atom other than nitrogen (N_2) as in NH_3, NO_3^-, and $(NH_4)_2CO_2$. The nitrogen compound produced by rhizobia is ammonia, which is used by plants to produce amino acids.

It is to be expected that soil with a sod cover, that is, with thick grass and roots, will have different characteristics from those of a soil that has few or no plants growing. These differences will be important in analyzing soil for specific components of concern.

4.3 MICROORGANISMS

Microorganisms are the most diverse group of organisms growing in soil. Here we define microorganisms as any organism that is only visible under a microscope. Algae, fungi, actinomycetes, bacteria, and even some worms, arthropods, ciliates, and other organisms are included in this group. This represents a truly diverse group of organisms capable of carrying out an immense diversity of physical, biological, and chemical changes in their environment.[1] Aerobic, anaerobic, heterotrophic, autotrophic, thermophilic, mesophilic, and cycrophilic are only some of the different types of microorganisms found in soil. Table 4.1 gives the characteristics of these different types of organisms. An indication of their power and importance is seen when it is noted that some of these organisms can create a whole new cell out of only carbon dioxide, light, and a mixture of inorganic compounds, including nitrogen and ions!

Figure 4.5 is a drawing of soil microorganisms; Figure 4.6 is a photomicrograph of common soil microorganisms, which are differentiated by their shape, size, and chemical activity. While the actinomycetes and fungi are both filamentous in growth pattern, actinomycetes are smaller and less branched, while fungi have larger, more highly branched mycelia. Cocci and bacteria can be found in virtually all environments, including boiling water and high acidity (pH 1), cold (0°C), basic (pH 12), and high salt concentrations. Fungi, being

---

[1] The three references to Web sites given in the Bibliography provide additional information and links to other relevant sites concerning microbial activity in soil.

TABLE 4.1. Designations of Different Types of Organisms Common in Soil

Organism	Characteristic
Aerobic	Lives in the presence of oxygen and carries out oxidation reactions using oxygen as an electron acceptor
Anaerobic	Lives in the absence of oxygen and uses electron acceptors, other than oxygen (i.e., nitrogen and carbon)
Heterotrophic	Requires preformed organic compounds for energy and cell production
Autotrophic	Obtains energy from oxidation of inorganic ions and obtains carbon from carbon dioxide
Thermophilic	Grows at elevated temperatures, may be any of the types of organisms above (>40 and <100°C)[a]
Mesophilic	Grows at moderate temperatures, may be any of the types of organisms above (>5 and <40°C)
Cryophilic	Grows at low temperatures, may be any of the types of organisms above (<5°C)[a]

[a]Organisms have been reported as growing beyond these temperature ranges.

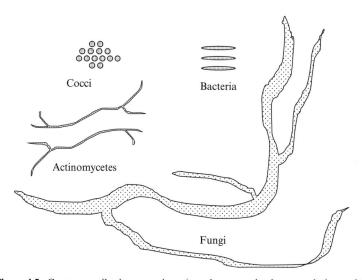

Figure 4.5. Common soil microorganisms (not drawn to absolute or relative scale).

aerobic, are somewhat more restricted in the environments they inhabit; however, some species can live in high-osmotic-potential environments, such as sugar syrups, as well as in or on hydrocarbon mixes, such as diesel fuels.

Although all microorganisms in soil are important, most attention is focused on bacteria (often, little or no distinction is made between bacteria and cocci, both being lumped together and simply referred to as "bacteria"). This is because they are extremely numerous and versatile in the reactions they carry out. In well-drained soils, they inhabit aerobic (oxidizing), anaerobic

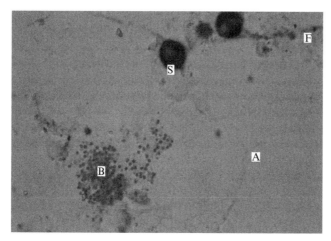

Figure 4.6. A view of soil microorganisms obtained after removing and staining a slide buried in soil for several days. Actinomycetes (A), bacteria (B), spore (S), and fungi (F) can be seen clearly.

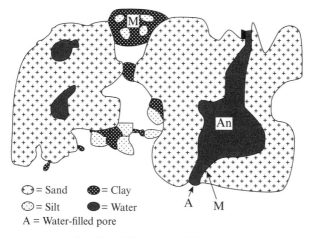

Figure 4.7. A representation of a soil ped showing aerobic areas around peds and at pore mouths (A) and anaerobic areas (An). The pore will be anaerobic in the middle, as indicated, and will not drain because of the small diameter of the pore mouths. The point labeled M will be microaerophilic.

(reducing), and microaerophilic zones in soil peds. Microaerophilic zones (**M**) are those with low concentrations of oxygen. In Figure 4.7, the area marked as **An** is anaerobic. The area just inside the mouths of the pore but before the inner water areas is microaerophilic, while all other areas will be either aerobic or anaerobic. Because of these environmentally different areas, a well-drained soil will have both oxidized and reduced forms of all components present at the same time. For example, carbon will be present in the oxidized form as carbon dioxide (CO_2) and in the reduced form as methane (CH_4). The common oxidized and reduced forms of sulfur are sulfate (SO_4^{2-}) and hydrogen sulfide (H_2S), respectively, and the oxidized and reduced form of iron are ferric (Fe^{3+}) and ferrous (Fe^{2+}), respectively.

In continually submerged soils, there is no oxygen. Thus, the entire environment is anaerobic and reducing. Under these conditions, there will be a predominance of the reduced forms mentioned earlier, namely, methane, hydrogen sulfide, ferrous iron, and so on.

Soil microorganisms play an extremely important role in the cycling of environmental elements such as carbon, nitrogen, and sulfur. The cycling of these elements and others is often represented as their respective cycle (carbon cycle, nitrogen cycle, sulfur cycle, etc.)[11]. Of these, the two most important are the carbon and nitrogen cycles. Organisms chiefly responsible for the carbon cycle are animals, plants, and microorganisms. They change carbon from carbon dioxide to plant and animal tissue and eventually back into carbon dioxide. A critical step in this process is the decomposition of organic matter by microorganisms. During this process, other elements important to life are either taken up or released to be used again. One of these other elements, nitrogen, is also affected by microorganisms. The nitrogen cycle is illustrated in Section 6.2.2.1.

There are many other cycles such as the phosphorous, potassium, halogen, and sulfur cycles. The latter is illustrated in Figure 4.8. All the transformations

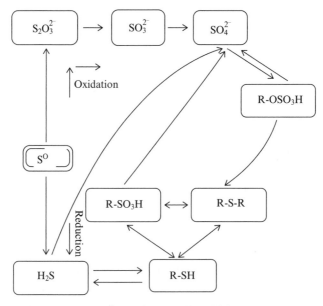

Figure 4.8. The sulfur cycle where S^0 is elemental sulfur, H_2S is hydrogen sulfide, $S_2O_3^{2-}$ is thiosulfate, SO_3^{2-} is sulfite, SO_4^{2-} is sulfate, $R\text{-}OSO_3H$ represents a sulfate ester, $R\text{-}SO_3H$ a sulfonic acid, $R\text{-}S\text{-}R$ a thioether, and $R\text{-}SH$ a thiol. (Adapted from Coyne MS. *Soil Microbiology: An Experimental Approach*. Boston: Delmar Publishers; 1999.)

illustrated are carried out by soil microorganisms. It is interesting that sulfur is converted from its elemental form to either fully oxidized or fully reduced forms by various microorganisms in soil. The starting point in the cycle may be either a reduction or an oxidation depending on the environment where the reactions are occurring. It should also be noted that there are a broad range of organic sulfur compounds (including linear, branched, cyclic, and aromatic) that, although not shown, can also occur as part of the sulfur cycle [11].

When conducting soil analyses, there are two significant reasons to recognize that microorganisms are present in the soil: (1) between taking soil samples and their analysis, these organisms can cause extensive changes in the chemistry and chemical composition of the soil sample, completely changing the species and amount of components found; and (2) almost all extraction procedures will cause the destruction of any cells (animal, plant, and microbial) found in soil with the consequential release of cellular constituents into the soil solution. Some constituents will be degraded by the extraction process; some enzymes may continue to function; and some cellular components will complex or form chelates with metallic components in the soil. These reactions will lead to a more complex mixture of components than one might expect. All of these eventualities can affect or change the analytical or instrumental results obtained [11–14].

BIOLOGICAL AND ORGANIC CHEMICALS OF SOIL

4.4   BIOCHEMICAL

There are two sources of biochemicals in soil; one is the cellular constituents released when cells are destroyed during the extraction process. It should be kept in mind that any handling of a soil sample will cause the destruction of some of the cells it contains. The simple act of sieving, air drying, and weighing soil will cause some lyses of cells and release of their contents. Extraction will typically cause complete destruction of all cells in soil, with the release of all their constituent parts. Some parts such as enzymes may continue to function after release from the cell and continue to change the makeup of soil components for some time.

The second source of biochemicals is molecules excreted from cells such as extracellular enzymes and other organic matter. A typical example is cellulase, which is excreted by fungi such as *Penicillium* in order to break down wood and woody material into sugars that can be used by the organisms. Other common extracellular enzymes found in soil are ureases and amylases. Often enzymes are associated with clay particles, and in such associations, their activity may be increased, decreased, unchanged, or completely destroyed [15].

All biomolecules tend to be large polymer or polymer-like combinations of individual molecular units that can be isolated by either acidic or basic hydrolysis. Lipids and fats are found as integral parts of membranes and cell walls.

Polysaccharides are used as structural units and as stored energy sources. Proteins are used to construct muscle and enzymes that also contain metals such as zinc, manganese, and iron. There are many other important biomolecules present at lower concentrations such as DNA and RNA, which are also released into the soil solution. All can be the source of smaller molecules in the soil solution.

The different groups of biomolecules, including fatty acids, triglycerides, polysaccharides, and proteins (illustrated in Figure 4.9) decompose at different rates depending on their composition. Lipids and fats are slower to decompose in soil because of their insolubility in water. Large polysaccharides are also insoluble in water but are more quickly decomposed than fats. Proteins and compounds such as DNA and RNA are more quickly decomposed in part

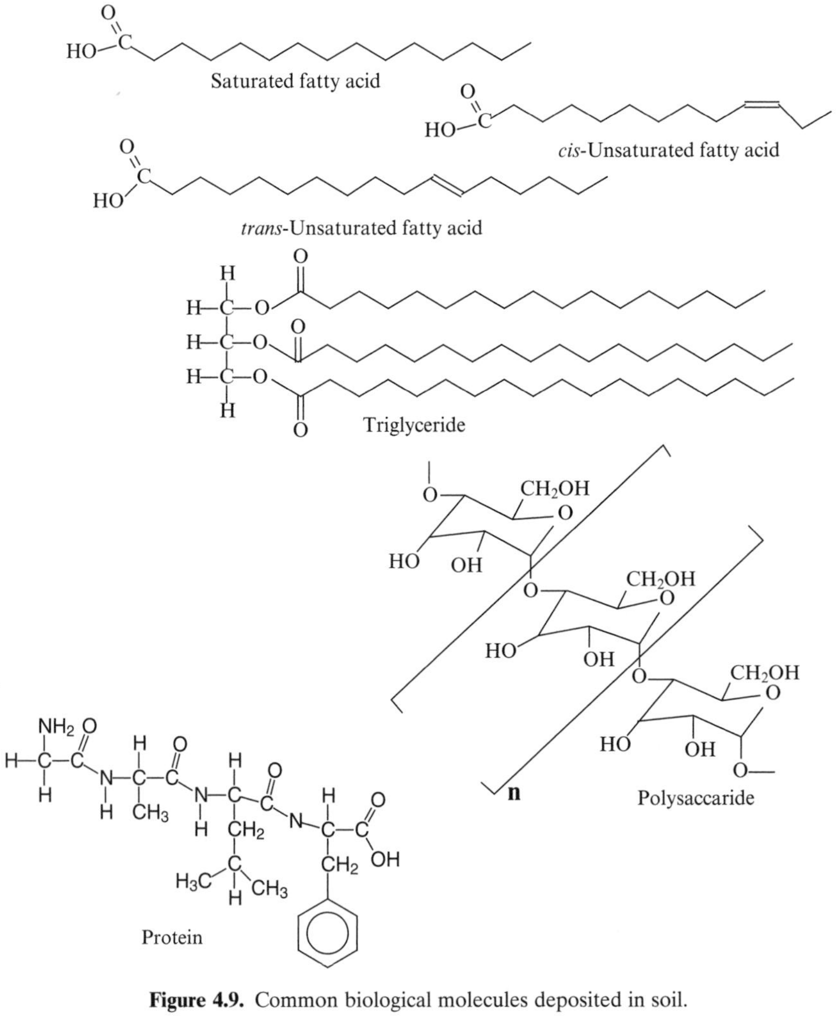

Figure 4.9. Common biological molecules deposited in soil.

because they contain nitrogen (fixed) that is often in short supply in the environment.

Biochemicals will be present in soil during any analysis and can react with components of interest, either organic or inorganic, including sand, silt, and clay particles. Possible reactions include chelation, decomposition, precipitation, solubilization, or dissolving such as dissolving in soil organic matter (humus). Several of these reactions will take place simultaneously and can lead to nondetection of the component of interest or an analytical result that is much lower than the true value [16].

4.5 BIOORGANIC

Bioorganic components in soil include those organic molecules that participate in biochemical reactions, initiate reactions, inhibit the action of other biochemical, or act as antibiotics. Bioorganic chemistry also uses synthesized molecules to study biological processes such as enzyme activity. Often these studies are undertaken to develop a mechanism for the reactions of interest. Bioorganic molecules will be present either as components of the synthesis chain or as part of the degradation products. Whenever a cell lyses, its compounds will be released into the soil solution.

Bioorganic molecules sometimes have the same or similar behavior as the analytes of interest in the analytical methods used to make chemical measurements of soil. Thus, they can falsely indicate the presence of more molecules than is actually the case, or even that an analyte or contaminant is present when it is not or has not been added from an external source. These compounds may also mask analytes of interest by reacting or simply associating with them (see Chapters 13 and 14 for a further discussion and specifics of this topic).

4.6 ORGANIC COMPOUNDS

Organic chemistry of soil mostly deals with relatively simple (when compared to biochemical compounds) organic compounds, which are found in all environments. Organic compounds are composed of carbon, hydrogen, oxygen, and nitrogen and may contain smaller amounts of other elements such as the halogens, phosphorus, and silicon. They can be straight-chained, branched, cyclic, and contain single, double, and triple bonds. For soil analysis considerations, there are nine principal organic functional groups as shown in Table 4.2. There are also four important functional groups derived from the acid functional group as shown in Table 4.3.

These functional groups, along with the general structure of the molecule, will determine its solubility and its ease of resistance to degradation. The solubility of organic molecules typically increases as the number of groups capable of forming hydrogen bonds with water increases. Conversely, the longer the straight-chain hydrocarbon portion of a molecule, the less soluble

TABLE 4.2. The Organic Functional Groups

Family Name	Composition	Structure and IUPAC[a] Name
Alkanes	Carbon, hydrogen, bonded with single bonds	$\begin{array}{c} \text{H} \quad \text{H} \quad \text{H} \\ \mid \quad \mid \quad \mid \\ \text{H}-\text{C}-\text{C}-\text{C}-\text{H} \\ \mid \quad \mid \quad \mid \\ \text{H} \quad \text{H} \quad \text{H} \end{array}$ Propane
Alkenes	Carbon, hydrogen, single bonds, and containing at least one double bond	$\begin{array}{c} \text{H} \quad \text{H} \quad \text{H} \\ \mid \quad \mid \quad \mid \\ \text{H}-\text{C}-\text{C}=\text{C}-\text{H} \\ \mid \\ \text{H} \end{array}$ Propene
Alkynes	Carbon, hydrogen, single bonds, and containing at least one triple bond	$\begin{array}{c} \text{H} \\ \mid \\ \text{H}-\text{C}-\text{C}\equiv\text{C}-\text{H} \\ \mid \\ \text{H} \end{array}$ Propyne
Alcohols	Carbon, hydrogen, single bonds, and at least one –OH (may contain double and triple bonds and other structures)	$\begin{array}{c} \text{H} \quad \text{H} \quad \text{H} \\ \mid \quad \mid \quad \mid \\ \text{H}-\text{C}-\text{C}-\text{C}-\text{OH} \\ \mid \quad \mid \quad \mid \\ \text{H} \quad \text{H} \quad \text{H} \end{array}$ 1-Propanol
Ethers	Carbon, hydrogen, single bonds, and at least one –C–O–C– in the molecule (may include double and triple bonds and other structures)	$\begin{array}{c} \text{H} \quad \text{H} \qquad \text{H} \quad \text{H} \\ \mid \quad \mid \qquad \mid \quad \mid \\ \text{H}-\text{C}-\text{C}-\text{O}-\text{C}-\text{C}-\text{H} \\ \mid \quad \mid \qquad \mid \quad \mid \\ \text{H} \quad \text{H} \qquad \text{H} \quad \text{H} \end{array}$ Diethyl ether
Aldehydes	Carbon, hydrogen, single bonds, and at least one –CHO in the molecule (may include double and triple bonds and other structures)	$\begin{array}{c} \text{H} \quad \text{H} \qquad \text{O} \\ \mid \quad \mid \qquad \nearrow \\ \text{H}-\text{C}-\text{C}-\text{C} \\ \mid \quad \mid \qquad \searrow \\ \text{H} \quad \text{H} \qquad \text{H} \end{array}$ Propanal
Ketones	Carbon, hydrogen, single bonds, and at least one $-\overset{\displaystyle O}{\underset{\displaystyle \|}{C}}-$ in the molecule (may include double and triple bonds and other structures)	$\begin{array}{c} \text{H} \quad \text{O} \quad \text{H} \\ \mid \quad \| \quad \mid \\ \text{H}-\text{C}-\text{C}-\text{C}-\text{H} \\ \mid \qquad \mid \\ \text{H} \qquad \text{H} \end{array}$ Propanone (acetone)
Acids	Carbon, hydrogen, single bonds, and at least one $-\text{C}\overset{O}{\underset{OH}{\diagup}}$ in the molecule (may include double and triple bonds and other structures)	$\begin{array}{c} \text{H} \quad \text{H} \qquad \text{O} \\ \mid \quad \mid \qquad \nearrow \\ \text{H}-\text{C}-\text{C}-\text{C} \\ \mid \quad \mid \qquad \searrow \\ \text{H} \quad \text{H} \qquad \text{OH} \end{array}$ Propanoic acid
Amines	Carbon, hydrogen, single bonds, and at least one $-\overset{\mid}{\underset{\mid}{C}}-\text{N}\big\langle$ in the molecule (may include double and triple bonds and other structures)	$\begin{array}{c} \text{H} \\ \mid \qquad \text{H} \\ \text{H}-\text{C}-\text{N}\big\langle \\ \mid \qquad \text{H} \\ \text{H} \end{array}$ Methyl amine

[a]IUPAC is the International Union of Pure and Applied Chemistry, which determines nomenclature for organic compounds.

TABLE 4.3. Functional Groups Derived from the Acid Functional Group

Family Name	Composition	Structure
Acid halides	Carbon, hydrogen, single bonds, and at least one –COX[a] in the molecule (may include double and triple bonds and other structures)	Propanoyl chloride
Anhydrides	Carbon, hydrogen, single bonds, and at least one –COOOC– in the molecule (may include double and triple bonds and other structures)	Ethanoic anhydride
Amides	Carbon, hydrogen, single bonds, and at least one –CONH$_2$[b] in the molecule (may include double and triple bonds and other structures)	Ethanamide
Esters	Carbon, hydrogen, single bonds, and at least one –COOC$^-$ in the molecule (may include double and triple bonds and other structures)	Methyl ethanoate

[a]X is always used as the general representation for any of the halogens.
[b]Hydrogens on the nitrogen can be substituted with alkyl groups.

it is. Low-molecular-weight alcohols, aldehydes, ketones, acids, and amines are soluble, while high-molecular-weight members of these families are insoluble. These represent a very large group of compounds with simple and complex structures, including cyclic and multicyclic compounds containing several functional groups and having a wide range of solubilities.

Cyclic organic structures are widespread and common. They can contain any of the functionalities listed in Table 4.2 and Table 4.3. Because of bonding angles and steric effects, some will be highly unstable, while others will be particularly stable and resistant to decomposition.

Any of the above-mentioned compounds can be found and even synthesized in soil. The simplest example, methane (CH_4), is commonly found in the soil atmosphere. It is produced during the decomposition of organic matter under anaerobic conditions, which can occur even in aerobic soils. It is interesting to note that methane can not only can be produced in aerobic soils but can also be oxidized by soil bacteria in the same soil.

In addition to methane, other simple organic compounds will be found in soil from two different sources. They can be either exuded from roots into the rhizosphere or derived from the decomposition of larger organic molecules.

During decomposition, organic matter is broken down into smaller and smaller organic molecules until it is completely converted into carbon dioxide, water, and humus.[2] The same is true for anaerobic decomposition except that one of the final decomposition products is methane. Thus, at any given time, intermediate decomposition products can be found in the soil solution.

Plant roots excrete a mixture of simple and complex compounds and materials, including fatty acids, amino acids, sugars, and phenolic compounds. Each of these groups can be composed of a complex mixture of compounds; thus, the most common amino acids can be found along with many common sugars. In addition, a high-molecular-weight compound called *mucigel* is secreted by root tip cells and is thought to lubricate root penetration into soil. Mucigel is somewhat slower to decompose but can also be a source of simple organic molecules during its decomposition. Because the compounds represent an energy source for microorganisms, their presence helps to determine the microorganisms present in the plant's rhizosphere.

No matter what the source of organic matter or the mechanism of its decomposition, an extremely important compound, humus, is formed during decomposition [6–10, 17].

4.6.1   Humus

Humus is the material synthesized or resynthesized during decomposition of organic matter by microorganisms. It is produced under both aerobic and anaerobic conditions and remains after decomposition of the original organic matter is complete. Humus is a complex molecule often described as being a polymer even though no mer[3] unit has ever been found. It is black or dark brown in color and has a high affinity for organic molecules, cations, and water. Organic molecules associate with humus via hydrogen bonding, van der Waals forces, and dipole–dipole interactions. Cations are held by cation exchange involving both acid and phenolic groups (see Figure 4.10 and Figure 4.11). A base extract of humus from an organic soil and a commercial sample of the sodium salt of humic acid are shown in Figure 4.12. Further information on the extraction of humus is given in Section 12.3.3, Procedure 12.5.

Many researchers have attempted to unravel the mystery of the structure of humus. One approach has been to isolate "fractions" by extracting humus using various extraction procedures. These procedures result in the isolation of three or more "fractions": humic acid, fulvic acid, and humin. Humic material is isolated from soil by treating it with alkali. The insoluble material remaining after this treatment is called *humin*. The alkali solution is acidified to a pH of 1.0 and the precipitate is called *humic acid*, while the soluble

---

[2] Although not often mentioned, this process is characterized by the release of energy that is utilized by the organism carrying out the decomposition.

[3] A mer is the individual repeating unit from which a poly**mer** is formed. Thus, polyethylene is made of ethylene units, the mer, bonded together.

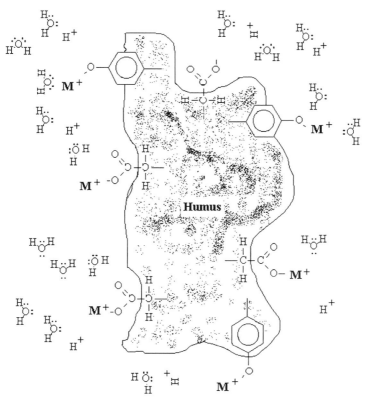

Figure 4.10. Humus with cation exchange sites created by the ionization of phenolic and acidic functional groups. The M$^+$ represent exchangeable cations.

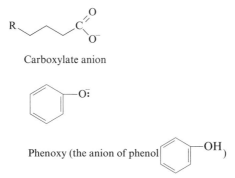

Carboxylate anion

Phenoxy (the anion of phenol)

Figure 4.11. The carboxylate anion (an ionized acid) and a phenoxy (the anion of phenol) groups. The R stands for the rest of the humus molecule to which these groups are attached.

Figure 4.12. A basic extract of an organic soil is shown on the left; a sample of the sodium salt of humic acid is seen on the right.

components are called *fulvic acid*. A lot is known about these components; for instance, they contain three carbon or propyl groups, aromatic moieties with various and usually multiple functional groups, and variable amounts of other components. Isolation of these components, however, has not generally brought us closer to a complete understanding the structure of humus. Applying other extraction procedures will allow the extraction and isolation of other "fractions" of humus; however, humin, humic acid, and fulvic acid are the three main components likely to be discussed in terms of humus.

What is known is that humus is an extremely important component of soil. Even small amounts can cause demonstrable differences in a soil's cation exchange capacity (CEC), and its other chemical and physical properties as well. It is active in binding soil particles together to form peds, increases the soil water holding capacity, and increases the sorptive capacity of soil for organic and inorganic constituents, both natural and synthetic. As an example of the importance of humus in terms of soil water, it could be noted that mineral soils absorb and hold 20–40% of their weight in water, while some organic soils can hold 10 times this amount. This is due, in part, to the fact that organic soils are less dense than mineral soils, but nevertheless, this increased water holding capacity is dramatic. Increased sorptive capacity is reflected in the increased need for herbicides on soils high in organic matter.

Different soils contain different amounts of humus. Some tropical African soils may contain less than 0.1% organic matter. At the other extreme, organic soils such as Histosols, generally must have 20% or more organic carbon in the upper 80 cm, although this will vary somewhat depending on the conditions of moisture, texture, and depth of the soil. Many agricultural soils contain 1–2% organic matter, although it is not unusual to find mineral soils containing 10% organic matter.

Humus and organic matter can have a dramatic effect on analytical results as shown in the work by Gerke [16]. Metal complexes, particularly aluminum and iron, were found to complex with phosphate. These complexes, which accounted for 50–80% of the phosphate present, were not detectable by standard phosphate analytical procedures. When developing a soil analytical method, it is essential that the method be either applicable to soils of all organic matter contents or that variations of the procedure applicable to soils of differing organic matter be developed [16–19].

4.7 ANALYSIS

There is a myriad of methods used for analyzing soil for animals, plants, and microorganisms. All are well developed and easily carried out; however, some are more useful when large numbers of samples are to be analyzed, while some are subject to significant error [20–22].

4.7.1 Analysis for Animals and Plants

Determination of the number of animals in a cubic meter of soil is usually done by simple isolation and counting. Plants are usually counted as the number per square meter, hectare, or acre. Determination of root numbers or mass is extremely difficult and is commonly done by assuming that there is a certain mass of roots associated with a plant's top, that is, the ratio of roots to top masses. Determination of microorganisms is usually done using standard microbiological techniques such as dilution plate counts. Because of the extreme diversity of soil microorganism, it is impossible to determine all the organisms present at one time by any one technique. In some cases, direct microscopic observation, as was done to produce Figure 4.6, is used to estimate the numbers of microorganisms present [2–4].

4.7.2 Determination of Soil Organic Matter

The analysis of soil for organic matter[4] is straightforward, involving oxidizing it to carbon dioxide and water. Oxidation can be accomplished in a number

---

[4] Specific directions for determination of soil organic matter are given in the bibliographic references.

of different ways, such as by applying air at high temperature, pure oxygen, or chemical oxidation using various common chemical oxidizing agents such as hydrogen peroxide, permanganate, and dichromate followed by titration. Of these, oxidation by hot dichromate followed by titration of unreduced dichromate is commonly used.

The real problem is the complete oxidation of all of the carbon present. As noted previously, soil organic matter can be associated with mineral surfaces or in pores with limited access. For these reasons, simple oxidation will not be sufficient and drastic oxidation procedures are necessary.

4.7.2.1 Direct Oxidation Using High Temperature and Atmospheric Air

The simplest and least expensive method of determining the organic matter in soil is high-temperature oxidation using air. In this case, a soil sample in a crucible is weighed, placed in an oven, and dried. Subsequently, after a prescribed time at an elevated temperature, it is removed, reweighed, and the difference taken as the amount of organic matter. This procedure is seemingly simple, straightforward, and easy to perform and does not involve the expense of waste disposal. While oxidation of organic matter with a consequent loss of weight occurs, so do other reactions, which also lead to a loss of weight. One such reaction is the loss of waters of hydration from soil minerals. Another is the decomposition of soil minerals. For instance, carbonates, which are very common soil minerals, decompose at high temperature to form the respective oxide and carbon dioxide, which is lost, resulting in a loss of weight in the sample.

For these and other reasons, direct oxidation of soil organic matter using high temperature and atmospheric air, although commonly used, is not the best procedure for determining soil organic matter content.

4.7.2.2 Oxidation Using Dichromate

The most common method used to determine soil organic matter content is oxidation using dichromate. In this procedure, a mixture of potassium or sodium dichromate with sulfuric acid is prepared and mixed with a soil sample. The heat released from mixing the soil with the dichromate and other accessory solutions heats the mixture to around 100°C for a period of time. Alternately, external heating may be used or required. After cooling, an indicator is added and the unreacted dichromate titrated. The amount of dichromate remaining can be related back to the amount of carbon and organic matter present originally (see also Section 10.3). Because of the relatively low temperatures and the reagents used, this is a reliable method that does not lead to loss of other components except carbonates. The loss of carbonate does not involve reduction of dichromate but can cause interferences that must be taken into account [12].

Caution: It should be noted that this procedure involves the use of hazardous chemicals. It also results in hazardous waste that must be disposed of carefully and properly.

4.7.2.3   Other Oxidative Procedures

Potentially, any oxidizing reagent can be used to determine soil organic matter. Of the many other possibilities, the use of catalytic oxidation in pure oxygen and or the application of hydrogen peroxide are two common methods. Any organic chemical can be oxidized to carbon dioxide and water using a catalyst, pure oxygen, and heating. A known amount of material is placed with a catalyst in a stream of oxygen and heated, producing carbon dioxide and water, which are trapped and weighed. From these masses, the amount of carbon and hydrogen in the original organic matter can be determined. As might be surmised, this method can also be used to determine the amount of nitrogen and sulfur in soil organic matter. This procedure is done on a one-by-one basis and is therefore slow and time-consuming. For these reasons, it is not used where a significant number of samples must be analyzed.

A soil sample can be mixed with hydrogen peroxide and heated to decompose organic matter. This procedure is often used when the objective is to remove organic matter and the amount present is not determined. Problems with this procedure include frothing due to decomposition of hydrogen peroxide by components in soil, and questions as to whether or not all the organic matter is oxidized. Directly measuring the amount of organic matter destroyed is difficult using this method.

There are still other methods of oxidizing soil organic matter. These are not generally or commonly used for a variety of reasons, which can be found by investigating them in the literature [22] (see also Chapter 10 for further details).

4.8   CONCLUSIONS

Soil biological components are important because of their role in, not only, mixing and decomposing both the inorganic and organic components therein but also in the addition of organic matter to soil. Organic matter is added by animals that live in the soil as well as on the soil, chiefly in the form of manure. Plants add organic matter from their tops and their roots. Microorganisms are involved chiefly in the decomposition of organic matter deposited by animals and plants. However, they also produce many different inorganic and organic compounds during the decomposition process. One of the most important of these components is humus. Analysis of total soil organic matter is accomplished by oxidizing the organic matter using various oxidative chemical procedures. Animals, plants, microorganisms, and organic matter in soil will have a pronounced effect on the soil characteristics and on procedures used to analyze it.

PROBLEMS

4.1. Name some common types of animals, plants, and microorganisms found in soil.

4.2. Explain how animals can change the physical characteristics of soil. Give some examples.

4.3. In soil microbiology, microorganisms are defined by which simple characteristic? Describe the variety of organisms that have this characteristic.

4.4. Explain how animals and plants add organic matter to soil and how this addition is different for these two different groups of organisms.

4.5. What is the rhizosphere and why is it important?

4.6. What is the difference between a biomolecule and an organic molecule?

4.7. Diagram the breakdown of organic matter in soil and give the products of this decomposition process.

4.8. Explain how humus is formed in soil and why it is important. Give some examples.

4.9. What features of organic molecules control their solubility in the soil solution?

4.10. Describe two ways in which soil organic matter is measured.

REFERENCES

1. Suzuki Y, Mastsubara T, Hoshino M. Breakdown of mineral grains by earthworms and beetle larvae. *Geoderma* 2003; **112**: 131–142.

2. Baker GH, Lee KE. Earthworms. In Carter MR (ed.), *Soil Sampling and Method of Analysis*. Ann Arbor, MI: Lewis Publishers; 1993, pp. 359–372.

3. Kimpinski J. Nematodes. In Carter MR (ed.), *Soil Sampling and Method of Analysis*. Ann Arbor, MI: Lewis Publishers; 1993, pp. 333–340.

4. Winter JP, Voroney RP. Microarthopods in soil and litter. In Carter MR (ed.), *Soil Sampling and Method of Analysis*. Ann Arbor, MI: Lewis Publishers; 1993, pp. 319–332.

5. Garner DT, Miller RW. *Soils in Our Environment*, 10th ed. Upper Saddle River, NJ: Pearson Educational Inc.; 2004, pp. 151–153.

6. Bais HP, Park S-W, Sternitz FR, Halligan KM, Vivanco JM. Exudation of fluorescent β-carbolines for *Oxalis tuberosa* L. roots. *Phytochemistry* 2002; **61**: 539–543.

7. Bruneau PMC, Ostle N, Davidson DA, Grieve IC, Fallick AE. Determination of rhizosphere ^{13}C pulse signals in soil thin sections by laser ablation isotope ratio mass spectrometry. *Rapid Commun. Mass Spectrom.* 2002; **16**: 2190–2194.

8. Dakora FD, Phillips DA. Root exudates as mediators of mineral acquisition in low-nutrient environments. *Plant Soil* 2002; **245**: 35–47.

9. Garcia JAL, Barbas C, Probanza A, Barrientos ML, Gutierrez MFJ. Low molecular weight organic acids and fatty acids in root exudates of two *Lupinus* cultivars at flowering and fruiting stages. *Phytochem. Anal.* 2001; **12**: 305–311.

10. Walker TS, Bais HP, Halligan KM, Stermitz FR, Vivanco JM. Metabolic profiling of root exudates of *Arabidopsis thaliana*. *J. Agric. Food Chem.* 2003; **51**: 2548–2554.

11. Tabatabai MA. Sulfur oxidation and reduction in soils. In Hickelson SH (ed.), *Methods of Soil Analysis Part 2: Microbiological and Biochemical Properties*. Madison, WI: Soil Science Society of America; 1994, pp. 1068–1078.

12. Bottomley PJ. Light microscopic methods for studying soil microorganisms. In Hickelson SH (ed.), *Methods of Soil Analysis Part 2: Microbiological and Biochemical Properties*. Madison, WI: Soil Science Society of America; 1994, pp. 81–104.

13. Germida JJ. Cultural methods for soil microorganisms. In Carter MR (ed.), *Soil Sampling and Method of Analysis*. Ann Arbor, MI: Lewis Publishers; 1993, pp. 263–276.

14. Zuberer DA. Recovery and enumeration of viable bacteria. In Hickelson SH (ed.), *Methods of Soil Analysis Part 2: Microbiological and Biochemical Properties*. Madison, WI: Soil Science Society of America; 1994, pp. 119–158.

15. Tabatabai MA. Soil enzymes. In Hickelson SH (ed.), *Methods of Soil Analysis Part 2: Microbiological and Biochemical Properties*. Madison, WI: Soil Science Society of America; 1994, pp. 778–834.

16. Gerke J. Humic (organic matter)-Al(Fe)-phosphate complexes: an underestimated phosphate form in soils and source of plant-available phosphate. *Soil Sci.* 2010; **175**: 417–425.

17. Buurman P, van Lagen B, Piccolo A. Increased in stability against thermal oxidation of soil humic substances as a result of self association. *Org. Geochem.* 2002; **33**: 367–381.

18. Sawhney BL. Extraction of organic chemicals. In Bartels JM (ed.), *Methods of Soil Analysis Part 3: Chemical Methods*. Madison, WI: Soil Science Society of America and American Agronomy Society of Agronomy; 1996, pp. 1071–1084.

19. Fujitake N, Kusumoto A, Tsukamoto M, Kawahigashi M, Suzuki T, Otsuka H. Properties of soil humic substances in fractions obtained by sequential extraction with pyrophosphate solutions at different pHs I. Yield and particle size distribution. *Soil Sci. Plant Nutr.* 1998; **44**: 253–260.

20. Swift RS. Organic matter characterization. In Bartels JM (ed.), *Methods of Soil Analysis Part 3: Chemical Methods*. Madison, WI: Soil Science Society of America and American Agronomy Society of Agronomy; 1996, pp. 1011–1070.

21. Wolf DC, Legg JO, Boutton TW. Isotopic methods for the study of soil organic matter dynamics. In Hickelson SH (ed.), *Methods of Soil Analysis Part 2: Microbiological and Biochemical Properties*. Madison, WI: Soil Science Society of America; 1994, pp. 865–906.

22. Tiessen H, Moir JO. Total and organic carbon. In Carter MR (ed.), *Soil Sampling and Method of Analysis*. Ann Arbor, MI: Lewis Publishers; 1993, pp. 187–200.

BIBLIOGRAPHY

Coyne MS. *Soil Microbiology: An Experimental Approach*. Boston: Delmar Publishers; 1999.

Huang PM, Schnitzer M, eds. *Interactions of Soil Minerals with Natural Organics and Microbes*. Madison, WI: Soil Science Society of America; 1986.

Linn DM, Carski TH, Brusseau ML, Chang FH, eds. *Sorption and Degradation of Pesticides and Organic Chemicals in Soil*. Madison, WI: Soil Science Society of America and American Agronomy Society of Agronomy; 1993.

Nardi JB. *The World beneath Our Feet: A Guide to Life in the Soil*. New York: Oxford University Press; 2003.

Pinton R, Varanini Z, Nannipieri P, eds. *The Rizosphere Biochemistry and Organic Substances at the Soil-Plant Interface*. New York: Marcel Dekker; 2001.

Sawhney BL. *Reactions and Movement of Organic Chemicals in Soils*, Brown K (ed.). Madison, WI: Soil Science Society of America and American Agronomy Society of Agronomy; 1989.

Stevenson FJ. *Humus Chemistry: Genesis, Composition, Reactions*. New York: Wiley; 1994.

WEB SITES

Loynachan T. Available at: http://www.public.iastate.edu/~teloynac/. 2009. Accessed June 3, 2013.

Skipper H. Available at: http://www.clemson.edu/cafls/safes/faculty_staff/skipper.html. Accessed June 3, 2013.

Sylvia D. Available at: http://ecosystems.psu.edu/directory/dms39. Accessed June 3, 2013.

SOIL BASICS PART IV
THE SOIL AIR AND SOIL SOLUTION

All soil includes air and water. Soil air is similar to atmospheric air in that it contains the same constituents but differs in that it has different concentrations of these constituents, and that the relative percentages vary over time. In nature, soil is never completely dry; it always has a layer of water surrounding the individual soil particles. The dust cloud behind a car driving down a sandy road may contain 1% water on a dry-weight basis, while the dust behind a car on a clayey road may contain 10% or more. In the field, water content is highly variable and can range from 1% to more than 100% in some organic soils (i.e., the mass of water is greater than the mass of the organic soil). A saturated soil will lose water through percolation, evaporation, and transpiration. As the water content decreases, two things happen. The air content increases and the water remaining in the soil is held more and more strongly. At some point,

Introduction to Soil Chemistry: Analysis and Instrumentation, Second Edition.
Alfred R. Conklin, Jr.
© 2014 John Wiley & Sons, Inc. Published 2014 by John Wiley & Sons, Inc.

which will depend on a soil's texture, structure, and organic matter content, water will be held so tightly that it is unavailable to plants.

Most components coming in contact with water will dissolve, if only to a minute degree, and thus soil water is not pure, but rather a complex solution of inorganic and organic ions, molecules, and gases that are constantly exchanging between the solid, liquid, gaseous, and biological phases. It is in this solution surrounding soil particles that reactions occur, and in which microorganisms live and function. The type and rate of reactions occurring in soil will depend on the soil air and on the soil solution composition, and thus soil air and water are intimately involved in all things happening in and to the soil.

5.1 SOIL AIR

Soil air is made up of the same basic constituents as atmospheric air; however, the ratios of various gases are different and more variable. First and most importantly, virtually all void volume of the soil can be occupied by either air or water. The amount of air in soil is thus inversely related to the amount of water present. When the air content is around 25% or more of the void volume, as shown in Figure 5.1, the soil is considered to be aerobic and oxidation reactions predominate. When all the void volume is occupied by water, the soil is anaerobic and reducing reactions predominate.

Oxidation reactions in soil, particularly those carried out by microorganisms, and plant roots increase the amount of carbon dioxide in soil air to 10 or more times the concentration in atmospheric air. The consequence of this is that the oxygen content is proportionally decreased. When the soil void volume is almost or completely filled with water, the remaining trapped and dissolved oxygen is quickly utilized by organisms and the oxygen content of any remaining gas is zero. The soil is then anaerobic and reducing conditions prevail (see Figure 5.1).

In addition to oxidation–reduction reactions occurring under various conditions, increased carbon dioxide will also affect soil pH. Carbon dioxide

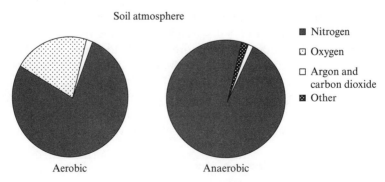

Figure 5.1. Soil atmosphere composition under aerobic and anaerobic conditions.

$$H_2O + CO_2 \rightarrow H_2CO_3 \tag{1}$$

$$H_2CO_3 \rightleftharpoons H^+ + HCO_3^- \quad pK_1 \; 6.35 \tag{2}$$

$$HCO_3^- \rightleftharpoons H^+ + CO_3^{2-} \quad pK_2 \; 10.33 \tag{3}$$

$$H^+ + CO_3^{2-} \rightleftharpoons HCO_3^- \tag{4}$$

$$Ca^{2+} + CO_2^{2-} \rightleftharpoons CaCO_3 \downarrow \tag{5}$$

Figure 5.2. Reactions of carbon dioxide and water, which illustrate its involvement in control of soil water pH.

dissolves in water and produces carbonic acid (H_2CO_3), which release protons into the soil solution as shown in the equations in Figure 5.2, which illustrate the involvement of CO_2 in control of soil water pH. In addition to providing protons, carbonic acid also leads to the generation of carbonate ions (CO_3^{2-}) that can react with cations to form insoluble precipitates. Both reactions can dramatically alter the composition of the soil solution and soil extracts.

The soil atmosphere also contains water vapor and, in most cases, is at 100% relative humidity. Water vapor evaporating from the soil surface is one mechanism by which water and dissolved components can move upward in a soil profile.

The gases discussed earlier are interesting and important but do not represent all the gases commonly found in analyses of the soil atmosphere. Even in aerobic soils, it is common to find reduced species such as methane. In addition, if ammonium is present in the soil solution, ammonia will be present in the soil atmosphere. Under oxidizing conditions, ammonium will be oxidized to nitrite and then to nitrate. Under reducing conditions, these species are converted to gaseous nitrogen oxides, which also occur in the soil atmosphere, and finally to nitrogen gas. Other gases, such as hydrogen and helium, can be found in the soil under some conditions and at some localities [1–4].

5.2 WATER

Water is a unique molecule, and when it is associated with soil, it is even more unique. The most frequently cited unique characteristics of water are its high melting and boiling points and its ability to dissolve a wide range of molecules and ions. A less often appreciated characteristic of water is its density, which decreases both above and below its freezing point with the maximum density actually occurring at a temperature between 3 and 4°C. These phenomena are related to hydrogen bonding in water discussed in Section 3.2.5, where it was pointed out that water is the prime example of this phenomenon.

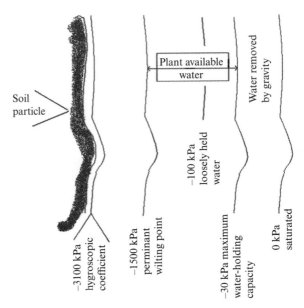

Figure 5.3. Potential (kPa) of "layers" of water surrounding a soil particle.

One way of thinking about water in soil is to envision it as a layer around and covering a soil particle (see Figure 5.3). As water is removed from outside layers, the remaining molecules are held more strongly. The outermost layers are held with a tension of 0 to −30 kPa,[1] and are removed by the pull of gravity. This is called gravitational water, and normally drains or percolates through the soil and into the groundwater. Soil containing gravitational water contains little or no air and because roots require air to function, this water is generally said to be unavailable to plants.

The next layer of water, held between −30 and −1500 kPa, is available to plants and is therefore called *plant available water*. The water present between −1500 and −3200 kPa is held in capillaries so tightly that it is not available to plants but can be lost by evaporation. The layer closest to the soil solid is held at more than −3200 kPa and is called *hygroscopic water*. A soil sample, heated in an oven for 24 hours at 105°C, and then left exposed to the air will adsorb water until a layer of hygroscopic water has been formed, illustrating the strong attraction of soil surfaces for water.

The unit kPa is in common use today and is part of the International System of Units (SI). However, it is also common to encounter the terms *bars* and atmospheres (*atm*) when reading about soil water. One bar is approximately equal to one atmosphere pressure, which is abbreviated atm, and a bar is equal to 100 kPa (−1 bar = −100 kPa).

---

[1] kPa is kilopascal and is 1000 Pa, where a pascal is a unit of pressure defined as 1 newton per meter squared (N/m^2).

The movement of water through soil is controlled by the diameter of soil pores and the surface tension of water. Water drains from larger pores and moves down through soil and into the groundwater. In small pores, the surface tension of water is strong enough to prevent movement of water into or out of pores. However, because of surface tension, pores will also draw water up from a free water surface until the surface tension is balanced by the pull of gravity. The smaller the pore, the higher the water will be raised. If water, moving down through soil, reaches a compacted zone, for example, having few and extremely small pores, it will move laterally along the dense layer. Thus, water can move down, up, and sideways in soil depending on the soil's pores.

Pores can control water movement in other ways related to size. To understand this control, soil pores can be grouped or classified simply as large, those that allow water to drain (move), and small, those that do not. Water in large pores will be drawn into smaller pores, but water in small pores cannot move into large pores unless energy is exerted. Thus, water will not move from a sandy horizon—smaller pores—into an underlying gravel layer—larger pores—unless the water exerts enough pressure to move the water from the small pores into the larger pores in the gravel. This relationship between pores and water movement is illustrated in Figure 5.4 using capillaries.

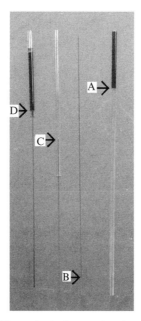

Figure 5.4. A large diameter capillary, tube A, draws colored water a shorter distance than a small diameter capillary (B). Colored water cannot move from the small diameter capillary to the large diameter capillary (C). Colored water can move from a larger diameter capillary to a smaller capillary (D).

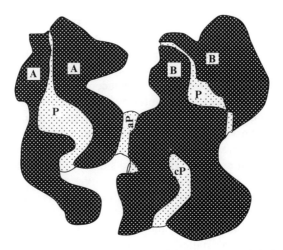

Figure 5.5. Two peds (AA and BB) each with a pore (P) that does not drain because the mouths are too small. Ped BB has a pore closed at one end, that is, cP. Between the peds is an apparent pore, aP, formed by the close proximity of the peds.

Some pores do not drain, including those that are simply too small and also those in which the interior is large but the entrances and exits are so small that they prevent drainage (see Figure 5.5). Both of these can be called *restricted pores*. Restricted pores can cause problems in any extraction and analysis procedure because they prevent the complete removal of solution from the soil. Exchange with extracting solutions is limited by the slow process of diffusion and can result in the component of interest occurring at low levels even in the last of multiple extractions [5]. An example of the importance of pores in relationship to the distribution of cations in soil is given in the work by Jassogne et al., which shows that the distribution of cations around pores is uneven [6].

5.3   SOLUBILITY

The solubility of components in the soil solution will be controlled by the innate solubility of the compound in question and the existing soil solution characteristics, particularly salts already present. High salt concentrations will result in salting out and precipitation of some components. Note here that salt concentration is not constant in that as the soil dries, the concentration of salt increases. Precipitation reactions may not be reversible when the soil water content is subsequently increased.

In addition to straightforward precipitation reactions, components may dissolve and react with components already present, including atoms on colloidal surfaces. For example, phosphate may dissolve from phosphate rock and react with iron present in the soil solution or on particle surfaces to form an iron phosphate that is insoluble.

5.4   ELEMENTS IN SOLUTION

Elements in the soil solution will be derived from the atmosphere, lithosphere, and biosphere. Thus, nitrogen, oxygen, and argon from the atmosphere will commonly be found dissolved in soil water. It is often surprising and sometimes disconcerting to find argon in soil air and water primarily because it is rarely mentioned as a component of air. Although its occurrence may be surprising, it does not represent an unusual situation.

Mercury and the noble metals are found in nature in their elemental forms; however, they are generally unreactive and so their occurrence in the soil solution is limited. Some elements, such as sulfur, can be reduced to their elemental state (see Figure 4.8) by soil microorganisms; however, they can also easily be both oxidized and the oxidized forms reduced and so are rarely found in their elemental form in soil.

5.5   DISSOLVED GASES

Molecular gases, the two most important of which are oxygen and carbon dioxide, dissolve in soil water. Dissolved oxygen results in the soil solution being oxidative, or aerobic, and thus tends to keep components in their highest oxidation state. Microorganisms present will be aerobic and dissolved, and suspended organic matter will undergo oxidative reactions.

Dissolved carbon dioxide produces carbonic acid, which ionizes to bicarbonate and carbonate ions, the reactions for which are shown in Figure 5.2 (equations 1–3). This reaction sequence is extremely important because bicarbonate is a counterion to many cations, is active in buffering the soil solution, and is involved either directly or indirectly in many soil chemical reactions. Bicarbonates are generally more soluble than carbonates, which are generally insoluble. Adding acid to carbonates or bicarbonates results in the release of carbon dioxide and the formation of the salt of the acid cation. The acid is thus neutralized.

At the other extreme, high levels of carbonate or high pH results in the formation of insoluble carbonates, frequently calcium and magnesium carbonates, thus removing them from the soil solution. These reactions limit the lower and upper pHs normally found in soil (Figure 5.2, equations 4 and 5). In addition to carbon dioxide, methane is another common molecular gas in the soil solution. It is largely insoluble in water but will be present because it is constantly being produced by methanogenic bacteria in anaerobic zones in soil. It may seem unlikely that methane would be found in an oxidizing environment; however, as noted earlier, some pores in soil do not drain and at some interior distance from their mouths form microaerobic and/or anaerobic zones (Figure 5.6). Conditions in these zones are conducive to the formation of methane, hydrogen, hydrogen disulfide, carbon monoxide, or other reduced species in the soil solution.

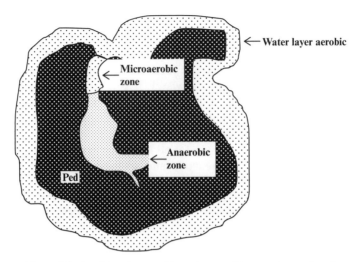

Figure 5.6. Aerobic, microaerobic, and anaerobic zones in a soil ped.

Any low-molecular-weight organic compounds are normally a gas at standard temperature and pressure (STP) may also be found in the soil solution. These compounds will be produced as a result of the decomposition of organic matter in soil. Many of these will be readily taken up and used by microorganisms and thus their life spans in the soil solution are short.

5.6   COMPOUNDS IN SOLUTION

The soil solution will contain numerous inorganic and organic compounds derived from the solid components making up the soil. Common compounds include oxides, particularly those of silicon, aluminum, iron, and titanium in low concentrations. These compounds move down the soil profile sometimes contributing to formations such as the spodic horizon, which can contain aluminum and iron oxides along with highly decomposed carbon.

A large variety of dissolved organic compounds released by decomposition of organic matter or from the activities of plants and animals are in soil. These compounds range from relatively simple molecules, such as acetic acid, to relatively complex materials such as enzymes, antibiotics, and various cellular components released upon cell lyses (see Chapter 4). These and intermediate products serve as sources of carbon and energy for organisms in soil and thus do not persist. However, it is possible for almost any relatively simple water-soluble organic compound to be found in the soil solution at any given time.

Other inorganic and organic compounds are brought into solution by the decomposition of their parent materials. Rocks and minerals will be decomposed by physical, biological, and chemical mechanisms. Enzymes released into the soil solution by microorganisms will decompose insoluble organic materials such as wood. Most of the organic material released will be taken up by the organism releasing the enzyme, but some will find its way into the soil solution.

Of particular interest to soil chemists are polycarbonyl and polyamino compounds that can form associations with metals. A multiplicity of polycarbonyl and polyamino compounds is extruded by plant roots, so solutions of these or similar compounds are used to mimic the extracting capabilities of plant roots. However, it is not possible to completely duplicate the extracting activities of roots.

Polycarbonyl and polyamino compounds include chelates, which are man-made compounds, and siderophores, which are biologically derived and more specialized. They can help dissolve inorganic ions, keep ions in one specific ionic state, and maintain them in biologically available forms. Because of their ability to bring ions into solution, they are often used to extract specific ions in specific forms from soil. Cationic micronutrients, particularly iron and zinc, are often applied in chelated form as a foliar spray to alleviate nutrient deficiencies. This is particularly true of plants growing in basic soil and is done even when plants do not show micronutrient deficiencies. Typically, both foliar and soil applications are effective in this regard.

Two examples of polycarbonyl and polyamino compounds are shown in Figure 5.7. Ethylenediaminetetraacetic acid (EDTA) is a synthetic ligand. The other, a siderophore, is biologically derived. Other polycyclic structures that coordinate with iron are also common.

The structures show the lone pairs of electrons that form bonds with metal cations. Both chelates and siderophores hold micronutrient cations in solution and in a biologically available form until the cations can be used by the

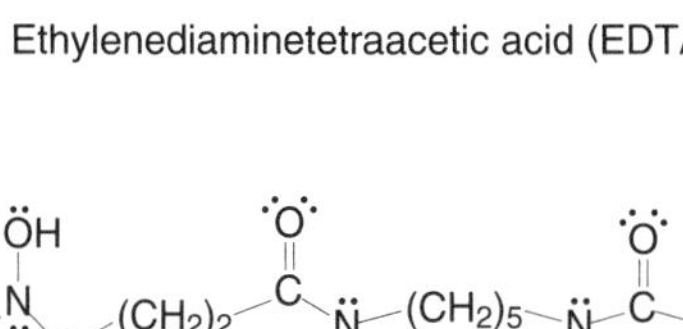

Ethylenediaminetetraacetic acid (EDTA)

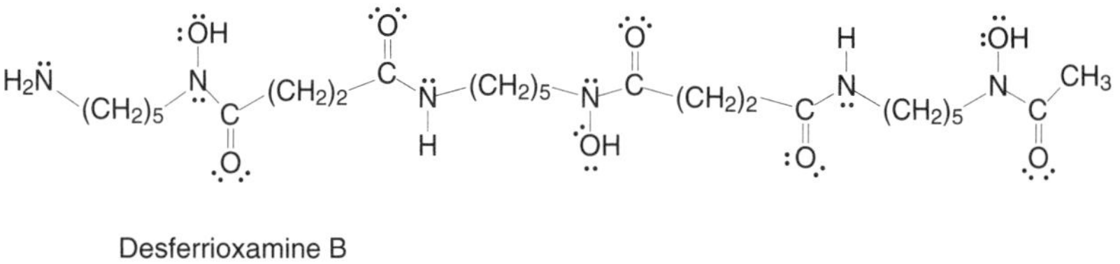

Desferrioxamine B

Figure 5.7. Example of a chelate (EDTA) and a siderosphore (desferrioxamine B).

organisms. Sidosphores are more specific than chelates in that they specifically coordinate with iron in the ferric, or Fe(III), state. This is important because Fe(III) is insoluble in water and thus biologically unavailable. When coordinated with a siderophore, the siderophore–iron combination is soluble in water because the iron is isolated from water. In this form, the iron can be taken into biological systems and used as needed.

5.7   INORGANIC IONS IN SOLUTION

Ionic species are generally more soluble in water than are neutral molecules. The partially positive hydrogen atoms in water associate with anions and the partially negative oxygen atoms with cations, forming a "shell" of water molecules around them as illustrated in Figure 5.8. Most cations in soil are simple metal ions, although some, such as iron, may be present in multiple oxidation states. An exception to this is molybdenum, which occurs not as a cation but as molybdate, an oxyanion. Two nonmetal cations in soil are ammonium and hydronium (hydrogen ions, protons, associated with water). In contrast, there are relatively few simple anions found in soil outside of the halogens, particularly chloride, Cl^-, and bromide, Br^-. Most anions in soil occur as complex oxyanions.

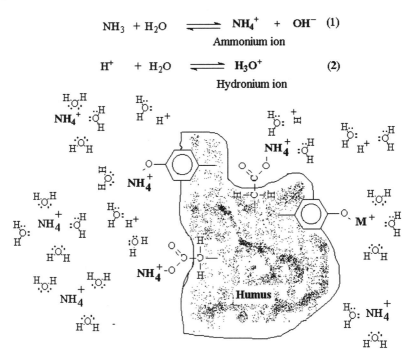

Figure 5.8. Reactions forming ammonium (1) and hydronium ions (2). Ammonium ions associated with exchange sites on soil humus.

5.7.1 Simple and Multi-Oxidation-State Cations

The most common simple cations in the soil solution are calcium (Ca^{2+}), magnesium (Mg^{2+}), potassium (K^+), and sodium (Na^+). Other alkali and alkaline-earth elements, when present, will be as simple cations also. Iron, aluminum, copper, zinc, cobalt, manganese, and nickel are also common in soil. Iron is present in both the ferrous (Fe^{2+}) and ferric (Fe^{3+}) states, while aluminum will be present as Al^{3+}. Copper, zinc, cobalt, and nickel can all be present in one or both of their oxidations states simultaneously. Manganese presents a completely different situation in that it can exist in several oxidation states simultaneously.

5.7.2 Multielement Cations

Ammonium and hydrogen ions (protons) are both present in the soil solution as multielement cations. Ammonia gas reacts with water to produce the ammonium cation, NH_4^+ (Figure 5.8, equation 1). Ammonium acts as a cation in all senses and will be attracted to cation exchange sites on soil particles. Ammonium in the soil solution and on exchange sites is available to plants.

The hydrogen ion (H^+) represents a very different situation. When hydrogen is released into the soil solution by ionization, it loses its electron. The naked proton (H^+) is naturally attracted to the partially negative oxygen of water and its lone pair of electrons (Figure 5.8, equation 2). The result of this interaction is the species H_3O^+, which is called a *hydronium ion*. This is the true species in the soil solution even though scientific papers and texts will use the simpler H^+ when writing equations.

It is important to remember that exchangeable cations (see Figure 5.8), including NH_4^+ and H_3O^+, when attached to exchange sites, cannot be measured directly; they must be brought into solution by displacing them from the site before analysis can be affected. Thus, extracting solutions must contain a cation capable of displacing all the cations of interest on the exchange sites of the soil. Once in solution, analysis can be carried out.

5.7.3 The Simple and Oxyanions in Soil

Simple anions in soil are the halogens, chlorine (Cl^-), and bromine (Br^-). If present, the other halogens—fluoride, F^-, and iodide, I^-—will also occur as simple anions. Because the compounds of these anions are generally soluble, they readily leach out of the soil and so are present at low concentrations. Exceptions occur in low-rainfall regions where significant, sometimes deleterious (to plants and animals), levels of simple anions can be found.

The two important oxyanions in soil are nitrate and phosphate. Nitrate (NO_3^-) is the predominant oxyanion of nitrogen; however, nitrite (NO_2^-) can also occur in the soil solution. Phosphate can exist as one of three species,

$H_2PO_4^-$, HPO_4^{2-}, and PO_4^{3-}, depending on the soil pH. Both nitrate and phosphate are important in plant nutrition and, because of contamination concerns, environmental work. Other important oxyanions are bicarbonate (HCO_3^-) and carbonate (CO_3^{2-}), and the various oxyanions of boron ($H_2BO_3^-$) and molybdate (MoO_4^{2-}).

It is reasonable to expect that because anions and most colloidal particles in temperate region soils have a negative charge, they will repel each other. The consequence is that anions will pass through soil and will not be adsorbed or even retarded. For the simple anions and some of the oxyanions, this is exactly what happens. All the halides, nitrite, nitrate, bicarbonate, and carbonate act in this fashion. However, there are some oxyanions that do not act as expected, and chief among them is phosphate.

Monobasic ($H_2PO_4^-$), dibasic (HPO_4^{2-}), and tribasic (PO_4^{3-}) phosphate react with iron and aluminum at low pH to form insoluble phosphates. In a similar fashion, calcium reacts with phosphate at high pH to form insoluble calcium phosphates. In addition, phosphates will react with clay minerals and organic matter to form insoluble compounds. For these reasons, phosphate seldom moves in soil. Exceptions occur when phosphate is associated with organic matter that is moving through the soil and when there is a large excess of phosphate such as in areas where phosphate is mined. Other oxyanions will also behave like phosphate, although the exact nature of the reactions leading to their attraction to soil particles and organic matter is not as well understood.

Some soils, particularly those in the tropics, have significant anion exchange capacity. For these soils, there is an attraction between soil colloids and the simple halogen and nitrate anions. Bringing these anions into solution for analysis requires an extraction, or replacing anions, just as does the analysis of exchangeable cations.

5.8   ORGANIC IONS IN SOLUTION

There are three types of organic functional groups—acid, phenolic, and amine—that are commonly ionized in soil (Figure 5.9). Whether the group is ionized or not will be determined by the pK_a of the acid or phenol or, in the case of amines, their pK_b. The solubility of organic molecules that contain ionizable groups is greatly increased once they become ionized. Ionized groups contributing to the cation exchange capacity (CEC) of organic matter are illustrated in Figure 5.8.

Once ionized, the acidic and phenolic molecules carry a negative charge and can thus attract cations and participate in the CEC of soil that are part of humus. The contribution of this source of negative charge will depend on the pH of the soil solution and will change as the pH changes. Thus, the CEC is pH dependent. For this reason, it is essential that the CEC values are measured at the same pH when comparing different soil samples.

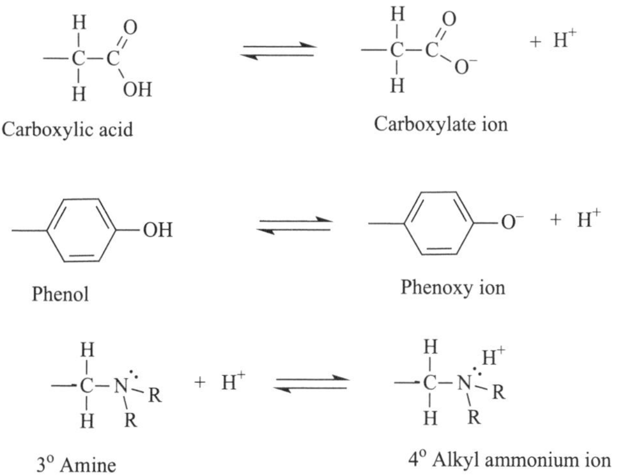

Carboxylic acid   Carboxylate ion

Phenol   Phenoxy ion

3° Amine   4° Alkyl ammonium ion

Figure 5.9. Reactions leading to the formation of charged organic species in soil. Note that the unsatisfied bond on the left is attached to some larger organic component in soil.

5.9 SOIL pH

In any aqueous solution, the pH is a measure of the hydrogen ion or proton activity. However, in many if not most cases, pH is treated as the concentration of protons in solution rather than their activity. Soil solutions are no different except that the measurement is much more complex. The complexity arises from two sources:

1. An electrical potential develops at all interfaces. In soil, there are interfaces between solids and solution, solution and suspension, suspension and the electrode surface, and between the reference electrode and all these interfaces.
2. The concept of activity is extremely important in soil. Protons or hydronium ions attracted to exchange sites or other components in the system will not be measured as part of the solution composition.

For these reasons, a standard method of measuring soil pH is chosen and all phenomena related to pH or involving pH are related to this "standardized" pH measurement. The most common method is to use a 1:1 ratio of soil to water, typically 10 mL of distilled water and 10 g of soil. In this method, the soil and water are mixed and allowed to stand 10 minutes and the pH determined using a pH meter.

This method does not measure exchangeable protons attached to cation exchange sites; therefore, it is also common to use a salt solution (either KCl or $CaCl_2$) instead of distilled water in determining soil pH. The K^+ or Ca^{2+} in the solution exchanges with exchangeable hydronium, thus bringing it into solution where it can be measured. These procedures therefore usually give a pH that is more acidic than that obtained using distilled water. The justification for this approach is that it is thought to more closely relate to the pH experienced by plant roots.

Many other methods of determining soil pH and exchangeable protons are described in the literature. Among these are a number of methods designed to determine what is called the *soil buffer pH*, which is determined by adding a highly buffered solution to the soil and measuring the change in its pH. This change can then be related back to what is called the *total acidity* in soil. It is also used as the basis for making liming recommendations in agriculture. See Section 9.3.1 for a detailed discussion of pH electrodes for measuring soil pH.

It is often tempting to design a new protocol or to change an existing method for pH determination. Keep in mind that when or if this is done, it will change all the analytical methods that use pH as an important component such as the determination of CEC as discussed previously. It will also change the interpretation of analytical results and the recommendations for applications of some agricultural amendments. Such a change is probably not advisable unless there is some highly significant improvement in the new method.

5.10   THE SOIL SOLUTION AROUND PARTICLES

Dissolved ions and molecules are evenly distributed throughout a pure solution. This is not true in soil because soil particles, particularly colloidal particles, have sorbed, exchangeable, and peripherally attracted components associated with them.

Cations attracted to colloid surfaces through their waters of hydration are said to be outer-sphere species, whereas those that interact directly with the oxygen atoms present on the surface are called inner-sphere species. Of the two, the latter species will be more strongly bonded and harder to extract than will be the outer-sphere species.

Cations form a diffuse layer of ions called the *diffuse double layer* or the *electrical double layer* around soil particles as depicted in Figure 5.10. The existence of the diffuse double layer means that the ions are not evenly distributed throughout the solution; rather, cations are more concentrated close to soil particle surfaces and are less concentrated further away. This phenomenon must be kept in mind, particularly when electrochemical analytical methods of analysis are developed [5, 7].

5.11   DISTRIBUTION BETWEEN SOIL SOLIDS AND SOIL SOLUTION

All components in the soil solution are to a greater or lesser extent distributed unevenly between the solid and liquid phases. Anions are generally only

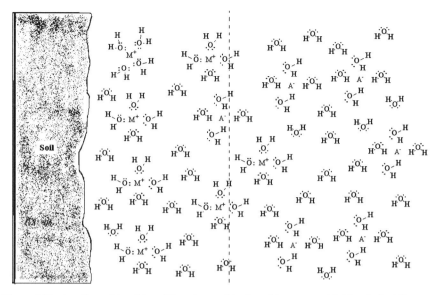

Figure 5.10. A soil particle with a diffuse layer of hydrated ions around it. The dashed line represents the boundary of the layer of tightly held cations. This diagram is not an exact representation of the diffuse double layer around a soil particle.

$$K_d = \frac{\text{mg of component/kg of soil}}{\text{mg of component/L of solution}}$$

$$K_{om} = \frac{\text{mg of component/kg of organic matter}}{\text{mg of component/L of solution}}$$

Figure 5.11. Equations for K_d and K_{om}; kg is kilogram; L is liter of the soil solution. For K_{om} applied to soil, the numerator would be kg organic matter in soil.

weakly attracted to soil solids, if at all. Cations are attracted to the soil colloids, while the interaction between soil solids and organic compounds is complex, depending on the structure and functional groups of the organic compound and the nature of the soil solids. However, measuring the attraction or having a measure of the distribution between the two phases is extremely important in understanding the movement of material in soil. It will also give invaluable information about the time and amount of extractant needed for an extraction procedure.

Two types of distribution coefficients are commonly measured and used in describing the distribution between solid and liquid phases. The first and simplest is the distribution between total solid and liquid phases. This can be represented by K_d, as given in the equation in Figure 5.11. Here, kg is kilogram and L is liter of soil solution.

Organic matter in soil has a much higher sorptive capacity than the inorganic component, and so sometimes it is more useful to describe the distribution between organic matter and a component of interest, particularly organic components. This can be done using a distribution coefficient called the K_{om}, which denotes the distribution between organic matter and water. The equation for this distribution is also given in Figure 5.11. When K_{oc}^2 is applied to soil, the numerator would be the kilogram of organic matter in soil.

These constants are determined by mixing a solution of known concentration measured in milligram per liter with a known amount of soil measured in kilograms. After a period of time, the solution and solid are separated and the amount of a given component in solution is measured. K_d can be calculated from this data. This procedure is often performed for various periods of time and at various concentrations of a target compound in solution. In the determination of K_{om}, the amount of organic matter in the solid phase, namely, soil, is determined (see Section 4.7.2), and this amount is used as the kilogram of organic matter.

In these two equations, the larger the amount of component sorbed to the solid phase, the larger the K_d or K_{om} and the less likely it is to move in the soil. Because K_d and K_{om} are determined using the compound in a pure solvent, usually water, and a soil suspension, these constants do not necessarily give information about how difficult the extraction of a component is likely to be when a specific extractant is used, such as mixed solvents or extractants using specific extracting or complexing agents (also see Chapters 11 and 12) [7, 8].

5.12   OXIDATIVE AND REDUCTIVE REACTIONS IN THE SOIL SOLUTION

Many chemical reactions occurring in soils are acid–base reactions; however, oxidation–reduction reactions are more frequently the cause of confusion and may complicate the interpretation of analytical results. During and after a rainfall, the soil will quickly become anaerobic because microorganisms and plant roots use up dissolved and trapped oxygen. Under these conditions, reduction of various constituents takes place. When the soil drains or dries, air replaces the water and there is movement of oxygen back into the pores. However, not all pores drain immediately, and some pores never drain at all. Both these situations lead to anaerobic zones. Even when soils become aerobic, the reactions leading to oxidation of the reduced species may not be fast enough to remove all reduced species before the next anaerobic event. Thus, the relative amounts of oxidized and reduced forms of some species changes over time, even if the total amount of the species remains the same.

Drying soil at an elevated temperature will result in the loss of water from these normally filled pores, thereby allowing reactions that would not

---

[2] A K_{oc}, where oc = organic carbon could be calculated if organic carbon is substituted for organic matter in the equation.

otherwise take place. Also, chemical reactions take place faster at higher temperatures, resulting in reactions taking place much more rapidly than they normally would in soil. This not only changes the amount of compounds in the soil sample but will also change the ratios of the components present. It may even lead to the formation of compounds not naturally present in soil at all. For these reasons, soil is not dried at elevated temperatures before analysis [9].

5.13 MEASURING SOIL WATER

Soil water can be measured using either laboratory or field methods. Common laboratory measurements include the percentage of water on a dry-weight basis and water content as a function of pressure. For the analyst, the determination of the percentage water on a dry-weight basis is the most important. Field methods include tensiometers, porous blocks (shown in Figure 5.12), psychrometers, time-domain reflectrometry (TDR), neutron scattering, and others primarily designed to determine plant available water.

Soil water content may be reported in a number of different ways but is most commonly reported as the amount of water in grams per gram or kilogram of oven dry soil (see equations in Figure 5.13). It may also be reported

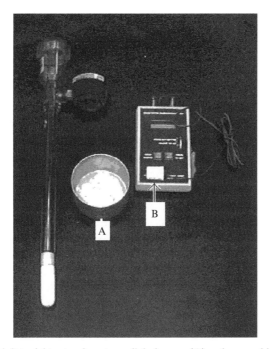

Figure 5.12. From left to right, a tensiometer, soil drying can (A) and porous block (B) and meter, used with B, for determination of soil water.

$$\% \text{ Water} = \frac{\text{soil wet weight} - \text{soil dry weight}}{\text{soil dry weight}} \times 100 \qquad (1)$$

$$\% \text{ Water} = \frac{[(\text{soil wet weight} - \text{can weight}) - (\text{soil dry weight} - \text{can weight})]}{(\text{soil dry weight} - \text{can weight})} \times 100 \qquad (2)$$

Figure 5.13. Determination of percent water on a dry-weight basis. Equation (1) gives the basic equation. In equation (2), soil is dried in a drying can, and thus the can weight must be subtracted to carry out the calculation (see also Figure 5.12).

as the volume of water per volume of soil, that is, the volumetric water content. However, for analytical purposes, it is most often simply presented as a percent of water on a mass basis. In field measurements, the water content of soil is reported as the kilopascals (kPa) of pressure holding the water in the soil. The importance of reporting soil water in this fashion can be seen in Figure 5.3.

More information about the soil solution in relationship to soil analysis is presented in Chapter 10.

5.13.1 Laboratory Methods

The most important laboratory measure of soil water is the percentage water on a dry-weight basis. In most cases, soil will be extracted or analyzed while still moist to minimize changes that occur during drying. To obtain comparable data from multiple analyses, the soil sample weight is corrected using the percent water on a dry-weight basis. Because the water content of soil is highly variable, the dry weight is used, as it is more constant. Soil is typically dried at 105–110°C for 24 hours in a drying cup as shown in Figure 5.13, and the amount of water lost is divided by the dry weight of the sample and multiplied by 100. The basic simplified equation is given in Figure 5.13, equation (1). An equation representing the actual calculation usually made (for soil dried in a drying can; see also Figure 5.12) is given in Figure 5.13, equation (2). Note that the can weight must be subtracted to carry out the calculation.

The percentage of water on a dry-weight basis is used to calculate the dry weight of soil taken for analysis as shown in Figure 5.14. First, a 50-g sample of soil is taken and dried. A second 25-g sample is taken and analyzed. The dried sample is found to weigh 48 g and thus lost 2.0 g of water representing 4.2%. The amount of water (i.e., 4.2/100) is used to determine the dry weight of sample taken for analysis as shown in equation (2), Figure 5.14.

The other common laboratory method uses pressure plates and membranes to measure the amount of water held by soil at various pressures (Figure 5.15). One advantage of this method is that it gives a pressure value that can be used in many calculations relevant to movement of water in soil. The plates in the apparatus shown are used in the pressure range of −10 to −30 kPa. Other

$$\% \text{ Water} = \frac{2}{48} \times 100 = 4.2\%. \tag{1}$$

The water in the moist sample, taken for analysis, is found using this percentage:

$$\frac{4.2}{100} = 0.042$$

$$\text{Dry weight sample analyzed} = \frac{\text{Weight of sample taken}}{1 + 0.042} = \frac{25}{1.042} = 24 \text{ g.} \tag{2}$$

Figure 5.14. Use of percent water on a dry-weight basis to determine the actual weight of soil analyzed.

Figure 5.15. Pressure apparatus for measuring the amount of water held in soil at different pressures; the pressure gage is to the left and a pressure plate is on top of the vessel.

similar pressure plate apparatus can be used to determine water at a pressure of -1500 kPa [10].

5.13.2   Field Methods

Field methods of measuring soil water are primarily designed to measure water in the range of -10 to -1500 kPa of pressure. However, different instruments have different ranges as shown in Table 5.1. Tensiometers, porous blocks, and thermocouple psychrometers are usually installed in the field and measurements are taken on a regular basis. Neutron probe and time domain reflectometry (TDR) equipment are usually carried to the field each time a

TABLE 5.1. Common Instruments for Measuring Soil Water Content in the Field

Instrument	Useful kPa Range	Characteristics
Tensiometers	0 to −86	Limited range in kPa and depth in soil
Thermocouple psychrometers	−50 to $< -10,000$	Wide range but limited accuracy
Porous blocks	−100 to $< -1,500$	Accuracy and range is limited
Neutron probe	0 to $< -1,500$	Cannot be used in high organic soils
Time-domain reflectometry (TDR)	0 to $< -10,000$	Accurate and can be installed at various depths

measurement is made. In addition, the neutron probe requires an access hole into which it is lowered to make a determination of water content. These methods provide data that are not generally used in soil analysis and will not be discussed further [1, 11–15]. Additional information about these methods can be found in the text by Nyle C. Brady and Ray R. Weil [5].

5.14 CONCLUSION

The compounds contained in soil air are basically the same as those in atmospheric air but are more variable. Also, the volume of soil occupied by air varies greatly. Water is a unique molecule in both its physical and chemical characteristics. It has higher than expected boiling and melting points and can dissolve a great variety of compounds. In the soil, it is even more unique in that it occurs in the liquid, gaseous, and solid (frozen) states. The water content of soil is highly variable, ranging from air dry, with as little as 1% water, to saturated, where all void spaces are filled with water. The soil solution contains many inorganic and organic compounds, ions, and gases, the concentrations of which change dramatically when soil water content increases or decreases.

In the laboratory, soil water content is measured by drying in the oven and with a pressure plate apparatus. Drying soil can change the form and species of components present, and for this reason, most soils are air dried carefully or at temperatures only slightly above room temperature before analysis. A number of different field measuring methods are used mostly to determine the amount of water available for plant use.

PROBLEMS

5.1. Describe the differences between atmospheric air and soil air.

5.2. Soil water can be thought of as existing as layers around soil particles. Explain how these layers are differentiated.

5.3. Explain how pores can affect the composition of both the soil atmosphere and the soil solution.

5.4. Using chemical equations, illustrate how soil carbon dioxide affects both the pH of soil and the ions present in the soil solution.

5.5. Give both general and specific examples of how organic compounds can lead to the formation of cation exchange sites in soil.

5.6. What role would K_d play in the extraction of a component from soil?

5.7. Using any available resources, find out how pressure plate methods for determining soil moisture levels work.

5.8. Explain why tensiometers and porous block methods are most useful in the field.

5.9. A soil sample is taken from a field. Half the sample, 50 g, is dried at 105°C for 24 hours after which it is found to weigh 45 g. What is the percent moisture, on a dry-weight basis, of this soil sample?

5.10. The undried soil sample in Problem 5.9 is extracted, analyzed for phosphate, and found to contain 5 µg of phosphate. What is the concentration of phosphate on a dry-weight basis?

REFERENCES

1. Brady NC, Weil RR. *The Nature and Properties of Soils*, 12th ed. Upper Saddle River, NJ: Prentice Hall; 1999, pp. 171–212.

2. Jacinthe PA, Lal R. Effects of soil cover and land-use on the relations flux-concentrations of trace gases. *Soil Sci.* 2004; **169**: 243–259.

3. Rolston DE. Gas flux. In Klute A (ed.), *Methods of Soil Analysis Part 1: Physical and Mineralogical Methods*, 2nd ed. Madison, WI: Soil Science Society of America and American Society of Agronomy; 1994, pp. 1103–1120.

4. Corey AT. Air permeability. In Klute A (ed.), *Methods of Soil Analysis, Part 1: Physical and Mineralogical Methods*, 2nd ed. Madison, WI: Soil Science Society of America and American Society of Agronomy; 1994, pp. 1121–1136.

5. Brady NC, Weil RR. *Elements of the Nature and Properties of Soils*, 3rd ed. Upper Saddle River, NJ: Prentice Hall; 2010, pp. 158–161, 188–191.

6. Jassogne L, Hettiarachchi G, Chittleborough D, McNeill A. Distribution of nutrient elements around micropores. *Soil Sci. Soc. Am. J.* 2009; **73**: 1319–1326.

7. Bohn HL, McNeal BL, O'Connor GA. *Soil Chemistry*, 3rd ed. New York: Wiley; 2004, pp. 217–229, 252.

8. Sparks DL. *Environmental Soil Chemistry*, 2nd ed. New York: Academic Press; 2003, pp. 115–132.

9. Bartlett RJ. Oxidation-reduction status of aerobic soils. In Dowdy RH (ed.), *Chemistry in the Soil Environment*. Madison, WI: American Society of Agronomy, Soil Science Society of America; 1981, pp. 77–102.

10. Klute A. Water retention: laboratory methods. In Klute A (ed.), *Methods of Soil Analysis Part 1: Physical and Mineralogical Methods*, 2nd ed. Madison, WI:

Soil Science Society of America and American Society of Agronomy; 1994, pp. 635–660.

11. Bruce RR, Luxmoore RJ. Water retention: field methods. In Klute A (ed.), *Methods of Soil Analysis Part 1: Physical and Mineralogical Methods*, 2nd ed. Madison, WI: Soil Science Society of America and American Society of Agronomy; 1994, pp. 663–684.

12. Campbell GS, Glendon GW. Water potential: miscellaneous methods. In Klute A (ed.), *Methods of Soil Analysis Part 1: Physical and Mineralogical Methods*, 2nd ed. Madison, WI: Soil Science Society of America and American Society of Agronomy; 1994, pp. 619–632.

13. Cassel DK, Klute A. Water potential: tensiometry. In Klute A (ed.), *Methods of Soil Analysis Part 1: Physical and Mineralogical Methods*, 2nd ed. Madison, WI: Soil Science Society of America and American Society of Agronomy; 1994, pp. 563–594.

14. Rawlins SL, Campbell GS. Water potential: thermocouple psychrometry. In Klute A (ed.), *Methods of Soil Analysis Part 1: Physical and Mineralogical Methods*, 2nd ed. Madison, WI: Soil Science Society of America and American Society of Agronomy; 1994, pp. 597–617.

15. Reeve RC. Water potential: piezometry. In Klute A (ed.), *Methods of Soil Analysis Part 1: Physical and Mineralogical Methods*, 2nd ed. Madison, WI: Soil Science Society of America and American Society of Agronomy; 1994, pp. 545–560.

BIBLIOGRAPHY

Brady NC, Weil RR. *The Nature and Properties of Soils*, 12th ed. Upper Saddle River, NJ: Prentice Hall; 1999.

Coyne, M. *Soil Microbiology: An Experimental Approach*. Cincinnati, OH: Delmar Publishers; 1999, p. 16.

Harper L, Sharpe RR, Scarbrough J, Thornton M, Caddis P, Hunter M, Miller C. Soil-Plant-Animal-Atmosphere Research. 2004. Available at: http://www.spcru.ars.usda.gov. Accessed June 3, 2013.

Hillel D. *Soil and Water Physical Principles and Processes*. New York: Academic Press; 1971.

Warrick AW. *Soil Water Dynamics*. New York: Oxford University Press; 2003.

CHAPTER

6

SPECIATION

For many people, the term *speciation* brings to mind oxidation states of metals and nonmetals; however, this is much too restrictive an idea of speciation, especially when applied to soil. *Speciation* can broadly refer to the form of an element or molecule present under a set of environmental conditions as illustrated in Figure 6.1. This would include metal cation, simple ion, oxyanion, oxidation states, cation and anion associations, ionization states of organic compounds, and organic and inorganic associations. It will also include association of metals, nonmetals, organic ions, and compounds with both inorganic and organic components of soil, especially colloids.

Several different types of species are illustrated in Figure 6.1. The potassium cation (K^+) at the top of the figure is separated from the soil surface by water molecules and would thus be considered an outer-sphere species. The potassium cation near the bottom of the figure is directly connected to the soil particle by an ionic charge and is therefore an inner-sphere species. Above this is an inner-sphere phosphate directly bonded to a soil surface aluminum. Also shown are potassium cations attached (inner sphere) to colloidal clay (CC) and colloidal soil organic matter (COM). Each of these is a different species.

Colloidal associations are particularly important in soil because they allow for movement of otherwise immobile or slowly mobile species. They can also lead to confusing analytical results in that atoms, ions, or molecules may appear to be soluble above their solubility limit when in reality the analytical results represent a very different species.

Introduction to Soil Chemistry: Analysis and Instrumentation, Second Edition.
Alfred R. Conklin, Jr.
© 2014 John Wiley & Sons, Inc. Published 2014 by John Wiley & Sons, Inc.

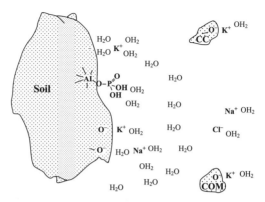

Figure 6.1. Species around soil particles. CC, clay colloid; COM, colloidal organic matter.

Organic compounds might also be thought of as having different species, although they are generally not discussed in this manner. Organic compounds are capable of existing in various conformations, some of which are easily changed, while others are more or less fixed by steric, hydrogen bonding, or other atomic and molecular interactions. In addition, compounds containing double bonds and chiral centers can exist as different optical isomers. For instance, fatty acids contain double bonds, which can either be *cis* or *trans*. Amino acids can be either S or R (D or L) optical isomers (see also Table 6.1). These various "species" are illustrated in Figure 6.2. It is well documented that the various conformations and optical isomers have dramatically differing biological activities. An especially vivid example of this is thalidomide, which exists in two optical isomers, one relieving nausea and the other causing birth defects.

Inorganic and metal-organic compounds also have conformations and isomers, including optical isomers. It is expected that these various conformers and isomers will also have varying biological activities.

There are many other reasons why speciation might be important, including plant nutrient availability, biological contamination by metals, human or animal toxicity, movement of species in the environment, and biological accumulation or amplification. Obtaining the data needed for modeling the chemistry of a species' role in the reactions and fates of various environment components, and development of a basic understanding of a species' chemistry are other reasons why understanding the behavior of different forms of ions and molecules is important. Although any one of these reasons by itself is important, often several of these reasons are important simultaneously [1].

Why might one be specifically interested in the chemistry of a particular species in soil? From a positive perspective, it is desirable to provide plants with nutrients in forms that are available and yet are not present in concentrations high enough to cause environmental harm. Potassium ions are found in the soil solution and on exchange sites. Both species are available to plants. It may also be part of the mineral structure of soil, as in muscovite,

TABLE 6.1. Common Important Ionic Species in Soil

Element/Compound	Cations[a]	Anions
Nitrogen	NH_4^+	NO_3^-, NO_2^{-b}
Phosphorus	None	$H_2PO_4^-$, HPO_4^{2-}, PO_3^{3-}
Potassium	K^+	None
Sulfur	None	S^{2-}, SO_4^{2-}
Chlorine	None	Cl^{-c}
Carbon	None	HCO_3^-, CO_3^{2-}
Iron	Fe^{2+}, $Fe(OH)_2^+$, $Fe(OH)^{2+}$, Fe^{3+}	None
Manganese	Mn^{2+}, Mn^{3+}, Mn^{4+d}	None
Copper	Cu^{2+}, $Cu(OH)^+$	None
Zinc	Zn^{2+}, $Zn(OH)^+$	None
Nickel	Ni^{2+}, Ni^{3+}	None
Boron	None	$H_2BO_3^-$
Cobalt	Co^{2+}	None
Molybdenum	None	MoO_4^{2-}, HMO_4^-

[a]In some cases, other species may exist in soil under certain conditions; however, they are not common.

[b]Nitrogen oxides and nitrogen gas from denitrification will also be present.

[c]Any halide found in soil solution will be present as its anion.

[d]Often found in mixed oxidation states.

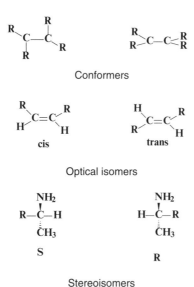

Figure 6.2. Common forms of organic compounds that have the same number and type of elements and bonds arranged in different orientations.

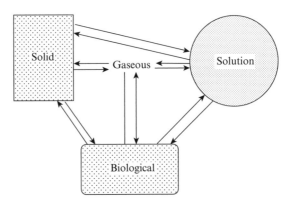

Figure 6.3. Soil compartments.

$KAl_2(AlSi_3O_{10})(F,OH)_2$, or trapped between clay layers, as occurs in illite. Both of these forms of potassium are unavailable to plants.

In addition to knowing about bioavailability of important nutrients, it is also important to recognize that some species are toxic, while others are benign. Some species accumulate in biological tissue; others do not. To fully understand the behavior of chemicals in the environment, it is important to know which of the above-mentioned categories it falls into.

A first step in deciding on an analytical procedure to use or a species to look for is to understand that the species of interest may be in one of four soil compartments (see Figure 6.3): the solid (both inorganic and organic), the liquid (soil solution), the gaseous (soil air), or the biological (living cells). It is also important to remember that molecules and ions can move between compartments and interconvert between species.

Once the compartment to be studied is selected, then the species to be analyzed for can be selected, as illustrated in Figure 6.3. In this case, it is important to consider all possible species that might be present in the compartment. It is also important to consider the changes a species is likely to undergo during the sampling and analytical procedures to which it will be subjected. A further discussion of this topic is given in Section 6.5 [2–4].

6.1 CATIONS

Any positively charged chemical species can be considered a cation or as being in a cationic form. Some will be present as relatively simple, single-oxidation-state cations such as sodium (Na^+). Others may be more complex in that they may have several oxidation states such as iron, in either the Fe^{2+} or Fe^{3+} states. Cations may also have oxygens or hydroxy groups associated with them, such as FeO^+.

In all cases, cations will associate with water molecules. Because of this, it is common to find them written as hydrated or hydroxide species such as

$FeOOH \cdot 0.4H_2O$. Such a representation more closely resembles the actual condition of cations in soil, particularly if the discussion is about cations in the soil solution. This introduces a troublesome situation. When a cation is removed (extracted) from a solid matrix into an extracting solution, a species change most likely occurs. In the solid matrix, the cation may not be associated with water molecules, while in solution it most certainly will be. For hydrated species, the amount of associated water and the activity will also change. This raises the question of what form of the species actually exists in the solid matrix. When analyzing for a specific cation in soil, it is important to keep this issue in mind.

It may seem natural to discuss molybdenum in this section because it is a metal. However, it is always present in soil as an oxyanion and so will be discussed in Section 6.2.2 [5–7].

6.1.1   Simple Cations in Soil

Simple cations are those that exist only in one oxidation state in soil and are mostly associated with water, although they may also be chelated and form other associations with inorganic and organic components.

The alkali and alkaline earth metals are examples of relatively simple cations that occur in only one oxidation state and are surrounded by water (see Figure 6.1). The most common of these ions in soil, in order of decreasing abundance, are calcium (Ca^{2+}), magnesium (Mg^{2+}), potassium (K^+), and sodium (Na^+). Sodium is typically present in very small amounts in high-rainfall areas, whereas it may have a relatively high concentration in low-rainfall areas.

Although simple, these cations can occur as a number of different species. They will be present as exchangeable or, in solution, as hydrated species. They can be part of the inorganic component structure, for example, as in isomorphous substitution (see Section 3.1.3), or can be trapped between clay layers (see previous description). They will also be associated with organic matter and with colloidal inorganic and organic matter, and as chelated species. In many cases, these cation species are further grouped into those available to plants and those that are unavailable. If a more detailed grouping is needed, species can be further divided regarding their availability to plants as into readily, slowly, very slowly, and nonavailable species.

Another simple cation occurring commonly in soil is ammonium (NH_4^+). Because of the unique role and chemistry of nitrogen and nitrogen species in soil, it is discussed separately in Section 6.3 [5,6].

6.1.2   Complex Cations in Soil

Complex cations are those that can occur in several oxidations states in soil and are often associated with oxygen and hydroxy groups while still carrying a positive charge. There are more than 65 metals that may be present as complex cations. Of these, only a few are common in soil. Even those that are not common in soil will, if present, occur predominately in one oxidation state under normal soil conditions. No attempt will be made to cover all these cations; rather, a limited number of the more important cations will be presented to illustrate the chemistry of complex cations in soil.

Examples of cations that are present in significantly lower concentrations than the simple cations are iron, manganese, zinc, copper, nickel, and cobalt. Except for cobalt, these have multiple oxidations states in soil as shown in Table 6.1. Because of their multiple oxidation states, they may be present as many more species than the simple cations. Typically, the higher oxidation states predominate under oxidizing conditions, while the lower oxidation states predominate under reducing conditions. However, it is common to find both or all oxidation states existing at the same time in either aerobic or anaerobic soil [7,8].

6.1.2.1 Iron

Iron and its various oxidation state species are common components of the environment. In addition to the oxides FeO and Fe_2O_3, it is found in minerals such as hematite, goethite, and ferrihydrite, and in a number of hydroxy and oxy compounds. Because of its common occurrence in the environment in general, and in soil in particular, the total iron content of soil is usually not a useful piece of information.

Most commonly, iron is discussed as being in either the ferrous (Fe^{2+}) or ferric (Fe^{3+}) state. Changes between these two depend on the soil's pH and Eh (where Eh is a measure of the oxidation–reduction potential of soil) as discussed in Chapter 9. Acid conditions and low Eh values tend to lead to the production of ferrous ion, while high pH and high Eh values result in the predominance of ferric ion. It should be noted that the ferrous ion is more soluble than the ferric ion and, thus, it will be more available to plants.

Iron cations in both the ferrous and ferric states can act as exchangeable cations; however, ferric ion is generally insoluble and thus not present on exchange sites as such. However, other iron cations that contain oxygen and/ or hydroxide groups may be exchangeable. There are still other iron cations bonded to other ligands (see siderophores as follows) that can exist. Any compound having atoms with electron pairs that can be shared with positive species will associate with iron cations. Any iron species may become attached to soil components such as sand, silt, clay, and inorganic and organic colloids to form still more species.

Because of its common occurrence and biological importance, it is an essential micronutrient for most organisms. Thus, a number of analytical procedures for the analysis of iron species have been developed and typically concentrate on biologically available species [9].

A specific unique chelation association occurs between iron and compounds called siderophores. These compounds usually have a mixture of α-hydroxy carboxylic acids and hydroxaminate functionalities (see Figure 5.7). Iron associates with these groups and thus remains available to plants. A large number of siderophores from both plant and microbial sources have been identified [10].

6.1.2.2 Manganese

Manganese in soil has many characteristics that are similar to iron; for instance, it exists in multiple oxidation states: Mn^{2+}, Mn^{3+}, and Mn^{4+}. Although manganese can exist in the laboratory in other oxidations states, from -3 to $+7$, the $+2$ to $+4$ species are the ones commonly found in soil. Manganese forms various oxide and hydroxide species and chelates with many soil components. Its low oxidation state (i.e., Mn^{2+}) is more soluble and more available than its high oxidation state (i.e., Mn^{4+}).

Manganese does, however, have some unusual characteristics. It is unusual to find soil situations where iron is toxic, whereas manganese toxicity is known. As noted earlier, iron is found in only two oxidation states, while manganese can have three common oxidation states in soil. However, the situation is found to be much more complex than this when soil is analyzed for the species of manganese present. A simple analysis for manganese might indicate an oxidation state of $+3.5$, indicating that the material analyzed contains an unknown mixture of the common oxidation states of soil manganese. This creates problems in understanding the chemistry of manganese and the reactions leading to its deficiency, toxicity, biological availability, and movement in the environment. This is an issue that has received significant attention and research [11].

Permanganate (MnO_4^-) is a strong oxidizing agent and is frequently used as a hoof treatment for farm animals. Because of this, it is not uncommon to find high levels of manganese in soils where animals are treated. Some African soils also naturally have significant concentrations of manganese.

6.1.2.3 Chromium

Chromium has numerous oxidation states, some of which are strongly oxidizing. The most highly oxidized species is Cr^{6+} and the reduced ion is Cr^{3+}. Because of its strong oxidizing characteristics coupled with the presence of species that are oxidizable, Cr^{6+} is rapidly reduced to Cr^{3+} in soil. Chromium species of intermediate oxidation states can exist in soil; however, the $+3$ state is the most common. As with the other metals, all the combinations of species with other components is possible and must be kept in mind when carrying out an analysis for chromium species in soil [12–14].

6.1.2.4 Mercury

Mercury is unusual in that it is found in the environment as both oxidized mercury ions and as reduced methyl mercury. The mercurous (Hg^+) ion is unstable and not likely to be found in soil, while mercuric (Hg^{2+}) ions and methyl mercury compounds are. Mercury ions can form the same types of interactions with soil constituents as those described for other multi-oxidation-state metals. Mercury in all its forms is toxic and thus of concern; however, methyl mercury, which can form in soil under anaerobic conditions, is particularly dangerous because of its extreme toxicity.

Mercury has several other characteristics that make it of particular environmental concern and make it likely to be found as many different species. It is a natural constituent of soil, although it occurs at low concentrations. It is widely used both in industry and in the laboratory, making it a common contaminant of reference soils. Metallic mercury has a relatively high vapor pressure, which means that it can occur in measurable amounts in the soil atmosphere. It has a high affinity for reduced sulfur compounds in soil solids and soluble organic matter that allows species to be present in the soil solution above mercury's solubility limit.

Analysis of mercury is difficult and specialized sampling and instrumental techniques are generally required to carry out an accurate analysis. Although atomic absorption is applicable, it requires specialized heating of the sample such as using a graphite furnace and other specialized sample handling [15–17].

6.1.2.5 *Aluminum*

Aluminum deserves special attention because, although it is present only in one oxidation state, Al^{3+}, it is commonly associated with both oxygen and hydroxy groups and is particularly important in acidic soils. Aluminum may also be bonded or associated with the colloidal inorganic and organic particles and the surfaces of other soil components. In solution, the aluminum ion, Al^{3+}, reacts with water, releasing protons into the soil solution. Thus, under acid conditions, aluminum is more soluble and thus some of its reactions, shown in Figure 6.4, lead to additional soil acidity. Although it is not toxic to most animals, it is toxic to most plants. Therefore, any increase in aluminum in soil is important.

Aluminum is always present in soil as it is a constituent of many soil minerals, particularly clay minerals. As the pH of soil decreases, aluminum from

$$Al^{3+} + H_2O \rightleftharpoons AlOH^{2+} + H^+$$

$$AlOH^{2+} + H_2O \rightleftharpoons Al(OH)_2{}^+ + H^+$$

$$Al(OH)_2{}^+ + H_2O \rightleftharpoons Al(OH)_3 + H^+$$

Figure 6.4. Reactions of aluminum ions in soil solution with the release of protons.

various sources is brought into solution. At very low pH levels, generally less than 3, the very fabric of soil begins to erode, which causes two things to happen:

1. Soil has very high buffering capacity at this point because the added acid is decomposing the inorganic components in soil. This means that a large amount of acid is needed to decrease the pH of soil to a point where all metals are solubilized and can be leached out.
2. The soil itself is destroyed and so at the end of the extraction process soil is no longer present, and what is left is a mixture of highly acidic salts that must be disposed of. It is for this reason that extraction or remediation methods that depend on the acidification of soil to low pH fail and should never be undertaken [18,19].

6.2 ANIONS

Simple anions are those that exist only in one oxidation state in soil and generally are only associated with water. Complex anions are typically oxyanions of nonmetals, although molybdenum occurs as an oxyanion.

6.2.1 Simple Anions in Soil

There is only one simple anion commonly found in soil, and that is chloride (Cl^-). Chloride is an essential nutrient for plants but is typically present in sufficiently high concentrations that deficiencies are never observed. If other halogens are present, they will also be present as simple anions. Most soils do contain small amounts of bromide as the second most common simple anion. In some cases, significant levels of fluoride and iodide may be present, although this is rare. These anions are generally soluble in water and tend to exist as the simple anion. However, they can combine with other components and exist as other species. For instance, halogens are present in organic compounds such as solvents, insecticides, and herbicides, which can be soil contaminants. There are also other nonionic species of these elements that may be present [20].

6.2.2 Complex Anions in Soil (Oxyanions)

Many important soil components are not present as simple cations or anions but as oxyanions that include both important metals and nonmetals. The most common and important metal oxyanion is molybdate (MoO_4^{2-}). The most common and important nonmetal oxyanions are those of carbon (e.g., bicarbonate [HCO_3^-] and carbonate [CO_3^{2-}]), nitrogen (e.g., nitrate [NO_3^-] and nitrite [NO_2^-]), and phosphorus (e.g., monobasic phosphate [$H_2PO_4^-$], dibasic

TABLE 6.2. Common Oxyanions in Soil, Their Chemical Characteristics and Mobility

Oxyanion	Chemical Characteristics	Mobility
Carbonate	Forms insoluble carbonates with cations under basic conditions	Precipitates from solution
Bicarbonate	Forms slightly soluble bicarbonates with cations	Mobile in solution
Nitrate	Available to plants, converted to N_2 gas under anaerobic conditions	Mobile
Nitrite	Readily oxidized to nitrate	Mobile
Monobasic phosphate	Stable at low pH	Considered immobile
Dibasic phosphate	Stable under neutral conditions	Considered immobile
Tribasic phosphate	Stable under basic conditions	Considered immobile
Molybdate	Reacts with various soil constituents	Some mobility
Borate	Acts as simple anion	Mobile

phosphate [HPO_4^{2-}], and tribasic phosphate [PO_4^{3-}]). Sulfate (SO_4^{2-}) and, to a much lesser extent, sulfite (SO_3^{2-}) also occur in soil but are typically at a much lower concentration than other oxyanions. The soil chemistry of oxyanions is complicated by the fact that some act as simple anions and move readily through soil, while others react with numerous soil constituents, forming insoluble immobile species. Common oxyanions in soil and their chemical characteristics and mobility are summarized in Table 6.2.

Molybdate, although present in small amounts in soil, is an essential nutrient for nitrogen fixation, specifically in the enzyme nitrogenase. It does not move readily through soil and is therefore considered to be of limited mobility.

Of the nonmetal oxyanions, those of carbon have a different role in soil than nitrogen and phosphorus. Bicarbonate and carbonate can act as counterions to cations to keep the soil electrically neutral. They are also important because all pH changes in soil tend to involve either carbonate or bicarbonate, and thus, both are involved in soil pH and buffering.

Both nitrogen and phosphorus oxyanions are important because they are a source of nutrients for plants but also are a source of pollution of water. Nitrite and nitrate are of great interest because they are readily formed in soil from organic matter and from inorganic nitrogen-containing compounds, particularly ammonia (NH_3). Soil must be moist but not saturated and at a temperature above 20°C for rapid oxidation of ammonia to nitrite and nitrate. Both oxyanions are mobile in soil and so can be leached into groundwater and find their way into lakes, ponds, and drinking water. Nitrate can move through soil with water and be taken up by plants. In saturated soil under reducing conditions and temperatures between 20 and 40°C, it is readily converted to nitrogen gas (N_2).

Phosphorus oxyanions are entirely different from nitrogen oxyanions. First, the oxyanion species present is controlled by the pH; also, phosphate oxyanions are generally not mobile in soil. However, sandy soils and soils high in phosphorus are exceptions to this rule. Any soil, though, can lose phosphate by erosion and this phosphate can cause environmental problems. Because of its unique chemistry, phosphorus will be discussed separately later.

Sulfur oxyanions are similar to nitrate and nitrite in that they are mobile in soil and can be converted to many different forms (see Chapter 4, Figure 4.8). However, they are different from the other oxyanions in that they are the source of the essential nutrient, sulfur, and are deposited on soil from the atmosphere. Sulfate species can be determined by X-ray fluoresence (XRF), making their determination easier than the other oxyanions [21,22]. Sulfur is discussed in more detail in Section 6.2.2.3.

Boron and arsenic are natural components of soil and are both present as oxyanions. Boron is present as boric acid or borate polymers, and arsenic is present as arsenate. While boron is weakly held by soil, arsenic is similar to phosphate in its interactions with soil constituents. Boron is an essential nutrient for plants; however, it is also toxic to plants at relatively low levels. Arsenic is toxic. The laboratory chemistry of both of these elements is well understood, but their environmental chemistry, speciation and movement, is less well understood [23–27].

Another oxyanion of interest to soil chemists is that of tungsten, which is in itself important but is also important because it forms polymers and reacts with both molybdenum and phosphorus oxyanions to form mixed polymers. Because tungsten species are toxic, its oxyanionic species, including polymers, are of interest [28].

6.2.2.1 Nitrogen

Nitrogen is unique because it is found in soil as both reduced and oxidized species, as part of organic molecules, as oxidized gaseous species, and in the elemental form (i.e., nitrogen gas). Many nitrogen species are soluble in water and thus tend to be found in the soil solution as opposed to attached to soil particles. There are, however, important exceptions to this rule. The ammonium ion has a positive charge and is thus attracted to soil cation exchange sites. In the fine-grained mica-type clays, ammonia can be trapped between clay sheets. In this condition, the ammonium is not available to plants and is seldom biologically or environmentally available.

Amino acids and proteins also contain nitrogen. While amino acids are soluble in water and thus are expected to move readily through soil, they are also zwitterions (having both positive and negative charges) and thus may interact with charges on soil particles, depending on the pH, and move slowly or not at all. Proteins are polymers of amino acid and may or may not be soluble but can have charges as do amino acids. There are also a large number of other types of nitrogen-containing compounds present in soil, for example,

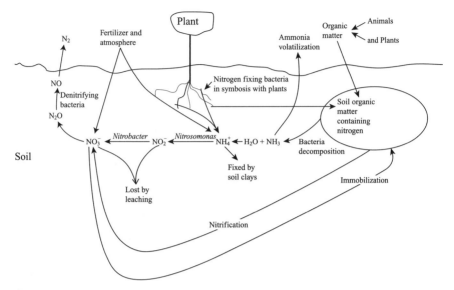

Figure 6.5. The nitrogen cycle showing some of the many species of nitrogen occurring in the soil environment.

amino sugars and nitrogen-containing lipids, and some are soluble while others are insoluble.

All positively charged nitrogen species and all organic nitrogen-containing compounds are readily converted to nitrite and then to nitrate in soil. Because of this ready conversion, all are sources of nitrogen for plants.

Except for those species mentioned previously, nitrogen moves between the various soil compartments easily, depending on the soil environmental conditions. These reactions and conversions are often outlined as the nitrogen cycle shown in Figure 6.5. The slowest and most energy-intensive part of the cycle is the "fixation" of nitrogen gas (N_2) forming ammonia (NH_3), various oxyanions, and amino acids. Once these initial compounds are formed, subsequent changes in nitrogen species is fairly rapidly and easily accomplished.

Ammonium (NH_4^+) may be added as a fertilizer or released during organic matter decomposition. It is the chief species present immediately after the injection of anhydrous ammonia into soil as a fertilizer. It is labile in soil in that, under aerobic, moist, moderate temperature conditions, it is rapidly oxidized to nitrite and nitrate. Ammonia may also be volatilized from the soil solution, particularly under basic soil conditions.

Oxidized species of nitrogen, chiefly nitrite and nitrate, occur in all soils and in the soil solution. Nitrite in the environment is of concern because of its toxicity. Its occurrence is usually limited because the oxidation of nitrite to nitrate is more rapid than the oxidation of ammonia to nitrite. Both nitrite and nitrate move readily in soil and nitrate is available to plants as a source of nitrogen and can move to plant roots with water.

As a result of these reactions and volatilization, concentrations of nitrogen species are expected to change during sampling and sample storage. This is particularly true if precautions are not taken to limit this eventuality [29–31].

6.2.2.2 Phosphorus

To help understand the occurrence of phosphate oxyanion species in soil, the titration of phosphoric acid is instructive. Phosphoric acid (H_3PO_4) is triprotic and each of the protons has a different pK_a. Thus, the titration curve has three inflection points reflecting the titration of each of these protons. At low pH (>2), the chief form is phosphoric acid. Between a pH of about 3 and 7, the primary form is $H_2PO_4^-$, which is called *monobasic phosphate*. At higher pHs (above 8 and less than 12), the chief form will be dibasic phosphate, HPO_4^{2-}, and finally at high pH (>13), the form will be tribasic[1] (PO_4^{3-}). It might be assumed that the species of phosphorus found in soil would be solely controlled by pH. This is not the case. While the inorganic phosphate species found is dependent in large measure on pH, organic species are not.

Phosphorus occurs almost exclusively as either monobasic or dibasic phosphate in the various soil compartments and in biological tissue. However, these two species react with a host of both inorganic and organic components, forming a multitude of other species (e.g., adenosine triphosphate, adenosine diphosphate, and adenosine monophosphate). Species that form when phosphate reacts with surface constituents on soil particles, including colloidal inorganic matter and organic particles, are often overlooked. Species containing phosphorus that is not in the form of phosphate are rare.

Organic phosphorus associations can occur under both acidic and basic conditions. Phosphorus can form esters with organic alcohol functional groups and can be associated with amine groups in various ways.

Phosphate reacts and forms insoluble compounds with many metals, particularly iron, aluminum, and calcium. Under acid soil conditions, both iron and aluminum become more soluble, and thus as soil pH decreases, its "phosphate fixing power" increases. This means that iron and aluminum react with phosphate to form insoluble species that are not available to plants. Under basic conditions, high concentrations of calcium exist and insoluble calcium phosphates form. Insoluble phosphate species are also formed with other metals that happen to be present; however, the three mentioned are generally present in the highest concentration, and so they represent the major reactants with phosphate. Iron, aluminum, and calcium phosphates can also occur as coatings on soil particles.

When analyzing soil for phosphorus, all these potential forms must be kept in mind. It is to be expected that all soluble forms of phosphorus will be

---

[1] The terms mono-, di-, and tribasic are used because the counterion is usually a base cation such as Na^+, K^+, Ca^+, or Mg^{++}, all of which form bases such as NaOH.

available to plants, while all insoluble forms will not. However, precipitation processes will also play a role in phosphorus availability. Initial precipitation results in small crystals with high surface area-to-mass ratio and thus greater reactivity and tendency to move into solution when the concentration of phosphorus in solution decreases as with plant uptake. On the other hand, as time passes, the crystals grow larger, thus decreasing their surface area-to-mass ratio as well as their reactivity and availability [32].

Phosphorus is one of the most important elements in soil chemistry because it is involved in numerous reactions with many different components. In addition to the species described earlier, phosphate will also form species with uranium, arsenic, and zinc. It also reacts with organic matter and with humic and fulvic acids to form environmentally important species [33–37].

6.2.2.3   *Sulfur*

Sulfur is an essential plant element occurring in soil primarily as sulfate, SO_4^{2-}, and in organic combinations. It can also exist in many other forms as shown in Figure 4.8. Sulfate is mobile in soil and easily extracted with water, but determining the quantity is not easy. There are four standard procedures used for its analysis: (1) oxiding all inorganic sulfur compounds to sulfate and analyzing; (2) reducing all inorganic sulfur compounds to hydrogen sulfide (H_2S), distilling, and analyzing; (3) converting inorganic compounds to sulfur dioxide (SO_2) and analyzing; and (4) decomposing organic sulfur compounds by methods similar to Kjeldahl analysis and analyzing by one of the inorganic methods mentioned earlier. Using these methods, information about the specific species present is lost. For this reason, new methods that preserve speciation information such as XRF and high-performance liquid chromatography–mass spectrometry (HPLC-MS) are preferred. For additional information on these methods, see Section 14.13 and Section 15.2.2.

Sulfur has four unique characteristics related to its occurrence and chemistry in soil. As sulfate, it is one of the principle counterions that keep the soil electrically neutral. Soil receives constant additions of sulfur through volcanic activity around the world and industrial pollution, usually in the form of acid rain. This means that soils usually have sufficient sulfur for plant growth. Lastly, plants can take and use sulfur dioxide from the air as a source of sulfur for growth [22,38].

6.3   ISOLATION OF SPECIES

To determine the various species and their concentrations in soil, many selective, semiselective, and sequential extraction methods have been developed. Species associated with various components in soil can be extracted with varying effectiveness (see Chapters 11 and 12). Thus, metal cations that are in solution, exchangeable, weakly held, or associated with carbonate and with

TABLE 6.3. BCR Extraction Sequence for 1 g of Soil. Extraction is at Room Temperature Unless Otherwise Noted[a]

	Components Extracted	Conditions
Step 1	Acid extractable	40 mL 0.11 M acetic acid, shaken 16 h, centrifuged
Step 2	Reducible	40 mL 0.1 M hydroxylamine chloride, pH 2, shaken 16 h, centrifuged
Step 3	Oxidizable	10 mL 8.8 M hydrogen peroxide 1 h, add 10 mL 8.8 M hydrogen peroxide heat 1 h 85°C, reduce volume, extract with 50 mL 1 M ammonium acetate, shaken 16 h, centrifuged
Step 4	Residual	7.5 mL 12 M HCl, 2.5 mL 16 M HNO$_3$, 16 h, 2 h reflux cool, filter

[a]Adapted from Nemati K, Baker NKA, Abas MRB, Sobhanzadeh E, Low KH. Comparison of unmodified and modified BCR sequential extraction schemes for the fractionation of heavy metals in shrimp aquaculture sludge from Selangor, Malaysia. *Environ. Mont. Assess.* 2011; **176**: 313–320.

organic matter can theoretically be removed selectively and analyzed. Also, oxides and reducible species can be extracted and determined. This is accomplished by using different extractants or a combination of extractants sequentially. Using a modified Community Bureau of Reference of the European Commission (BCR) sequential extraction method, acid extractable, reducible, oxidizable, and residual metals can be determined in soil.

A sequential listing of the various steps involved in a modified BCR method is given in Table 6.3.

Analyzing Table 6.3, it is obvious that this procedure harkens back to the work originally done by Liebig, Laws, and Way in the 1600s. The second thing that is striking is that the total procedure would take 68 hours or nearly 3 days for the extractions alone, not considering centrifugation and analysis of the extracts.

There have been some questions raised about the validity of results of extensive and sequential extraction methods. There is the possibility that species of an analyte may change during the extraction process. It is also possible that a species may be liberated and then reabsorbed during extraction or subsequent isolation. The same or similar question could be raised about various initial sampling and subsequent storage methodologies as described next. In spite of these questions, the BCR methodology has been widely accepted [39–41].

6.4   SAMPLING, SAMPLE STORAGE, AND SPECIATION

The problem of species changes during sampling and storage can be ameliorated in two ways. First, the soil component of interest can be measured *in situ*, thus removing problems associated with sampling and extraction. This approach, while optimal, is feasible in only a few instances. The second approach is to make the analysis as quickly after sampling as possible, preferably in the field so as to eliminate storage problems. Keeping the time between sampling and analysis to minutes or seconds is desirable.

When the above-mentioned approaches are not possible, steps must be taken to account for changes occurring during sampling and storage. These can be in either direction; that is, the concentration of a species may increase or decrease or be converted to an entirely different species. It is not possible to give a prescription for sampling or storage that will guarantee the stability of a soil species over a period of time. Some species will be sensitive to oxidation, while others will be changed by the lack of oxygen. In some cases, species will be stable in air dry soil, while others are more stable in moist soil. In some cases, refrigeration at 0°C is best, while other species will require storage at –40°C or even lower temperatures. In yet other cases, storage at room temperature will be sufficient. The solution to sampling and storage problems is to study the stability of the species of interest under various sampling and storage conditions and thus determine which is best. The method of standard additions can be used under these conditions to investigate potential changes and take these into account [42].

Some generalized sampling and storage recommendations, however, can be made. During sampling and storage, samples must be kept under the same oxygen concentration, temperature, and pressure conditions that exist in the field if species integrity is to be maintained. Samples taken from anaerobic, low-, or high-temperature or high-pressure conditions should be maintained under these conditions during sampling, storage, and until actual analysis is carried out. For example, a sample taken from the bottom of a lake will be under higher-pressure and lower-oxygen conditions than is surface water and should be kept under these same conditions to maintain sample integrity [2–4].

6.5   CONCLUSIONS

This discussion is not intended to be a comprehensive discussion of all the elements, cations, anions, and organic species present in soil. It gives a range of species that illustrate the common situations related to the analysis of these species. Different species have different biological reactivity and availability and it is important to know which species is present. Cations have water molecules surrounding them and some cations react with water, forming oxy and hydroxy cations. Other cations can be present in multiple oxidation states and in mixtures of these oxidation states. Some species added to soil will rapidly be converted to other species, such as the conversion of Cr^{6+} to Cr^{3+}, such that analysis for the original species may be fruitless, especially if the soil sample has been stored for any length of time. All species may be associated in various ways with the inorganic, organic, and colloidal components of soil to form species of interest. Some oxyanions are soluble in the soil solution and readily move in soil, whereas others are immobile. In some cases, species of interest may be combined into new compounds such as to be readily, slowly, or non-available to plants.

PROBLEMS

6.1. Give some examples of anions including oxyanions that are common in soil.

6.2. Give examples of cations including oxy and hydroxy cations common in soil.

6.3. Write reactions that illustrate how metal cations can lead to the release of protons into the soil solution.

6.4. Illustrate solution, cation exchange, and outer- and inner-sphere species around a soil particle.

6.5. Compare the reactivity and movement of nitrogen and phosphate oxy-anions in soil.

6.6. Explain why the occurrence of colloidal species is important in the analysis of soil for other species.

6.7. Compare the characteristics of manganese and iron in soil and describe their similarities and differences.

6.8. What special roles do carbonate and bicarbonate play in the chemistry of soil?

6.9. Explain why analysis of soil for Cr^{6+} is generally not an enlightening analysis.

6.10. Diagram the compartments in the soil environment and the movement of species between these compartments.

REFERENCES

1. Nolan AL, Lombi E, McLaughlin MJ. Metal bioaccumulation and toxicity in soils— why bother with speciation? *Aust. J. Chem.* (online) 2003; **56**: 77–91. http:// www.publish.csiro.au/journals.ajc. Accessed May 31, 2013.

2. Conklin AR, Jr. *Field Sampling Principles and Practices in Environmental Analysis.* New York: Marcel Dekker; 2004, pp. 224–245.

3. Weber G, Messerschmidt J, von Bohlen A, Alt F. Effect of extraction pH on metal speciation in plant root extracts. *Fresenius. J. Anal. Chem.* 2001; **371**: 921–926.

4. Göttlein A, Matzner E. Microscale heterogeneity of acidity related stress-parameters in the soil solution of a forested cambic podzol. *Plant Soil* 1997; **192**: 95–105.

5. Helmke PA, Sparks DL. Lithium, sodium, potassium, rubidium, and cesium. In Bartels JM (ed.), *Methods of Soil Analysis Part 3: Chemical Methods.* Madison, WI:

Soil Science Society of America and American Society of Agronomy; 1996, pp. 551–574.

6. Suarze DL. Beryllium, magnesium, calcium, strontium, and barium. In Bartels JM (ed.), *Methods of Soil Analysis Part 3: Chemical Methods*. Madison, WI: Soil Science Society of America and American Society of Agronomy; 1996, pp. 575–601.

7. Wells ML, Kozelka PB, Bruland KW. The complexation of "dissolved" Cu, Zn, Cd and Pb by soluble and colloidal organic matter in Narragansett Bay RI. *Mar. Chem.* 1998; **62**: 203–217.

8. Marin A, López-Gonzálvez A, Barbas C. Development and validation of extraction methods for determination of zinc and arsenic speciation in soils using focused ultrasound: application to heavy metal study in mud and soils. *Anal. Chim. Acta* 2001; **442**: 305–318.

9. Loeppert RH, Insheep WP. Iron. In Bartels JM (ed.), *Methods of Soil Analysis Part 3: Chemical Methods*. Madison, WI: Soil Science Society of America and American Society of Agronomy; 1996, pp. 639–664.

10. Sandy M, Butler A. Microbial iron acquisition: marine and terrestrial siderophores. *Chem. Rev.* 2009; **109**: 4580–4595.

11. Gambrell RP. Manganese. In Bartels JM (ed.), *Methods of Soil Analysis Part 3: Chemical Methods*. Madison, WI: Soil Science Society of America and American Society of Agronomy; 1996, pp. 665–682.

12. Bartlett RJ, James BR. Chromium. In Bartels JM (ed.), *Methods of Soil Analysis Part 3: Chemical Methods*. Madison, WI: Soil Science Society of America and American Society of Agronomy; 1996, pp. 683–701.

13. Howe JA, Loeppert RH, DeRose VJ, Hunter DB, Bertsch PM. Localization and speciation of chromium in subterranean clover using XRF, XANES, and EPR spectroscopy. *Environ. Sci. Technol.* 2003; **37**: 4091–4097.

14. Wang J, Ashley K, Marlow D, England EC, Carlton G. Field method for the determination of hexavalent chromium by ultrasonication and strong anion-exchange solid-phase extraction. *Anal. Chem.* 1999; **71**: 1027–1032.

15. Hempel M, Wilken RD, Miess R, Hertwich J, Beyer K. Mercury contamination sites — behavior of mercury and its species in lysimeter experiments. *Water Air Soil Pollut.* 1995; **80**: 1089–1098.

16. Crock JG. Mercury. In Bartels JM (ed.), *Methods of Soil Analysis Part: 3 Chemical Methods*. Madison, WI: Soil Science Society of America and American Society of Agronomy; 1996, pp. 769–791.

17. Ravichandran M. Interactions between mercury and dissolved organic matter — a review. *Chemosphere* 2004; **55**: 319–331.

18. Bertsch PM, Bloom PR. Aluminum. In Bartels JM (ed.), *Methods of Soil Analysis Part 3: Chemical Methods*. Madison, WI: Soil Science Society of America and American Society of Agronomy; 1996, pp. 517–550.

19. Kerven GL, Ostatek-Boczynski Z, Edwards DG, Asher CJ, Oweczkin J. Chromatographic techniques for the separation of Al and associated organic ligands present in soil solution. *Plant Soil* 1995; **171**: 29–34.

20. Frankenberger WT, Jr., Tabatabai MA, Adriano DC, Doner HE. Bromine, chlorine, fluorine. In Bartels JM (ed.), *Methods of Soil Analysis Part 3: Chemical Methods*.

Madison, WI: Soil Science Society of America and American Society of Agronomy; 1996, pp. 833–867.

21. Prietzel J, Thieme J, Neuhäusler U, Susini J, Kögel-Knabner I. Speciation of sulphur in soils and soil particles by X-ray spectromicroscopy. *Eur. J. Soil Sci.* 2003; **54**: 423–433.

22. Boye K, Almkvist G, Nilsson SI, Eriksen J, Persson I. Quantification of chemical sulphur species in bulk soil and organic sulphur fractions by S K-edge XANES spectroscopy. *Eur. J. Soil Sci.* 2011; **62**: 874–881.

23. Keren R. Boron. In Bartels JM (ed.), *Methods of Soil Analysis Part 3: Chemical Methods*. Madison, WI: Soil Science Society of America and American Society of Agronomy; 1996, pp. 603–626.

24. Vaughan PJ, Suarez DL. Constant capacitance model computation of boron speciation for varying soil water content. *Vadose Zone J.* 2003; **2**: 253–258.

25. Tu C, Ma LQ. Effects of arsenic concentrations and forms on arsenic uptake by the hyperaccumulator ladder brake. *J. Environ. Qual.* 2002; **31**: 641–647.

26. Roy P, Saha A. Metabolic and toxicity of arsenic: a human carcinogen. *Curr. Sci.* 2002; **82**: 38–45.

27. Sims JL. Molybdenum and cobalt. In Bartels JM (ed.), *Methods of Soil Analysis Part 3: Chemical Methods*. Madison, WI: Soil Science Society of America and American Society of Agronomy; 1996, pp. 723–737.

28. Bednar AJ, Mirecki JE, Inouye LS, Winfield LE, Larson SL, Ringelberg DB. The determination of tungsten, molybdenum, and phosphorus oxyanions by high performance liquid chromatography inductively coupled plasma mass spectrometry. *Talanta* 2007; **72**: 1828–1832.

29. Bremner JM. Nitrogen — total. In Bartels JM (ed.), *Methods of Soil Analysis Part 3: Chemical Methods*. Madison, WI: Soil Science Society of America and American Society of Agronomy; 1996, pp. 1085–1121.

30. Stevenson FJ. Nitrogen — organic forms. In Bartels JM (ed.), *Methods of Soil Analysis Part 3: Chemical Methods*. Madison, WI: Soil Science Society of America and American Society of Agronomy; 1996, pp. 1185–1200.

31. Mulvaney RL. Nitrogen — inorganic forms. In Bartels JM (ed.), *Methods of Soil Analysis Part 3: Chemical Methods*. Madison, WI: Soil Science Society of America and American Society of Agronomy; 1996, pp. 1123–1184.

32. Kuo S. Phosphorus. In Bartels JM (ed.), *Methods of Soil Analysis Part 3: Chemical Methods*. Madison, WI: Soil Science Society of America and American Society of Agronomy; 1996, pp. 869–919.

33. Negrin MA, Espino-Mesa M, Hernandez-Moreno JM. Effect of water : soil ratio on phosphate release: P, aluminium and fulvic acid associations in water extracts from Andisols and Andic soils. *Eur. J. Soil Sci.* 1996; **47**: 385–393.

34. Doolette AL, Smernik RJ. Soil organic phosphorus speciation using spectroscopic techniques. In Bunemann EK, Oberson A, Frossard E (eds.), *Phosphorus in Action: Biological Processes in Soil Phosphorus Cycling Book Series: Soil Biology*, Vol. 26. New York: Springer; 2011, pp. 3–36.

35. Gerke J. Humic (organic matter)-Al(Fe)-phosphate complexes: an underestimated phosphate form in soils and source of plant-available phosphate. *Soil Sci.* 2010; **175**: 417–425.

36. Van Moorleghem C, Six L, Degryse F, Smolders E, Merckx R. Effect of organic P forms and P present in inorganic colloids on the determination of dissolved P in environmental samples by the diffusive gradient in thin films technique, ion chromatography, and colorimetry. *Anal. Chem.* 2011; **83**: 5317–5323.

37. Dou Z, Ramberg CF, Toth JD, Wang Y, Sharpley AN, Boyd SE, Chen CR, Williams D, Xu ZH. Phosphorus speciation and sorption-desorption characteristics in heavily manured soils. *Soil Sci. Soc. Am. J.* 2009; **73**: 93–101.

38. Tabatabai MA. Sulfur. In Bartels JM (ed.), *Methods of Soil Analysis Part 3: Chemical Methods.* Madison, WI: Soil Science Society of America and American Society of Agronomy; 1996, pp. 921–960.

39. Whalley C, Grant A. Assessment of the phase selectivity of the European Community Bureau of Reference (BCR) sequential extraction procedure for metals in sediment. *Anal. Chim. Acta* 1994; **291**: 287–295.

40. Fernández E, Jiménez R, Lallena AM, Aguilar J. Evaluation of the BCR sequential extraction procedure applied for two unpolluted Spanish soils. *Environ. Pollut.* 2004; **131**: 355–364.

41. Sheppard MI, Stephenson M. *Evaluation critique des methodes d'extraction selectives pour sols et sediments Contaminated Soils.* 3. International Conference on the Biogeochemistry of Trace Elements, Paris (France), 15–19 Mai 1995 INRA, Paris (France) publication Date, 1997.

42. Saxberg BEH, Kowalski BR. General standard addition method. *Anal. Chem.* 1979; **51**: 1031–1038.

BIBLIOGRAPHY

Allen HE, Huang CP, Bailey GW, Bowers AR. *Metal Speciation and Contamination of Soil.* Boca Raton, FL: CRC Press; 1994.

Bohn HL, Myer RA, O'Connor GA. *Soil Chemistry*, 3rd ed. New York: John Wiley; 2001.

Coyne M. *Soil Microbiology: An Exploratory Approach.* Cincinnati, OH: Delmar Publishers; 1999.

Gimpel J, Zhang H, Davison W, Edwards AC. In situ trace metal speciation in lake waters using DGT, dialysis and filtration. *Environ. Sci. Technol.* 2003; **37**: 138–146.

Lindsay WL. Inorganic phase equilibria of micronutrients in soil. In Mortvedt JJ, Giordano PM, Lindsay WL (eds.), *Micronutrients in Agriculture.* Madison, WI: Soil Science Society of America; 1972.

Ritchie GSP, Sposito G. Speciation in soils. In Ure AM, Davidson CM (eds.), *Chemical Speciation in the Environment*, 2nd ed. Malden, MA: Wiley-Blackwell Science, Inc.; 2002, pp. 237–263.

Ure AM, Davidson CM. Chemical speciation in soils and related materials by selective chemical extraction. In Ure AM, Davidson CM (eds.), *Chemical Speciation in the Environment*, 2nd ed. Malden, MA: Wiley-Blackwell Science, Inc.; 2002, pp. 265–299.

SOIL AND SOIL SOLUTION SAMPLING, SAMPLE TRANSPORT, AND STORAGE

When sampling, local rules apply. No single, simple set of guidelines is applicable to all sampling situations. General sampling principles will need to be adapted for each field and sampling situation.

ALFRED R. CONKLIN, JR. 2004 [1]

The emphasis in this chapter is on obtaining representative soil samples from the field for subsequent laboratory analysis. The term field is used generally to represent any soil area to be sampled whether it is a small contaminated area or a multihectare agricultural area. It might also be an area partially or wholly submerged under water some part of the year. Soil solution sampling as part of a field sampling exercise will also be discussed. Various methods of

_____

Introduction to Soil Chemistry: Analysis and Instrumentation, Second Edition.
Alfred R. Conklin, Jr.

obtaining chemical information about a field that do not involve physically taking a sample to the laboratory will be touched upon.

There are two phases to soil sampling, a field sampling phase and a laboratory sampling phase. Of these two, field sampling will always be the major source of variation and inaccuracy. Soil solution sampling will be subject to the same variations and inaccuracies as field and laboratory samplings.

Field sampling, sample transport, and laboratory sampling are the three steps that must be carried out before sample analysis in the laboratory. Not getting a representative sample in the field, transport, and storage under nonideal conditions, and improper sampling in the laboratory can all cause dramatic changes in the results of an analytical procedure and thus alter its accuracy. The effect of these factors on variation in the data obtained is always larger than the inherent accuracy of the actual chemical procedure.

Because of the heterogeneity of soils in general and fields in particular, statistics alone cannot provide the best guidance regarding sampling for all situations. Instead some random samples can be taken and the results used in combination with statistical approaches to guide the selection of additional sampling sites.

7.1   FIELD SAMPLING

The sample taken from the field is extremely small compared to the area it is intended to represent. A typical agricultural field will have a composite sample taken for every 5 or more acres it contains. At this point, let us consider simply 1 acre or hectare (abbreviated as "ha") as the size being sampled. The sample represents both the area of the field and a certain depth of soil, usually 15 cm (6 in.).[1] Although soil bulk density varies considerably, it is generally assumed that a soil area of 1 ha, 15 cm deep, contains 2,000,000 kg (an acre 6 in. deep is assumed to contain 2,000,000 lb) of soil. Even if a kilogram of soil is taken as the sample, which would be a large sample, it still only represents 0.00005% of the tillable soil in that field. Then, in most cases, in the laboratory, a 1-g subsample of this kilogram will actually be analyzed, representing just 0.1% of the kilogram sample taken from the field, or 0.00000005% of the tillable soil in a hectare.

Considering a soil profile (see Figure 2.1, Figure 2.2, Figure 2.3, Figure 2.4, Figure 2.5, Figure 2.6, Figure 2.7, and Figure 2.8) and samples taken to a 15-cm depth, the kilogram sample of soil represents an even smaller percentage of the total soil. Assuming a soil depth of 1 m, it would contain 13,000,000 kg of soil, so the kilogram represents an even smaller percentage of the mass of soil. Considering a deep soil and subsoil down to contact with rock or to the water table, all soil samples are very small compared to the volume or mass of soil they are to represent [2].

---

[1] This depth is taken from the common plowing depth.

This sample is like using a snapshot of you at one particular instant in your life and saying it is a valid representation of your whole life. Although this analogy is not mathematically rigorous, it puts in perspective the difficulty of obtaining a good, representative soil sample.

With this in mind, it is obvious that field sampling is the most important component of soil analysis, especially in terms of obtaining a truly good indication of the average conditions of that field and determining its average chemical composition. It is common to come to the conclusion that the more samples taken, the better. Taken to its logical conclusion, this would mean digging up the whole field and somehow analyzing all of it. This is impractical and counterproductive, particularly if the soil is intended to be used for some purpose. Thus, compromises must be made in order to get a soil sample that is representative of the field and yet small enough to analyze.

There are two general types of field sampling. In many cases, tracking the fate of a particular analyte that originates from a point source is of concern. This would be the case of a leaking pipe or a spill. In this case, sampling is somewhat simpler in that sampling can start at the source and radiate outward in all directions until the analyte is found to be at background level. At this point, the area of interest has been found and can be treated appropriately. Figure 7.1 shows a leaking pipe as the point source, located on a downhill slope with areas of contamination around the leak.

Millions of soil samples per year are analyzed for nonpoint source analytes such as potassium, phosphorous, insecticides, and herbicides. It is necessary to obtain an idea of the level of these components throughout the field and to take corrective action as necessary. Thus, a more random sampling methodology is called for. Figure 7.2 shows a field that includes both different soil types (letters) and topographical features (e.g., a ditch) that are to be sampled. Also, a transect line and locations of sample sites along it are indicated. Transect lines and sampling sites are discussed later.

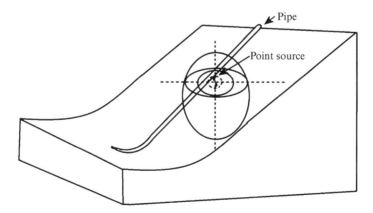

Figure 7.1. Point source of contamination. Circles represent areas of different levels of contamination.

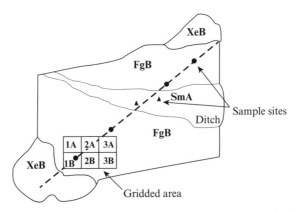

Figure 7.2. A field surface with transect, a gridded section (grid in lower left) and different topographic features. If the whole field were to be sampled, the whole field would be gridded. FgB, SnA, and XeB are standard abbreviations for different types of soil.

In either of the above-mentioned cases, two types of information need to be obtained before sampling and analysis are performed. First, information about the background level of the component of interest is needed. It might be assumed that transect samples can be taken in the area starting from one place that is contamination free and continuing across the contaminated area to another point where there is no contamination. This would delineate the area of interest or concern.

However, this might not be possible. The analyte of interest may very well be naturally present in the soil. Thus, trying to find the point where the analyte does not exist in the soil would be futile. In many cases, samples need to be taken from a point where only background or natural levels of the component of interest are present to another point where the same conditions exist. Knowledge of the background levels of components of interest is thus essential in all soil sampling and analysis.

Information about the soil in its natural state is thus an essential part of any soil sampling. This information may be available from previous soil analysis, that is, before contamination took place. This type of information is commonly available from the state's land grant university and the areas' soil survey. Another option would be to obtain historical soil samples. Historical soil samples are samples of the soil taken before contamination has occurred and thus can be used to ascertain the natural levels of components of interest in that soil. These contain information that may not be readily available otherwise. Caution must be used because storage of soil samples can change analyte composition including the most prominent species present (see Table 7.1 and Reference 2).

7.1.1 Before Sampling

It is always beneficial to obtain as much information as possible about a field to be sampled before actual sampling begins. Important questions about the

TABLE 7.1. Container, Preservative, and Storage Times for Various Soil Pollutants

Sample Type	Container	Preservative	Storage Time (Days)
Soil for plant nutrient analysis	Plastic, paper, or fabric	None	1 day before air drying
Soil for contaminant analysis (see below)	As below	As below	As below
Semivolatiles			
Solid samples	Varies with sample	Varies with method and sample	7
Acrolein/acrylonitrile	Glass with PTEF[a] liner	$\leq 6°$, pH 4–5	7
Water samples	Glass with PTEF	$\leq 6°$, pH 2	14
Inorganics			
Chloride	Plastic or glass	None	28
pH	Plastic or glass	None	As soon as possible
Nitrate	Plastic or glass	$\leq 6°$	2
Sulfate	Plastic or glass	$\leq 6°$	28
Metals			
Chromium VI	Plastic or glass	$< 6°$	1
Mercury	Plastic or glass	HNO_3, pH < 2	28
Hexane extractables			
Oils and greases	Glass	$\leq 6°$	28

[a]Polytetrafluoroethylene (Teflon).
Sample Preservation, Storage, Handling and Documentation. http://denr.sd.gov/des/gw/spills/Handbook/SOP7.pdf 04/26/2011.

area include the types of agricultural or industrial activities that have taken place in the area and their duration. In some instances, information gained from remote sensing and geographic information systems (GISs) may be available and valuable in guiding a chemical investigation.

In places where there are concerns about buried materials or sources of contamination, ground penetrating radar (GPR) surveys may be helpful in determining the best sampling approach. GPR can also detect coarse textured subsoils and water relationships in soils having these types of horizons. However, there are some limits as to where GPR can be used and the equipment for doing a GPR survey is expensive [3].

The field and any subportions of the field to be sampled must have specific unique labels. These labels are typically recorded in a field or laboratory notebook. The notebook will also contain their exact location using a global positioning system (GPS). The position of any samples taken in any part of the area is also recorded using its GPS location. This is the only way one can be certain were a sample was taken. Flags or other markers are always subject to being lost or moved.

It is important to determine and record the accuracy of the GPS equipment being used in the particular sampling situation. Despite frequently discussed accuracies of GPS equipment, many times, the accuracy and precision of GPS equipment is not up to advertised levels.

All rules pertaining to scientific notebooks are applicable to field notebooks and must be followed. Every entry must be dated and each sample labeled with their associated description in the book. No removed pages, no erasures, and no sections whited out are allowed. Sections may be crossed out if necessary, but in a manner in which the words can still be read.

7.1.1.1   Sampler Composition

Because soil is abrasive, it can be expected that small amounts of the sampler itself may be transferred from the sampler to the sample during sampling. Many soil samplers are constructed of metal and pose no problems for most samples. Figure 7.3 shows two of many types of metal samplers available. Plastic samplers of similar design are also available. If low concentrations of metals are an important component of the analysis, then knowing the sampler composition may be important.

For obtaining deeper samples or large numbers of samples, it may be practical to use a mechanical sampler. Truck-mounted samplers are available and can be used to take samples meters deep. Other more robust samplers may be necessary if very deep samples must be taken.

7.1.1.2   Containers

Container composition and type are a critical consideration in soil sample collection and transport. Plastic, cloth, and paper bags are commonly used for containing, transporting, and storing soil samples (see Figure 7.4). Several aspects of the container must be addressed if accurate analyses are to be obtained. Paper bags can allow samples to lose analytes and water. They also allow samples to absorb contaminants, particularly gases and volatile organics. Plastic will retain soil water but will probably be subject to gas exchange. Plastics may also contribute contaminates to the sample. For example,

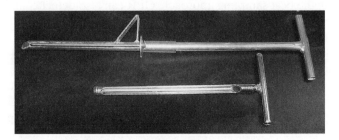

Figure 7.3. Typical soil samplers.

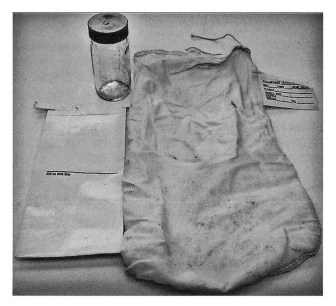

Figure 7.4. Common soil sample containers.

bisphenols may leach into soil samples during transport and storage. Such contaminants can have wide-ranging effects on analytical results. Cloth bags are the most porous soil containers and allow soil to dry out during transport and storage. As with paper, contamination from environmental sources is possible.

Glass jars make good containers because they are more likely to isolate soil samples from possible contamination. They can, however, break. It is important to remember that the composition of lid liners may be important in maintaining soil sample integrity. Liners can have the same drawbacks as plastic or paper containers.

7.1.1.3   Analyte Variation

In addition to sampler composition and container type, the analyte of interest will affect sampling. Table 7.1 and Table 7.2 give a partial list of considerations relative to sample container and storage that are determined by the analyte of interest.

7.1.2   Grab Sample

There is a tendency to "grab" a sample when first visiting a site. While this shows enthusiasm, it does little else. There is little or no attention paid to where the sample is taken or whether or not the sample is representative of the field or some portion of the field. Because of this, any analysis done on the sample

TABLE 7.2. Analyte Variation Requiring Differential Field Activities

Variant	Field Activity
Contaminant concentration	Different remediation intensities or remediation methods. High concentrations may require intensive leaching or even soil removal. Low concentrations may best be handled by phytoremediation.
Plant nutrient level	Appropriate fertilizer levels applied to minimize fertilizer cost and loss.
pH	Varying levels of pH modifier applied, calcium carbonate for acidic soils and sulfur for basic soils.
Pest concentration	Different pesticides and pesticide application rates, herbicides insecticides, and others, depending on pest and pest population.

is not useful or can even be detrimental. For instance, if the area is contaminated, this could lead to contamination of personnel or equipment. Furthermore, because sampling equipment and proper containment and storage are probably not available when the sample is "grabbed," the validity of the sample is unknown. Obtaining "grab" samples is to be discouraged at all times.

7.1.3 Point Source Sampling

In some cases, the source of contamination may be known, such as a point source, and sampling is simplified somewhat in this case (see Figure 7.5). Leaking gasoline, other fuel, or waste water pipes, as indicated in the figure, are common point sources of contamination. Other common sources are underground or aboveground storage tanks or other storage facilities. This type of contamination is the simplest to sample and analyze for two reasons. First, the area of contamination is usually well defined. Second, the analyte of interest is usually well known and methods of extraction and analysis are well developed.

Only the area around the source of contamination need be sampled. As a first step, transect sampling, at random points along dashed lines in Figure 7.5 (also see Figure 7.2), can be done to find the general extent of the contamination, that is, from a point where background levels of contamination are present on both sides of the contamination.

Transect sampling provides information on the extent of contamination, the intensity of contamination (level), and the area that exceeds the actionable levels[2] such that remediation is required. The solid circular lines in Figure 7.5 indicate various levels of contaminant around the leak.

---

[2] A level of analyte enough above the background level to be of concern and require remediation. In agriculture, it might be a low level of nutrient requiring additional fertilizer.

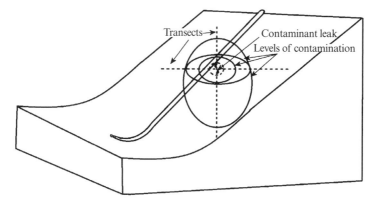

Figure 7.5. Point source contamination with transects and varying levels of contamination.

With point sources, particularly when they occur on areas with significant slope, there is a tendency to assume that the contamination is all downslope. Most of it will be, but contamination both upslope and to the sides is also likely. Transects that take these areas into account must be made.

7.1.3.1 Transect Sampling

Transect sampling is used to obtain initial basic information about a field and the inherent variability in that field. An imaginary line is drawn across the area in such a way as to intersect as many obvious variations as possible. Samples are then taken along this line using the method and instrumentation that will be used in the final sampling. Samples are taken randomly along the transect as indicated by the circles. Additional samples may be taken in unusual portions of the field as indicated by the triangles. If the area is extremely variable or on a slope, it may be advisable to sample along another transect 90° to the original transect.

If contamination is the reason for sampling, samples at depth along the transect line will also be required. Depending on the soil depth, depth to the water table, the severity of contamination, and the type of soil, deep samples may need to be taken at each location. As with the transect, depth samples must be taken until the contaminant or analyte of interest is found to be at background levels. Thus, several depths may need to be taken at each location along the transect.

7.1.3.2 Grid Sampling

After transect sampling, a more detailed sampling of the contaminated area can be undertaken. A first approach would be to take random samples from the whole area. This could be done by randomly picking GPS coordinates that occur in the field and randomly sampling these coordinates. Although

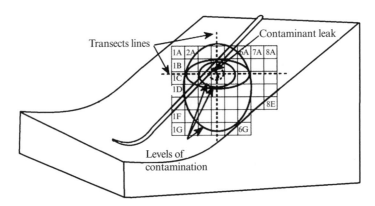

Figure 7.6. Point source with transect and grid sampling.

this is statistically valid, sampling is usually not done this way. For both environmental and economic reasons, different parts of the area will probably be treated separately as indicated by the different contamination levels shown in Figure 7.6. Areas with lower contamination may be remediated in place, whereas soil in areas of higher concentration areas may need to be removed for remediation.

The area is gridded and each grid area given a specific designation as shown in Figure 7.6. Note that the grid can be placed in any orientation desired to obtain the needed samples and information. Different methods of grid labeling can be used, but the one shown is common and simple. Note that areas outside the indicated contaminated area are also gridded. It is important to sample these areas to verify that the contamination has not spread outside the delineated areas. Samples from these areas are also used as reference areas showing the native levels of soil components.

Each grid area is typically sampled by taking three to four cores such as would be obtained using the sampler shown in Figure 7.3. The cores are combined and mixed to obtain the final sample for analysis.

In point source sampling, consideration of the topography, in this case (Figure 7.6) the slope and soil type, will be important. Discussion of soil types and sampling is presented in the next section.

7.1.4   Nonpoint Source Sampling

Nonpoint source sampling occurs where the analyte of interest is dispersed over a large area such that a specific point of origin cannot be ascertained. The innate occurrence of analytes of interest would be an example of a nonpoint source. The occurrence of plant nutrients, either naturally occurring or from fertilization, is an example of a nonpoint source of agricultural analytes. Herbicides, insecticides, and pest-control agents are, once applied on a field scale, also potential nonpoint sources of analytes. It is common to think of crop

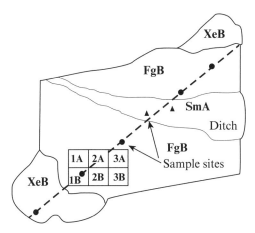

Figure 7.7. Transect sampling for a nonpoint source analyte.

production fields as nonpoint sources; however, golf courses, forests, and parks can also be nonpoint sources of components of environmental concern because they also have amendments, fertilizers, herbicides, and insecticides applied to them. As with point source sampling, Table 7.1 and Table 7.2 apply.

7.1.4.1   Transect Sampling

Transect sampling for nonpoint sources is done the same way and with the same purpose as it is in point source sampling. Figure 7.7 shows a transect across an area to be sampled. Different markers indicate that different sampling will be accomplished at these sites. These would also be sites for depth sampling if it is called for.

7.1.4.2   Grid Sampling

After transect sampling, a more detailed sampling of the field can be undertaken. A first approach is to take random samples from the whole area (see Section 7.1.3.2 for further discussion of this approach). In addition, there may be different soil types indicated by the letters and numbers, as can be seen in Figure 7.7. Even if the surface soil has a more or less uniform color, the soil types may be very different. Mollisols, Vertisols, and Andisols (see Chapter 2) can all have dark surface horizons and yet be very different in their physical and chemical characteristics, including sorption characteristics. For these reasons, fields are usually sampled using a grid or similar sampling methods. In this method, an area to be sampled is partitioned into a grid and each segment of the grid is sampled individually. In most instances, it is best to have each grid encompass an individual soil type, although in some cases, the entire area may be one soil type with only small areas of other soils included. In this case, the variations caused by these inclusions may be ignored.

Figure 7.7 shows an area of the field that has been gridded for sampling. Note that the gridding is accomplished in the same way it was as for point source sampling and is done for the same reasons. Each grid is given either just a number or a number and letter designation as discussed previously. Each area is sampled for the analyte of interest. It is common to take three to four cores from each area and to combine them to make one composite sample of the area. For areas where contamination exists or is suspected, samples at various depths must be taken and can be taken following the general plan used for surface samples.

The topography of an area must also be taken into account when gridding it for nonpoint sampling [4,5]. The tops of slopes will have more erosion of contaminates and soil. Thus, they will generally be lower in contaminant concentration. Lower areas, particularly where water ponds or accumulates, will have higher concentrations of contaminants. Soil color is a valuable characteristic to observe and record in these situations. The tops of slopes will be lighter and typically redder in color, while lower areas will have darker soil colors.

Even if the area of deposition of contaminants is not obvious by other means, it is observable by soil color. As the soil color becomes darker, the deposition of material, including organic material, is higher. These areas thus need special attention when sampling. The dark color is due to increased organic matter in most cases. The higher organic matter content of these soils affects analytical procedures and results and thus is important to note when sampling [6].

Samples at Depth. Samples must also be taken at various depths during both the transect and gridded sampling when contamination is suspected. These samples are similar to transect samples in that one must determine the depth to which the contamination has leached. If, however, contamination is found in wells around the contaminated area, it can be assumed that contamination is most probably throughout the soil profile. Even in this case, soil samples must be taken at various depths to ascertain the level of contaminants. Soils with prominent B horizons high in clay (illustrated in Figure 2.2) may have a buildup of contaminant in the B horizon with lesser amounts both above and below it. Soils with coarse B and C horizons (see Figure 2.2 and Figure 2.5) may have high concentrations of contamination in the A horizon and low levels in the B and C horizons because of their high sand content.

Sampling the surface to a depth of 15–20 cm is the most common practice in agricultural areas. Attention to soil type is critical in this kind of sampling. The best analytical results will be obtained when all samples are from the same soil type, although even here, large variations can and do occur. Two aspects of soil that are important to observe in the field are color and topography. Soils are linked in that soil movement in the field will be controlled by topography and erosion or deposition of soil will change its color. This will also affect the depth of penetration of the analyte of interest. Low-lying soils are, in general, darker and have more contaminants and plant nutrients both at the surface and at depth.

7.2 SAMPLING CROPPED LAND

Sampling cropped land is similar to the sampling described previously in most respects. Agricultural chemicals are spread more or less evenly over large areas and are thus considered nonpoint sources of contamination. However, when sampling, it is important to keep in mind that on a sampling scale, agricultural chemicals, particularly insecticides and fertilizers, are applied in rows along with the seed. Thus, sampling in the rows will give different results than sampling between the rows. In addition, these chemicals are applied at one time of the year, and so their concentration changes with the time between application and sampling. When crop rows are discernible, it is recommended that samples be taken between the rows to avoid having unnaturally high levels of agricultural chemical in the samples (see Figure 7.8).

Usually, soil sampling will be concentrated during one particular time of the year. If analytical results are to be compared with previous work, samples should be taken during the time period that the previous samples were taken. Temperature, moisture, freezing, and thawing can all have an effect on analytical results.

In addition to obtaining soil samples and transporting them to the laboratory, there has been research into on-the-go soil analysis. This is accomplished using sensors attached to implements in contact with the soil. A soil sample is taken, analyzed, and then returned to the field over a short period of time. In addition to specific sensors, reflection and conductance are other approaches used to obtain information about soil. Electrical measurements of soil are further discussed in Chapter 9.

On-the-go measurements are accomplished as the equipment moves across the field. GPS location and the duration of time of sampling, measuring, and recording must be accurate to make sure that the data refer to the place where

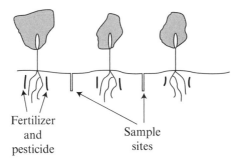

Figure 7.8. Sampling between rows of a crop planted field.

the sample is taken because the equipment is in a different part of the field by the time the data are recorded. Using this approach, individual sample accuracy is lower than laboratory results, but taking hundreds of samples in place of only one results in greater location accuracy. Accuracy and precision of this type of sampling is questionable although improving all the time [7].

7.3   ENVIRONMENTAL SAMPLING

Environmental sampling is undertaken to ascertain the levels of contaminants in soil. The general outline for environmental sampling given in Sections 7.1 and 7.2 can be applied depending on the situation. In the United States, the United States Environmental Protection Agency (USEPA) is responsible for rules and regulations regarding sampling of hazardous sites, such as Superfund areas. These sampling and test methods can be found in Reference 8, with specific reference to SW-846 Methods. Other sources of sampling recommendations related to specific topics can be found at the American Society for Testing Materials (ASTM) and the Occupational Safety & Health Administration (OSHA) Web sites. State-specific sampling rules and regulations may also be applicable. See the state's EPA sites for these [2,8].

7.4   OTHER ENVIRONMENTAL SAMPLING SITUATIONS

Terms used, methods of sampling, and sample handling will be different for water and submerged solid and semisolid samples. The methods described earlier (i.e., transect and grid sampling) are also applicable to these situations. However, semisolid samples, such as those obtained from lake bottoms, require a special sampler.

7.5   SAMPLE TRANSPORT AND STORAGE

Samples must be carefully labeled as they are taken. Labels must tell where, when, and how they were taken and include reference to the field notebook page where they are described. One other effective method of labeling is to use bar codes for each sample. This facilitates sample handling and decreases misidentification of samples. Bar codes can be generated in the field and attached to samples before shipment.

If there is a question of either loss of sample or contamination during transport, a quality control sample can be included with the shipment. A sample containing a known amount of the analyte or one of its related compounds can be included with the samples during transport. If the analyte or related material is changed in any way, then the samples are not acceptable for further investigation.

No single sample storage condition is applicable to all samples. This is particularly true when considering the type of analysis to be carried out and the analyte of interest. A general indication of maximum holding times for various analyses is given at the Environmental Protection Agency (EPA Web site http://www.epa.gov), SW-846 Chapter 2, Test Methods for Evaluating Solid Waste, Physical/Chemical, pp. 61–65, Tables 2-40(A) and 2-40(B)). Table 7.2 gives the general idea of the contents of this site. This site does not cover all analytes nor all storage conditions. This is not a copy of Tables 40(A) and 40(B) but rather an interpretation of the tables.

Changes in analyte concentration can easily occur during transport and storage. These changes can be brought about by changes in temperature, moisture content, and exposure to fumes. This might result in loss of contaminant by evaporation or the absorption of contamination if transport conditions are not carefully controlled. In addition, all soils contain millions of microorganisms that will be active during transport and storage. Thus, significant changes in analyte concentration can occur. Common examples would be the oxidation of ammonia to nitrate and the oxidation of hydrocarbons.

It would be logical to assume that sample composition would be best preserved by storage at low temperature since this would slow or stop chemical and biological reactions. However, microorganisms can be active at or below freezing temperatures [9] and storage at 4°C can have a dramatic effect on microbial populations [10].

Typically, it is best to move the sample from the field to the laboratory as quickly as possible and to minimize storage time in the laboratory. A container, often insulated, is commonly used for containing samples during transport. Using an insulated container allows one to maintain a desired temperature between field and laboratory and to isolate the samples from possible contamination. Table 7.1 shows the container, preservative, and storage times used for various types of common samples.

7.6 LABORATORY SAMPLING

The most common first step in laboratory sampling is air drying the sample unless the analyte or method specifically calls for other handling and storage. This is done for most agricultural samples where the analytes of most interest are phosphorus, potassium, calcium, magnesium, and micronutrients. Drying samples at significantly higher or lower than room temperature can significantly alter analytical results [11]. The same holds for environmental samples. Drying can significantly alter the results of analysis of inorganic components. Any drying, even air drying at room temperature can significantly alter organic contaminants in soil, particularly if these components have a significant vapor pressure.

Once dry, the second most common step is to sieve and mix the individual soil sample cores from a single grid. In some cases, samples may need to be ground before mixing and sieving. This can be accomplished with a mechanical grinder or a simple mortar and pestle. Grinding must not pulverize rocks present in the sample, heat the sample, or add components to the sample, such as metals from the grinder.

Samples, assuming they do not need to be ground, are at a minimum sieved through a number 10 sieve that has openings of 2 mm. This is done because the upper limit of sand, for consideration as part of soil, is 2 mm in average diameter. Components larger than this have a relatively low surface-to-volume ratio and very little sorptive capacity. Sieved samples can be further mixed and then divided into subsamples.

Subsampling can be done by hand, although several machines for subsampling are available. Another approach is to have an analytical method or methodologies that allow analysis of many different analytes in one sample. In this way, there is no question of variability between subsamples. This is not always possible, but minimizing steps in any procedure will decrease variability and uncertainty in the analytical results.

At this point, the samples are ready for analysis. However, before analysis, it is important to determine the moisture content of the air-dried soil. This is reported as the percent water on a dry-weight basis. Air dry soil contains water, usually 1–2%, although as organic matter increases, so does the air dry water content. In extreme cases, limited to organic or high organic matter soils, the percent water may approach 100%.

As stated previously, even low-temperature oven drying changes analytical results—sometimes dramatically [12]. The effect of drying a soil at 23–27°C vs. 93.3°C is shown in Figure 7.9. Soil samples were dried at the indicated temperatures and analyzed by standard soil analysis methods designed to determine biologically available forms of nutrients. Some results were only slightly different, as in the case of pH, while others were significantly changed by drying. Differences ranged from 16% for potassium to 47% for sulfur as sulfate.

A more recent study of the effect of drying on various species of phosphate in soil showed that drying had little effect on some species while having a pronounced effect on others [13]. In nature, soil is never heated to the temperatures used in oven drying, and thus the results of analysis of oven-dried soils are not considered representative of the soils' natural conditions.

Other considerations aside, such as changes in mass, heating causes changes in the species present. This results from loss of water, which causes changes in solubility, loss of water of hydration, precipitation, and reactions with other species present. The analytical results thus do not represent the true amount of biologically available species present and thus do not represent their environmental impact. Considering the importance of these nutrients, these differences can make a significant difference in the assessment of their environmental effect and fertilizer recommendations. Thus, analysis is usually carried out on air dry or field moist soil. The purpose of oven drying soil is to be able to correct analytical results on a soil oven dry basis by correcting for the sample's air dry water content, which allows comparison of analytical results over a wide range of environmental conditions.

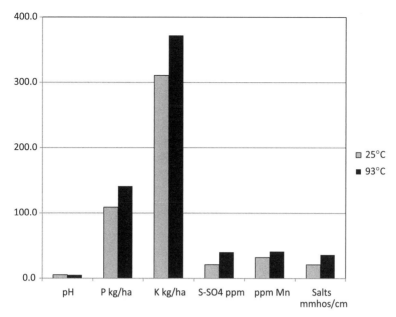

Figure 7.9. Effect of drying temperature on the analysis of a Brookston silty loam developed from glacial till and eolian silt. kg/ha is kilogram per hectare and ppm is parts per million (selected data adapted from Molloy and Lockman [12]).

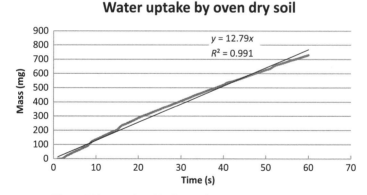

Figure 7.10. Uptake of hydroscopic water by oven dry soil.

Soil is hygroscopic. The water content of air dry soil will increase and decrease as the relative humidity of the air it is exposed to changes. The uptake of water by oven dry soil is shown in Figure 7.10. Thus, determination of water content just before analysis is critical.

Because soil contains pores that tightly hold water, it is not sufficient to simply dry soil at 100°C. On the other hand, heating too much above 100°C will cause changes in soil mass due to loss of various constituents other than water. The compromise is to heat the soil slightly above 100–105°C for 24

Figure 7.11. Drying cup for determination of soil moisture.

hours. Ideally, the sample should be dried until a constant mass is obtained; however, this is not usually done in practice.

Drying is often accomplished in a cup such as the aluminum cup shown in Figure 7.11. Similar cups made of iron are available, and although they quickly become rusty, they are still quite serviceable in this state.

Method 7.1 gives the general equation used for calculating the percentage of soil moisture on a dry-weight basis. Note that because the percent moisture is small, a relatively large sample, compared to other samples taken for analysis, is usually taken for this analysis. The results give the percentage moisture on a mass-to-mass basis, which is the most common way to report soil moisture.

Method 7.1. Determination of Moisture in Air Dry Soil on a Mass Basis

A 50- to 100-g sample of air-dried soil is placed in a drying can of known weight and placed in an oven at 105°C for 24 hours. It is removed and weighed and the percentage of moisture on a dry-weight basis determined using the following equation:

$$\% \text{ Moisture} = \frac{(\text{wet weight of soil} - \text{can weight})^1 - (\text{dry weight} - \text{can weight})^1}{(\text{dry weight} - \text{can weight})^1} \times 100.^2$$

1. Wet weight and dry weight of soil refer to wet weight and dry weight while still in the drying can.
2. To make this measurement truly scientific, the procedure would be repeated until a constant weight is obtained.

In some analyses, it is important or relevant to report the percentage moisture on a volume basis. In doing this, it is common to assume that the density of water is 1 g/mL (at room temperature, the density of water is a little less than 1). Knowing or determining the bulk density of the soil being used will allow the grams of soil to be converted to volume and, thus, a percentage water on a volume basis can be determined. The general form of the operations needed to make this calculation is given in Method 7.2.

Method 7.2. Conversion of Percentage of Moisture on a Mass Basis-to-Volume Basis Using Inverse Density

$$\frac{cc}{gram\ soil} = \frac{1}{bulk\ density}$$

$$grams\ soil \times \frac{cc}{gram} = volume\ soil$$

$$\frac{volume\ water}{volume\ soil} \times 100 = \%\ moisture\ on\ volume\ basis$$

For this calculation, $1\ cc = 1\ mL = 1\ cm^3 = 1$ g water.

7.7 SAMPLING THE SOIL SOLUTION

Although sampling and subsequently extracting analytes of interest is the standard method of soil analysis, sometimes it is advantageous to be able to sample and analyze the soil solution without addition of extractant. To do so, the soil solution can be collected in the field by collecting water that has percolated through a soil profile. Alternately, soil samples can be taken and the water isolated in the laboratory. In either case, the question as to the validity of the results of analytical analysis will depend on the status of the soil and the questions being asked.

7.7.1 The Lysimeter

Lysimeters [14] allow the collection of soil solution that has percolated through a largely undisturbed soil column under natural conditions. A soil monolith is isolated from surrounding soil by a metal cylinder (see Figure 7.12). The bottom of the soil is removed and replaced by filter material, sand in some cases, and the bottom sealed but with a drain that allows for the collection of water percolating through the soil column. The monolith may have various sampling probes placed in it that allow for the measurement of various parameters and sampling of water.

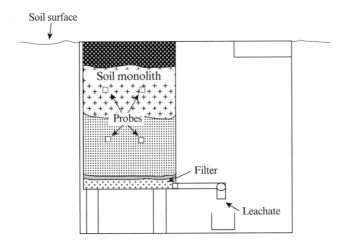

Figure 7.12. A typical setup for a lysimeter.

It might be argued that the same information could be obtained by analyzing groundwater, and to a certain extent, it can. But the lysimeter allows for two types of measurements that cannot be obtained otherwise. First, the rate of movement and fate of solutes within the monolith can be followed. Second, the fate of toxic elements and compounds that cannot be released into the environment can be ascertained.

Porous cups can be used as probes in the monolith and suction can be applied to them to move water out of the soil and into a sample container under unsaturated conditions. Passive samplers that do not involve pressure or vacuum are also available [11].

7.7.2 Water-Saturated Soil

Not everyone has access to a lysimeter, which means that one often needs to obtain the soil solution from a soil sample that has been collected in the field. It is easiest to remove the soil solution when the soil is saturated. Thus, experiments are sometimes carried out using soil suspensions. In these cases, questions regarding oxidation–reduction potential and its effect on the species or compound under investigation come in to play. For example, are reduced species that are observed under these conditions common in the field and do they play a significant role in the chemistry of that particular soil, or are they artifacts of the experimental conditions? This question must be both asked and answered for the results to be useful.

Soil solution samples from saturated soils can be obtained by simple filtration. Simple gravity filtration is preferable to vacuum filtration methods because vacuum filtration can lead to distortions in the composition of analyte composition in filtrates. Syringe filters are usually not capable of handling soil and so are not recommended. Also, some filters can retain analytes of interest.

If a syringe filter is to be used, it must be checked to make sure that it is not retaining the analyte of interest or, if it is, how much is being retained.

One place where soil solution from saturated soils is routinely used is in the determination of salts in soil. Salts can build up in soil in low-rainfall areas. A measure of the salt content is essential for these soils. Typically, a saturated paste of soil is made and the water filtered by vacuum filtration. The filtrate can then be analyzed for electrical conductivity and this measurement, together with calcium, magnesium, and sodium content, can be used to calculate various measures of salt content and potential detrimental effects of salts on crops and water quality.

Of particular concern in this analysis is sodium because it destroys soil structure, is associated with increased soil pH, and can be toxic to plants. Sodium can easily be determined by atomic absorption spectroscopy (AAS), flame ionization spectroscopy (FIS), and inductively coupled plasma (ICP) methods. Soil structure is discussed in Chapter 2 and the various spectroscopic methods discussed in Chapter 14.

7.7.3 Unsaturated Soil

If it is necessary or desirable, water can be extracted from unsaturated soils in the laboratory. This requires either pressure or suction to move water from the soil. A common laboratory method for removing water from unsaturated soils is the pressure plate (see Figure 7.13). Plates for this type of extractor can be used for extraction of water at field capacity ~-33 kPa and at permanent wilting point ~-1500 kPa.[3]

Other methods such as those used in lysimeters may also be used in laboratory sampling situations, when soil is not saturated with water.

7.8 CONCLUSIONS

The major source of variation in soil analysis occurs during field sampling as a result of the inherent variability in soil. This variability must be accounted for by using a sampling strategy that takes these differences into account. Two types of situations relative to soil analyte content are the point source and nonpoint source of analyte of interest. Two general types of sampling are commonly used: transect sampling and grid sampling. Both are applied to point and nonpoint sources of analytes. In the laboratory, soil samples are commonly air-dried, sieved, and the percentage of moisture on a dry-weight basis determined before analysis. Soil moisture from either saturated or unsaturated soil can be extracted and analyzed for contaminant levels if necessary or desirable.

---

[3] Different sources will give different values for field capacity and permanent wilting points; however, these are the values most commonly used.

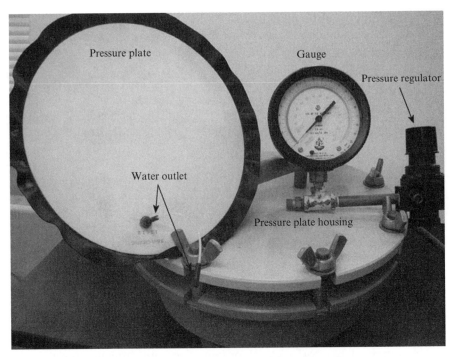

Figure 7.13. Pressure plate for removing water from soil at various pressures.

PROBLEMS

7.1. Soils are highly variable. Using information in this chapter and in previous chapters explain where this variability comes from.

7.2. Describe the differences in sampling procedures that would be used when sampling point source and nonpoint source contaminations. Are there any methods of analysis that would be common to both situations?

7.3. Explain how and why storage and transport of soil samples might affect analytical results.

7.4. Soil is hygroscopic. What does this mean and how might it affect analysis?

7.5. After air drying, a subsample is taken and the percentage of moisture on a dry-weight basis is determined. What might this information be used for?

7.6. Considering chemical reactions only, what, in general, happens when the temperature of a chemical reaction is increased?

7.7. Using the information in Method 7.1, complete the following calculation. If a 75-g sample were taken in Problem 7.5 and was found to weight 74 g after drying at 105°C for 24 hours, what is the percent moisture of this soil sample on a dry-weight basis?

7.8. Water that can easily be removed from soil is called gravitational water. Some of the water retained against the pull of gravity is called plant available water. Which of these two would be expected to have more soluble salts? Why?

7.9. Describe two situations where a lysimeter would be invaluable for studying the chemistry of a contaminant in soil.

7.10. Normally, mass measurements are not made on hot objects. Waiting for an oven-dried soil sample to cool in air will have what effect on its mass?

REFERENCES

1. Conklin AR, Jr. *Field Sampling Principles and Practices of Environmental Sampling*. New York: Marcel Dekker; 2004, p. 120.

2. Gerlach RW, Nocerino JM. http://www.epa.gov/osp/hstl/tsc/Gerlach2003.pdf. 2003. Accessed June 3, 2013.

3. Doolittle JA, Jenkinson B, Hopkins D, Ulmer M, Tuttle W. Hydropedological investigations with ground-penetrating radar (GPR): estimating water-table depths and local ground-water flow pattern in areas of course-textured soils. *Geoderma* 2006; **131**: 317–329.

4. Cavigelli MA, Lengnick LL, Buyer JS, Fravel D, Handoo Z, McCarty G, Millner P, Sikora L, Wright S, Vinyard B, Rabenhorst M. Landscape level variation in soil resources and microbial properties in a no-till corn field. *Appl. Soil Ecol.* 2005; **29**: 99–123.

5. Khalili-Rad M, Nourbakhsh F, Jalalian A, Eghbal MK. The effects of slope position on soil biological properties in an eroded toposequence. *Arid Land Res. Manag.* 2011; **25**: 308–312.

6. Lombnaes P, Chang AC, Singh BR. Organic ligand, competing cation, and pH effects on dissolution of zinc in soils. *Pedosphere* 2008; **18**: 92–101.

7. Adamchuk VI, Rossel RAV, Marx DB, Samal AK. Using targeted sampling to process multivariate soil sensing data. *Geoderma* 2011; **163**: 63–73.

8. http://www.epa.gov/epawaste/hazard/testmethods/index.htm. 2013. Accessed June 3, 2013.

9. Dortz SH, Sparrman T, Nilsson MB, Schleucher J, Oquist MG. Both catabolic and anabolic heterotrophic microbial activity proceed in frozen soil. *Proc. Natl. Acad. Sci. U.S.A.* 2010; **107**: 21046–21051.

10. Nuñez-Reguerira L, Barros N, Barja I. Effect of storage of soil at 4°C on the microbial activity studied by microcalorimetry. *J. Therm. Anal.* 1994; **41**: 1379–1383.

11. Koopmans GF, Groenenberg JE. Effects of soil oven-drying on concentrations and speciation of trace metals and dissolved organic matter in soil solution extracts of sandy soils. *Geoderma* 2011; **161**: 147–158.

12. Molloy MG, Lockman RB. Soil analysis as affected by drying temperatures. *Commun. Soil Sci. Plant Anal.* 1979; **10**: 545–550.

13. Achat DL, Augusto L, Gallet-Budynek A, Bakker MR. Drying-induced changes in phosphorus status of soils with contrasting soil organic matter contents—implications for laboratory approaches. *Geoderma* 2012; **187–188**: 41–48.

14. Meissner R, Seyfarth M *Measuring Water and Solute Balance with new Lysimeter Techniques*. The Regional Institute ISBN 1 920842 268, SuperSoil, Regional Institute, Australia, 2004.

BIBLIOGRAPHY

Allen R, Howell T, Pruitt W, Walter I, Jensen M, eds. *Lysimeters for Evapotranspiration and Environmental Measurements*. Proceedings of the International Symposium on Lysimetry. Honolulu, Hawaii. July 23–25, 1991.

Conklin AR, Jr. *Field Sampling Principles and Practices in Environmental Analysis*. New York: Marcel Dekker; 2004.

Fuhr F, Hance JR, Plimmer JR, Nelson JO, eds. *The Lysimeter Concept Environmental Behavior of Pesticides*. American Chemical Society Symposium Series, No. 699. September 10, 1998.

Jol HM. *Ground Penetrating Radar Theory and Applications*. Amsterdam, Holland: Elsevier Science; 2009.

Kaplan ED, Hegarty C, eds. *Understanding GPS: Principles and Applications*, 2nd ed. London, England: Artech House; 2005.

Kennedy M. *Introducing Geographic Information Systems with ArcGIS: Workbook Approach to Learning GIS*, 2nd ed. New York: Wiley; 2009.

DIRECT AND INDIRECT MEASUREMENT IN SOIL ANALYSIS

Direct and indirect measurements are important in understanding the complex chemistry of soil and the soil environment. For a full understanding of its chemistry, all measurements of all types must be synthesized into a whole picture of that soil. The various specific types of measurements will be discussed in greater detail in subsequent chapters. In this chapter, only a general outline of the basic concepts and tools used to analyze different soil components will be covered. A flowchart of methods of analyzing soil is given in Figure 8.1.

The ideal situation is to directly measure soil chemistry as it is happening, in its native, undisturbed state. We would like to be able to do this without changing the soil's inorganic or organic components, air, water, or biology. In some cases, for some limited types of analysis, this ideal can be met or nearly met. In the case of a pollutant, it may be necessary to know not only how it will react with inorganic, organic, and biological components in soil but also

Introduction to Soil Chemistry: Analysis and Instrumentation, Second Edition.
Alfred R. Conklin, Jr.
© 2014 John Wiley & Sons, Inc. Published 2014 by John Wiley & Sons, Inc.

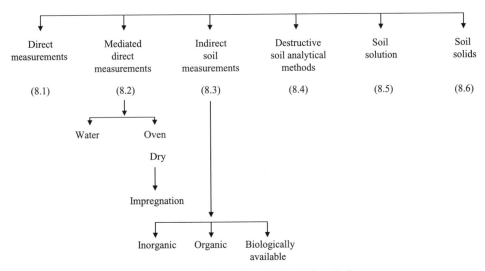

Figure 8.1. Flowchart of methods of soil analysis.

how it will physically move or otherwise change. This information is best obtained by observing it in the soil.

In most cases, this ideal cannot be met and will not yield the desired understanding of the chemistry of interest. This can sometimes be overcome by simple mediation of soil, such as bringing the soil sample to some standard moisture state before making certain measurements such as pH or salt concentration.

However, many analyses will require extraction of the component of interest before analysis is carried out. It may be essential to remove it from the soil matrix and/or interfering components before reliable analyses can be made. Extraction may be designed to extract only a portion of the analyte of interest. For example, extraction of a particular oxidation state may be important because oxidation states can affect the mobility and biological availability of chemicals. It may also be necessary to concentrate the extract after isolation.

A combination of direct observation and extraction may be carried out. The whole soil may be analyzed by various methods and then specific components sequentially extracted and measured. This approach has been used in the investigation of the speciation of metals in soil under various conditions. Using X-ray spectroscopy, metals and their various ionic forms in soil can be directly identified [1].

8.1 DIRECT MEASUREMENTS

Direct measurement of soil is most often carried out on air-dried soil and involves spectroscopic instruments and methods. For example, X-ray dispersion (XRD), X-ray fluorescence (XRF), infrared (IR) spectroscopy, near-infrared (NIR) spectroscopy, and mid-infrared (MIR) spectroscopy, solid-state nuclear magnetic resonance (NMR) spectroscopy, advanced spectroscopy using synchrotron radiation, neutron activation, visible (Vis) microscopy, fluorescence microscopy, and electron microscopy are all used to make direct measurement of the physical and chemical characteristics of soil.

In some techniques, including XRD, some IR spectroscopic techniques, and neutron activation, the surface of the soil sample is analyzed using radiation reflected or emitted from the sample. In the other types of spectroscopy, such as NMR, information is obtained from radiation passing through the sample. Infrared spectroscopy (both NIR and MIR) can be used in both transmission and reflection analyses.

8.1.1 X-Ray Methods

XRD and XRF are powerful tools used to determine both the mineralogy and elemental composition of soil. In XRD, a beam of X-rays is focused on a soil sample and the dispersion of the beam is used to determine the type of clay or clays present. Different types of clays have different numbers and spacing of layers as discussed in Section 3.1.3. These layers disperse X-rays at different angles. By measuring the angles, the type of clay present can be determined. Clays have large surface areas, are chemically very active, sometimes acting as catalysts, and have interlayer spaces that can isolate components from the soil solution and biological processes. They are thus important in understanding the chemistry of soil [2].

In XRF, an X-ray beam or gamma rays are used to displace electrons from the inner orbitals of elements. When electrons fall into these orbitals, replacing the removed electrons, photons of specific wavelengths and energy are emitted, detected, and measured to determine which elements are present. The X-rays used in XRF do not penetrate deeply and so elements on the surface of the sample are measured, while those in the interior may not be detected [3].

8.1.2 Spectroscopic Methods

Ultraviolet (UV) and Vis spectroscopy has not found appreciable application to direct measurements of soil. They have, however, found considerable use in identifying and measuring components extracted from soil (see Section 8.3). A more detailed explanation of spectroscopic methods applied to soil is given in Chapter 14.

IR spectroscopy, unlike ultraviolet-visible (UV-Vis) spectroscopy, can be used to directly measure both inorganic and organic components in soil, although it is more commonly used to identify organic compounds. It is carried out in two different wavelength ranges, the NIR, which is from 0.8 to 2.5 μm (800–2500 nm), and the MIR, which is from 2.5 to 25 μm (4000–400 cm^{-1}).[1]

---

[1] Wave length ranges may be different depending on the instrument manufacturer.

Both ranges are used in the analysis of soils, particularly soil organic compounds and their functional groups.

Making a direct measurement of soil organic matter requires that the soil have a relatively large amount of organic matter. It also requires that the soil sample have low water content as water absorbs strongly in the MIR. In spite of these limitations, IR spectroscopy can be used to identify both inorganic and organic components and molecules present in soil and can yield valuable data about the chemistry of soil.

Direct NIR or MIR measurements of whole soil can be made using attenuated total reflectance (ATR) and diffuse reflectance infrared Fourier transform spectroscopy (DRIFTS) samplers. This type of measurement detects only components on the surface and so has severe limitations when information about the bulk soil is needed [4].

Transmission spectra of soil components can also be made. This is most frequently accomplished by grinding soil with potassium bromide (KBr) and pressing the combination to make a transparent KBr pellet (KBr is transparent in the IR region of the spectrum). IR radiation will pass through the KBr and a spectrum of the soil components will be obtained. This will give a more complete picture of the soil components present. However, while no chemical change occurs, this process involves physical changes in terms of the grinding process necessary to produce the pellet.

8.1.3   Magnetic Resonance Spectroscopy

Solid-state NMR spectroscopy has not found as wide an application in soils as it has in other fields. The great advantage of NMR is that it is specific for specific elements; that is, it is tuned to a specific element and other elements are not detected. NMR spectroscopy shows the environment or multiple environments in which a particular element exists. For example, in a proton NMR spectrum, primary, secondary, and tertiary protons can be differentiated, as can protons attached to oxygen, nitrogen, and other atoms.

While many elements can be studied using NMR spectroscopy, only those elements at the top of the periodic chart are commonly measured by soil chemists, and of those, carbon-13 (^{13}C) and phosphorus-31 (^{31}P) are the most frequently investigated. Phosphorus is of particular interest because it is an essential nutrient for both plants and animals.

NMR spectra of solids, and thus soil, are obtained by what is called magic angle spinning. The spectra obtained have broader absorption features than NMR spectra of components in solution or liquids. Numerous NMR experiments such as ^{1}H–^{13}C heteronuclear chemical shift correlation (HETCOR), which identifies which hydrogen atoms are attached to which carbon atoms, can also be carried out on solid samples. A great deal of useful information about the structure of components in soil can thus be obtained from NMR investigations [5,6].

There has been concern that iron, manganese, and perhaps other magnetic or paramagnetic components in soil might interfere in obtaining usable NMR spectra. Although this appears not to be the case, it is still important to consider this possibility when carrying out an NMR investigation of soil or soil components [7].

8.1.4 Neutron Activation Analysis

Two different types of information can be obtained by bombarding soil with neutrons. Fast neutrons are slowed when they interact with water and thus can be used to measure the amount of water present. This type of analysis is most often conducted in the field rather than in the laboratory. Figure 8.2 illustrates the use of a fast neutron source and a slow neutron detector to measure the moisture content of soil. This method depends on the interaction of neutrons with hydrogen and so it is not as useful in soils with significant or highly variable organic matter contents.

In the second type of analysis, high-energy bombardment of soils with neutrons, on the other hand, leads to what is called neutron activation. Reemission of radiation from neutron-activated soil allows for the identification of the elements present. This type of analysis typically requires a source of high-energy neutrons and so requires special equipment [8].

8.1.5 Soil Air

Soil air is most often measured directly because it is easily removed from soil and can be directly analyzed by gas chromatography (GC). This can be done by simply sticking a syringe in soil, withdrawing the plunger, and then injecting the collected gas into a gas chromatograph. Soil gases can also be trapped as they exit the soil by placing a cover over the soil of interest. The gases can be concentrated by cooling if needed and subsequently analyzed by GC. In either

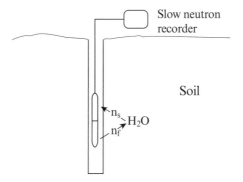

Figure 8.2. Neutron probe for determining soil moisture, where n_f is fast neutrons and n_s is slow neutrons.

case, the gas is not substantially changed before analysis and thus this is a direct analysis.

The soil atmosphere can also be passed into a long-path IR gas sampling cell. Because of the long path length, gases that might otherwise not be detectable are measured and identified. This should not be confused with IR used in remote sensing, which is covered in Section 14.14.

In some cases, the gas can be trapped, such as with sodium hydroxide, and changes in the mass of trapping material used as a measure of the amount of gas. Trapping agents may also trap water which must be accounted for. Other trapping agents and methods are also available.

Of the gases common in soil, carbon dioxide and the oxides of nitrogen, generated by denitrification, are perhaps the most interesting and most studied. Of particular interest are the gaseous oxides of nitrogen, such as NO and N_2O. These two gases result from the denitrification of nitrate under anaerobic soil conditions [9]. Because these products are part of the nitrogen cycle and result when nitrogen fertilizers undergo denitrification, they are of particular interest to both agriculturalists and environmentalists.

Contamination of soil often leads to a situation where the contaminants are not only in the soil but are also in the soil air. These can be analyzed using the same or similar methodologies that are used to analyze natural soil air constituents.

8.1.6   Other Direct Measurements

Other direct measurements of soil are common, although they are not frequently used to obtain chemical information. They may involve frequencies of the electromagnetic spectrum not involved in the analyses described previously. For instance, microscopy using light, including UV and polarized light, can be used to obtain information about the mineralogy of soils [10]. However, today, these methods are not commonly used in studying soil chemistry.

Electron microscopy, both transmission and scanning, is used to obtain chemical information about soil. Transmission electron microscopy (TEM) is used to identify the elements in a soil sample under investigation. Scanning electron microscopy (SEM) is also used to investigate soil chemistry, although it is a little less powerful than transmission microscopy [11].

Fluorescence spectroscopic studies of soil organic matter can yield important information about its formation and composition. This is different from XRF in that organic molecules are excited with UV light, and reemitted UV or Vis light is measured. In Figure 8.3, an electron, commonly in a conjugated π system, is promoted to a higher energy level by incident light (absorbance). From this position, the electron loses energy, eventually falling back to the ground state, leading to fluorescence. The excited electron then loses energy by nonradiative interactions and is finally at the lowest energy in the S_1 state. Because the energy of the emitted light is less than that absorbed, the wavelength of the emitted light is longer than that absorbed. This is a standard

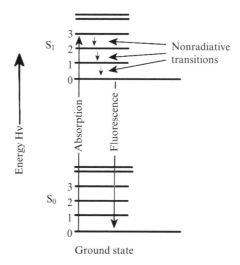

Figure 8.3. Electron excitation and relaxation in fluorescence.

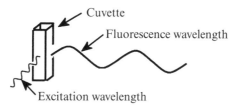

Figure 8.4. Fluorescence measurement showing excitation and fluorescence radiation.

representation but does not illustrate all the possible transitions of electrons when they absorb light energy.

To conduct fluorescence measurements, the sample is contained in a cuvette, all sides of which are transparent to UV and Vis light. As shown in Figure 8.4, excitation light enters the cell on one side and emitted light exits at right angles through the other side to be detected. This type of spectrophotometric measurement yields particular insight into the degree of unsaturation and aromaticity of soil organic matter during its synthesis [12,13].

Another interesting method of investigating soil, although it is not normally used to study soil chemistry, is ground-penetrating radar (GPR). This tool can penetrate soil up to several meters depending on the type and condition of the soil. From this, the identity of various inclusions in the field that could affect the results of sample analysis can be discovered. Thus, it can be used to guide soil sampling. GPR can identify discontinuities in soil such as buried pipes, wires, voids, foundations, and cement. When unusual analytical results are obtained, it can help identify the causes of these results without taking additional samples from the area [14].

8.2 MEDIATED DIRECT MEASUREMENT

There are two general situations where some amendment, which does not change the soil significantly, is made to soil prior to analysis. This is true with pH and salt measurements and, to a lesser extent, with Eh (electron activity) determinations. In some microscopic investigations, preparation of thin sections of soil may be used. Soil is, in these cases, modified before an analysis or measurement is made. Nothing is removed, the original soil matrix is intact, and minimal chemical changes in the soil occur during preparation and analysis.

8.2.1 Soil pH and Eh Measurements

Soil pH is perhaps the most critical and common soil measurement where a definite amount of water is added before a measurement is made. Soil pH is a particularly complicated measurement because the proton can and does exist as a hydronium ion in the soil solution, as an exchangeable ion on the cation exchange sites, and bonded to various soil constituents. Because of these complexities, a soil sample is usually brought to a standard moisture content before a pH measurement is made. By bringing different soils to a common moisture content, they can be compared and analytical results from different laboratories will be comparable. Although there is a number of ways to measure soil pH, typically it is carried out using a pH meter and a pH electrode.

Protons in soil are associated with soil water molecules to form hydronium ions, which can also be on an exchange site in soil (see Figure 8.5). Additionally, protons are bonded with different strengths to both organic and inorganic components in soil. All of these are potential sources of protons for chemical reactions. They also produce buffering of soil pH such that the pH of soil is hard to change.

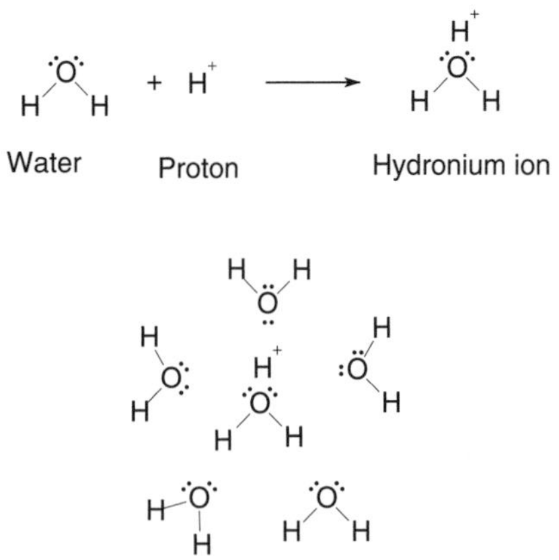

Figure 8.5. Top: formation of hydronium ion. Bottom: hydronium ion surrounded by water molecules.

It is important to take into account not only soluble hydronium ions but also exchangeable and bonded protons when studying soil pH. Different additions to soil are made to differentiate active protons (hydronium ions) and exchangeable protons. Bonded protons that might become available during changes in soil pH can be ascertained by titrating soil. Exchangeable protons are especially important because they are in equilibrium with protons in solution. This relationship can be illustrated by defining exchangeable protons as H_e^+, solution protons as H_s^+, and bonded protons as H_b^+ (see Figure 8.6). It is important to keep in mind that, in all cases, the protons are either in the form of hydronium ion, H_3O^+, or are associated with water molecules in some manner (see Figure 8.5). Figure 8.7 illustrates the relationship between the species of protons and the soil surface.

In a manner similar to pH, one can describe the availability or concentration of electrons, abbreviated as Eh, in an environment. This then is the negative log of the electron concentration. As with pH, it is really a measure of the electron activity rather than the concentration and is a measure of the oxidation–reduction potential (often referred to as "redox" potential) of the soil environment. Aerobic conditions represent electron-losing or oxidizing environments, and anaerobic conditions represent electron-gaining or

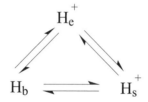

Figure 8.6. Equilibrium between species of protons in soil.

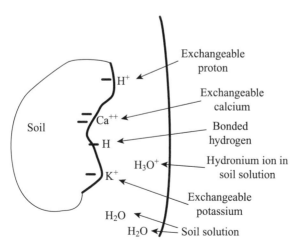

Figure 8.7. Diagram of exchangeable protons around a soil particle.

reducing conditions. Unsaturated soil is aerobic and soil saturated with water is anaerobic. The Eh measurement can thus indicate the predominant ionic state, oxidation state, or species of components of interest in a soil under a specific redox situation. These types of measurements are usually made using a pH meter equipped with platinum and reference electrodes (rather than pH electrodes). Eh values are reported in millivolts (mv).

Soil solutions can also be titrated to obtain information about both their pH status and their buffering capacity. Chapters 9 and 10 give a more detailed discussion of electrical and titration methods applied to soil.

8.2.2   Soil Impregnation

The relationship between soil solids, their composition, and chemistry can also be studied in soil thin sections. This is a mediated measurement because the soil sample must be impregnated with a resin or frozen for a thin section to be prepared for study.

A soil sample is typically impregnated with a liquid resin that is then cured (made solid) by heating. A diamond saw can then be used to cut the sample into thin sections suitable for observation using light microscopy and X-ray methods. A thin section allows one to visualize the relationship between the solid components and voids in a soil. Some alteration of the soil solids occurs during sectioning; however, these are generally obvious and can be discounted when making observations.

Thin sections can also be obtained from frozen soil. This type of thin section is best suited to soils that are naturally frozen at least part of the time. In these cases, the soil thin section must be kept frozen during the subsequent examination. Sectioning of frozen soil is generally more difficult than sectioning of resin-impregnated solids [10].

8.3   INDIRECT SOIL MEASUREMENTS

Indirect measurement of soil and its constituents is necessary in a majority of cases. Extracting solutions have been devised for inorganic and organic constituents of soil such that they can be removed from the soil matrix, in some cases purified and concentrated, and subsequently measured (see Chapters 11 and 12).

One of the questions that arises in soil analysis is whether a determination of the total amount of a component in soil is desired or if just the biologically available amount is more relevant. Related to this is the question of which species of the component is present. In some cases, speciation is of utmost importance. For instance, chromium can be present as Cr(III) or Cr(VI). Cr(VI) is more toxic and thus of greater concern [15]. This concern is also related to the biological availability of a specific species. In this case, while knowing the total chromium content (i.e., the sum of Cr(III) and Cr(IV) concentrations) may be useful, it may be even more important to know the Cr(VI) level specifically.

All chromatographic methods (including GC,[2] high-performance liquid chromatography [HPLC],[3] and thin-layer chromatography [TLC]), capillary, electrophoresis, and spectroscopic methods (including UV, Vis, IR, NMR X-ray, fluorescence, atomic absorption [AA or AAS], and inductively coupled plasma [ICP]), and mass spectrometry (MS) methods are widely used to analyze soil extracts. All offer the soil chemist a variety of methodologies that have different strengths and weaknesses. All are powerful tools for isolation and identification of elements, molecules, and compounds obtained from soil.

8.3.1   Chromatographic Methods

Chromatography is a powerful method for separating and quantifying the components of complex mixtures, including mixtures of organic and inorganic components commonly found in soil. Of the common chromatographic methods, GC and HPLC are the most commonly used. GC is favored because it is fast, relatively easy to use, and can be easily connected to various spectroscopic methods.

GC is particularly useful in soil analysis because of both the wide variety of columns available for separating specific mixtures and because of the variety of very sensitive detectors available. There are specific columns for separation of gases commonly found in the soil atmosphere and specific columns for the analysis of herbicides, insecticides, and pollutants.

Detectors range from the universal, but less sensitive, to the very sensitive but limited to a particular class of compounds. The thermal conductivity detector (TCD) is the least sensitive but responds to all classes of compounds. Another common detector is the flame ionization detector (FID), which is very sensitive but can only detect organic compounds. Another common and very sensitive detector is called electron capture. This detector is particularly sensitive to halogenated compounds, which can be particularly important when analyzing pollutants such as dichlorodiphenyltrichloroethane (DDT) and polychlorobiphenyl (PCB) compounds. Chapter 13 provides more specific information about chromatographic methods applied to soil analysis.

8.3.2   Spectroscopic Methods

Spectroscopic methods are used almost exclusively for the identification of pure components isolated from soil. Of these, AAS and ICP in their various

---

[2] Separation and spectrophotometric methods are commonly referred to by their abbreviations or acronyms. It is thus important to be familiar with these common abbreviations.

[3] The P in HPLC is sometimes referred to as pressure, as in "high-pressure liquid chromatography," and at other times performance.

forms and MS are the most frequently used. Metals extracted from soil are generally analyzed by ICP because this type of instrumentation can be used to determine multiple elements simultaneously more easily than AAS. In addition, modern ICP instruments can determine the concentration of several different metals at the same time, thus decreasing the total analysis time.

Because MS identifies components on the basis of their atomic or molecular mass and, in the case of organic compounds, fragmentation pattern, it is a particularly useful and powerful detector. The molecular mass, which is commonly referred to as the molecular weight, is a particularly useful piece of information when trying to identify a compound. For elements, atomic mass determination can also determine the isotope present, which in turn can be used to identify the source and movement of an element through the environment.

MS is the method used in stable isotope analysis. By enriching samples with a particular isotope, that sample or isotope can be followed as it passes through the environment. This technique has been particularly useful in studying the fate of nitrogen as ^{15}N in environmental studies.

MS requires that the components in the sample are driven into the gas phase. All gas chromatographic separations also require the sample components to be in the gaseous state. Thus, the two methods are compatible. It is possible, however, to do an MS analysis on samples without the MS being attached to a gas chromatograph (see Chapter 14).

8.4 DESTRUCTIVE SOIL ANALYSIS METHODS

Destructive methods, often called pyrolysis, are also used to determine components in soil. This commonly involves heating soil to a high temperature and analyzing the components given off. Often, analysis is by GC, mass spectrometry, or a combination of the two, although spectroscopic analysis is also possible [16].

Another approach is called either differential thermal analysis (DTA) or thermogravimetric analysis (TGA). A sample of soil is gradually heated and changes in temperature or mass are measured. A change in either quantity indicates a transition or the loss of components. The temperature at which components are lost can indicate the type and amount of that component present in the soil sample. However, in many cases, it is more useful to actually identify the components lost [17]. The material lost during either DTA or TGA is commonly analyzed and identified using either IR spectroscopy or mass spectrometry. Because the components are in the gaseous state, they are easy to sample and analyze by either method [17–19]. Although these methods can lead to information about soil and it constituents, information about the source of the products and their relationship to soil composition is lost.

There is one case where total destruction of soil organic matter and its constituents is standard practice and that is in the determination of nitrogen

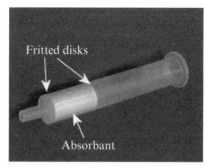

Figure 8.8. Commercial extraction tube used for isolating, purifying, and concentrating soil extractants.

in soil by the Kjeldahl method. This method involves digesting soil with sulfuric acid that has a catalyst and salts added to it. Using variations on the basic method, organic and inorganic nitrogen in soil can be determined individually. In addition, the amount of both nitrate and nitrite can be found, and thus, all the common forms of nitrogen in soil, NH_4^+, NO_2^-, and NO_3^-, can be determined (also see Section 10.5) [20].

8.5   SOIL SOLUTION

The soil solution can be extracted from soil as described in Section 7.7. It can then be measured directly or concentrated. When extracted or concentrated, it would be considered an indirect measure. Soil solution can be passed through an absorptive column that absorbs the component of interest (see Figure 8.8). Two things happen during this process: purification and concentration. The component of interest is absorbed to the column packing and contaminants are washed out of the column. When the component of interest is eluted from the column using a smaller amount of eluant compared with the original solution, it is concentrated. Concentration of the analyte of interest is common even if the original purpose was simply cleanup. At this point, this would become an indirect measurement of the soil solution. Column cleanup of soil extracts is commonly used in cases where the soil has been contaminated. This is particularly true if the contaminant is a liquid.

8.6   SOIL SOLIDS

Soil solids are either inorganic or organic, and each requires different analyses. Direct measurements on whole soil are not as common as indirect measurements. Direct measurements are limited to X-ray, IR, and NMR analysis. X-ray analysis is commonly used to investigate mineral structure, while IR is used to investigate organic constituents, and NMR has been used most commonly to investigate phosphate in soil.

Both organic and inorganic solids are commonly extracted to obtain specific components and species that are then analyzed. Whole soils can be extracted with dilute acid and chelating solutions for measurement of inorganic components. Organic components can be isolated using organic solvents. However, aqueous solutions of bases such as NaOH are usually used as the first extractant in organic matter investigations, particularly investigation of humus. These solutions are then analyzed for the analyte of interest, thus making them an indirect analysis.

8.7   CONCLUSIONS

Both direct and indirect methods are used in studying soil chemistry. While in all cases direct methods are preferable, it is not always possible to make direct observations of all the chemical species, and physical and chemical changes of interest. Thus, it is often necessary to modify the soil before analysis. In many cases, it is essential to extract components before analysis can be carried out. It is also possible to obtain valuable information about the chemistry of soil by carrying out analyses that destroy all or a part of the soil matrix. A summary of analysis types and instruments commonly used in soil analysis is given in Table 8.1.

TABLE 8.1. Summary of Analysis Types and Instrumentation Used in Soil Analysis

Analysis Type	Section	Analytical Methods
Direct analysis	7.1	XRD, XRF, infrared spectroscopy (NIR and MIR), solid-state nuclear magnetic resonance (NMR), advanced spectroscopy using synchrotron radiation, neutron activation, fluorescence, and visible and electron microscopy
Mediated analysis	7.2	pH, Eh, thin sections

Extraction	7.3	Separation methods
		    GC, HPLC, thin layer, column, capillary, electrophoresis
		Identification methods
		    UV, visible, IR, NMR, X-ray, fluorescence, AAS, ICP, MS

Destruction	7.4	Pyrolysis, DTA, TGA, Kjeldahl

PROBLEMS

8.1. Explain the difference between direct, mediated direct, and indirect measurement of soil components.

8.2. Give an example of each of the measurements given in Problem 8.1.

8.3. Give both the benefits and the drawbacks of each of the types of measurements given in Problem 8.1.

8.4. What two types of information can be obtained from soil by bombarding it with neutrons?

8.5. GPR does not identify soil chemical components, so how can it be a beneficial measurement in soil chemical investigations of soil?

8.6. Give both the benefits and the drawbacks of preparing and observing thin sections of soil.

8.7. Explain how and why NMR spectroscopy is a powerful tool for investigating elements in soil.

8.8. Explain how indirect measurement of soil chemistry is different from direct measurement.

8.9. How are AAS and ICP different and how are they the same?

8.10. Pyrolysis of soil can provide information about its composition, particularly with reference to its organic constituents. Using the reference given, describe some types of information available from this type of analysis (use information from other chapters to answer this question).

REFERENCES

1. Scheinost AC, Kretzschmar R, Pfister S, Roberts DR. Combining selective sequential extractions, X-ray absorption spectroscopy, and principal component analysis for quantitative zinc speciation in soil. *Environ. Sci. Technol.* 2002; **36**: 5021–5028.
2. Drever JI. Preparation of oriented clay mineral specimens for X-ray diffraction analysis by a filter-membrane peel technique. *Am. Mineral.* 1973; **58**: 553–554.
3. Pyle SM, Nocerino JM, Deming SN, Palasota JA, Palasota JM, Miller EL, Hillman DC, Kuharic CA, Cole WH, Fitzpatrick PM, Watson MA, Nichols KD. Comparison of AAS, ICP-AES, PSA, and XRF in determining lead and cadmium in soil. *Environ. Sci. Technol.* 1996; **30**: 204–213.
4. McCarty GW, Reeves JB III, Reeves VB, Follett RF, Kimble JM. Mid-infrared and near-infrared diffuse reflectance spectroscopy for soil carbon measurement. *Soil Sci. Soc. Am. J.* 2002; **66**: 640–646.
5. Baldock JA, Oades JM, Waters AG, Peng X, Vassallo AM, Wilson MA. Aspects of the chemical-structure of soil organic materials as revealed by solid-state C-13 NMR spectroscopy. *Biogeochemistry* 1992; **16**: 1–42.
6. Cao X, Olk D, Chappell M, Cambardella CA, Miller LF, Mao J. Solid-state NMR analysis of soil organic matter fractions from integrated physical-chemical extraction. *Soil Sci. Soc. Am. J.* 2011; **75**: 1374–1384.

7. Shand CA, Cheshire MV, Bedrock CN, Chapman PJ, Fraser AR, Chudek JA. Solid-phase P-13 NMR spectra of peat and mineral soils, humic acids and soil solution components: influence of iron and manganese. *Plant Soil* 1999; **214**: 153–163.

8. Raven KP, Loeppert RH. Trace element composition of fertilizers and soil amendments. *J. Environ. Qual.* 1997; **26**: 551–557.

9. Kim KR, Craig H. Nitrogen-15 and oxygen-18 characteristics of nitrous oxide-a global perspective. *Science* 1993; **262**: 1855–1857.

10. Miralles-Mellado I, Canton Y, Sole-Benet A. Two-dimensional porosity of crusted silty soils: indicators of soil quality in semiarid rangelands? *Soil Sci. Soc. Am. J.* 2011; **75**: 1330–1342.

11. Leroy JM. *Soil Chemical Analysis Advanced Course.* Upper Saddle River, NJ: Prentice-Hall; 1965, p. 471.

12. Houck MM, Siegel JA. *Fundamentals of Forensic Science*, 2nd ed. New York: Academic Press; 2010, p. 409.

13. Yu H, Xi B, Ma W, Li D, He X. Fluorescence spectroscopic properties of dissolved fulvic acids from salined flavo-aquic soils around Wuliangsuhai in Hetao Irrigation District, China. *Soil Sci. Soc. Am. J.* 2011; **75**: 1385–1393.

14. Doolittle JA, Collins ME. Use of soil information to determine application of ground penetrating radar. *J. Appl. Geophys.* 1995; **33**: 101–108.

15. Katz SA, Salem H. The toxicology of chromium with respect to its chemical speciation—a review. *J. Appl. Toxicol.* 1993; **13**: 217–224.

16. Schnitzer M, Schulten HR. The analysis of soil organic-matter by pyrolysis field-ionization mass-spectrometry. *Soil Sci. Soc. Am. J.* 1992; **56**: 1811–1817.

17. Marinari S, Masciandaro G, Ceccanti B, Grego S. Evolution of soil organic matter changes using pyrolysis and metabolic indices: a comparison between organic and mineral fertilization. *Bioresour. Technol.* 2007; **98**: 2495–2502.

18. Dean WE, Jr. Determination of carbonate and organic matter in calcareous sediments and sedimentary rocks by loss on ignition-comparison with other methods. *J. Sediment. Petrol.* 1974; **44**: 242–248.

19. Shurygina EA, Larina NK, Chubarova MA, Kononova MM. Differential thermal analysis (DTA) and thermogravimetry (TG) of soil humus substances. *Geoderma* 1971; **6**: 169–177.

20. Bremner JM. Determination of nitrogen in soil by the Kjeldahl method. *J. Agric. Sci.* 1960; **55**: 11–33.

BIBLIOGRAPHY

Joi HM. *Ground Penetrating Radar Theory and Applications*, 1st ed. New York: Elsevier Science; 2009.

Stoops G, Vepraskas MJ. *Guidelines for Analysis and Description of Soil and Regolith Thin Sections.* Madison, WI: Soil Science Society of America; 2003.

Tovey NK, Smart S. *Electron Microscopy of Soil and Sediments: Examples.* New York: Oxford University Press; 1981.

CHAPTER

9

ELECTRICAL MEASUREMENTS

In making measurements using electricity, all its components—voltage, amperage, resistance, capacitance, frequency, and dielectric characteristics—can be used singly or in combination to obtain information about various conditions in the medium through which electrons are moving. All soils contain many ions of different complexities: the simpler hydrated K^+; more complex NO_3^-; organic ions, such as ionized acids; and charged solids, such as clays. All contribute to the electrical characteristics of a soil and its solution.

One easily demonstrated electrical characteristic of moist soil is seen in the production of electricity when two different metals, namely, copper and zinc, are inserted into it. This is not unexpected because any salt-containing solution adsorbed in media, such as paper or cloth, and placed between these same two electrodes will cause a spontaneous reaction that produces electricity. The source of this flow of electrons is an oxidation–reduction reaction, represented as two half-reactions as shown in Figure 9.1 for copper and zinc.

This was the original method used to produce electricity (i.e., an electrical current) for scientific experiments. Thus, if copper and zinc electrodes are

Introduction to Soil Chemistry: Analysis and Instrumentation, Second Edition.
Alfred R. Conklin, Jr.
© 2014 John Wiley & Sons, Inc. Published 2014 by John Wiley & Sons, Inc.

$$Zn^0 \rightarrow Zn^{++} + 2e^- \quad \text{Oxidation}$$

$$Cu^{++} + 2e^- \rightarrow Cu^0 \quad \text{Reduction}$$

Figure 9.1. Oxidation–reduction reactions that, when coupled, produce an electric current.

Figure 9.2. Zinc and copper strips inserted into wet soil and producing a voltage as shown on the voltmeter.

inserted into wet soil, electricity will be produced by the same process. Figure 9.2 shows copper and zinc strips inserted into moist soil and connected to a voltmeter, which displays the resulting voltage, demonstrating the existence of salts in the soil. This simple setup is not, however, used to measure soil characteristics such as salt content or pH.

The electrical characteristics, along with the salts, their movement through soil, and the diffuse double layer must be kept in mind when making any soil measurement using electricity or electrodes.

Although all characteristics of electricity have been used to investigate soil and its properties, only a limited number are used routinely. The most common are those used for the determination of pH, salt content, and soil water content. Of these three, pH is the most common measurement and frequently the first measurement made prior to all other determinations. Although pH can be determined by many methods, for soil, the most common is to use a pH meter and electrode. Conductivity or resistance is used to measure soil salt content, while several different electrical characteristics of soil are used to determine

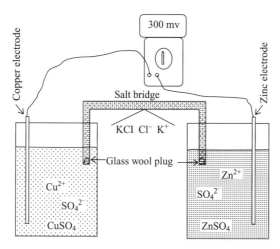

Figure 9.3. A typical electrochemical cell.

its water content. In addition to their inherent importance, the pH, salt, and water content are important in determining how analyses must be carried out and in determining the effects or interferences that these properties have on other analytical methods, particularly chromatography and spectroscopy.

9.1 THE BASIC ELECTROCHEMICAL CELL

The terms cathode and anode used to describe electrodes often cause confusion, and to a certain extent, the characteristics of an electrode depend on whether they are being looked at from the outside in or the inside out. One way to keep the terms straight is that an anode begins with a vowel, as does an electron. Likewise, cathode begins with a consonant as does the word "cation" or that both cathode and cation begin with a **c**.

A basic electrochemical cell is depicted in Figure 9.3 and is made of a copper wire in one container with a solution of copper sulfate and a zinc rod in a different container with a zinc sulfate solution. There is a salt "bridge" containing a stationary saturated KCl solution between the two containers. Electrons flow freely in the salt bridge in order to maintain electrical neutrality. A wire is connected to each rod and then to a measuring device such as a voltmeter to complete the cell.

9.2 ELECTRICITY GENERATION IN SOIL

The generation of small electrical currents in soil is possible and may affect any electrical measurement made therein. Recent theoretical and experimental work has shown that charged particles, including ions, passing through microchannels under low pressure (i.e., 30 cm of water), can generate small amperages. In addition, amperages from multiple channels are additive and might be a useful method of generating electricity or of making a new type of battery [1,2].

Soil solids have channels with charged surfaces (such as those in clays) and an accompanying *electrical double layer*. Salts in these soils can move through the channels and generate a current. Although generation of electrical currents under standard extraction conditions may not be of concern, they may be in other soil analytical procedures. *In situ* measurements, taken with electrodes buried in soil, may record or be susceptible to interference from such currents. This will be particularly true for electrodes buried in fields during rainfall or irrigation events and the subsequent percolation of water through the soil profile. It will be more pronounced in soils such as *Aridisols*, which naturally contain more salts, as well as in soil with more clays, that is, with Bt horizons, particularly for high-activity clays with high cation exchange capacity and high surface area.

9.3   POTENTIOMETRY (ELECTRODES IN SOIL MEASUREMENTS)

Numerous types of electrodes are used for soil analysis. Simple elemental electrodes such as platinum, mercury, and carbon are the most frequently used, while other unreactive metals such as gold and silver and the more reactive copper and zinc and others have also been used. Both stirred and unstirred solutions and suspensions are used during analysis. In some cases, electrodes are rotated or, in the case of mercury, dropped into the solution being analyzed.

Soil scientists represent the electrode potential as Eh and the standard potential as Eh^0,[1] which is measured in millivolts or volts. All (non-oven-dried) soil contains water, which limits the possible upper and lower potentials. The upper potential is controlled by the oxidation of water. Under highly oxidizing conditions, oxygen is oxidized to oxygen gas, namely, O_2. The lower potential is limited by the reduction of hydrogen or protons, specifically H^+, which leads to the formation of hydrogen gas, H_2. Figure 9.4 shows the electrode potentials for the oxidation of oxygen and the reduction of hydrogen at common soil pHs.

Although the reduction of hydrogen and the oxidation of oxygen determine the lower and upper limits of electrochemical redox measurements in soil, it is still possible to have both hydrogen and oxygen produced by biological processes. Photosynthetic production of oxygen by algae and phototrophic

---

[1] The standard electrode potential is that potential generated by an electrode when compared to a standard hydrogen electrode under standard conditions, which typically include a temperature of 298 K (kelvin).

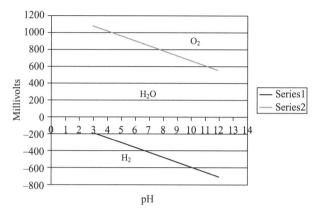

Figure 9.4. Electrode potentials for the reduction of hydrogen, series 1, and the oxidation of oxygen, series 2, at a gaseous pressure of 1 atm and pHs common in soil.

bacteria occurs in all soils. Likewise, the biological production of hydrogen gas occurs under anaerobic conditions. Thus, it is possible to find oxygen and/or hydrogen gas being produced in soil.

In soil analysis, pH, specific ion, oxidation–reduction (redox), electrical conductivity (EC) cells, and oxygen electrodes are commonly used. For each of these measurements, a different specific electrode along with a separate or integral reference electrode is needed. In some cases, with extended use or long exposure to soil or soil–water suspensions, electrodes may become polarized. When this happens, erroneous results will be obtained and depolarization will need to be carried out using the electrode manufacturers' directions [3].

9.3.1   pH

In soil, one can conceive of the presence of four "types" of protons. The naked proton not associated with any group or species would be the first "type." However, protons in solution are associated with water molecules, forming hydronium ions (the second type), and are the only measurable protons. The third type is those that are associated with cation exchange sites, are exchangeable, and contribute to soil buffering. They cannot be measured directly but can be exchanged with cations from salts or buffer solutions, and once in solution, they can be measured. The fourth type of proton is bonded to either inorganic or organic soil components and normally will be thought of as being covalently bonded. However, they may be part of a functional group from which they may be easily removed and thus become part of the protons in solution. Both organic acid and phenolic groups are examples of compounds that have protons that fall into this category.

A combination pH electrode is the most commonly used electrode to determine soil pH. It is illustrated in Figure 9.5 and shown in Figure 9.6 (D), which shows the pH sensing bulb and reference side. As an alternative to the

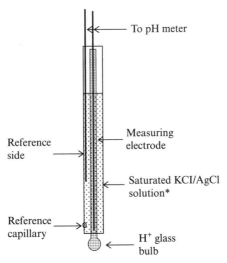

Figure 9.5. Typical setup for a combination pH electrode. Note (*) that the saturated KCl/AgCl solution can be different depending on the electrode manufacturer.

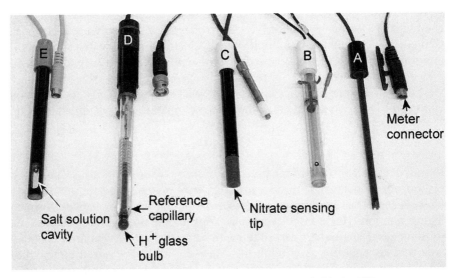

Figure 9.6. A—temperature electrode; B—reference electrode suitable for ISE measurements; C—nitrate ISE electrode; D—combination electrode; E—conductivity electrode.

combination electrode, two separate electrodes, one pH sensing (i.e., the H^+ glass bulb) and the other a reference electrode (Figure 9.6 [B]), may be used and may be best in cases where fouling of the reference electrode is a particular problem. While some pH electrodes have robust pH sensing electrodes, others are quite delicate. Therefore, care must always be taken to avoid scratching or breaking the pH sensing bulb when making a pH measurement. Even

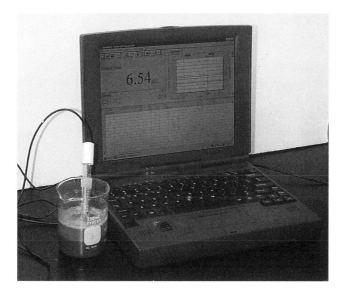

Figure 9.7. A pH electrode in a soil suspension. The electrode is connected to a computer that shows the readout.

robust electrodes must be treated with care and kept wet with the appropriate solution as directed by the manufacturer. Figure 9.7 shows a pH electrode in a soil suspension. The electrode is connected through a card, which is a pH meter, to a laptop computer that provides data output.

The pH meter can be analog, digital, or, as mentioned earlier, attached to a computer. As shown in Figure 9.6, each electrode has a different connector illustrating the need to match electrode connector with meter. Regardless of the make, pH meters are usually robust and should need little maintenance.

Standardization of pH meters is essential and must be done on a regular basis. The starting point in standardization is to adjust the meter and electrodes using a standard buffer solution of pH 7.00 and adjusting the meter to this pH. The second step is to set a second point, which is in the range of pH levels expected to occur in the measurements. Thus, if the pH values are all expected to be acidic, the second point will commonly be set using a buffer of 4.01; if they are expected to be basic, then a pH 10.00 buffer will be used. If both acidic and basic pH values are likely, the meter can be standardized using all three buffers.

Standard buffers can be purchased as prepared solutions or as powders that are dissolved in distilled or deionized water. In the latter case, the powder is typically dissolved in 100 mL of distilled or deionized water for use. Buffers may be color-coded such as green or yellow for pH 7.00, pink for pH 4.01, and blue for pH 10.00.

Because standardization is an essential component of soil pH measurement, care must be taken to have good buffers. If the buffers show any indication of contamination, such as material floating in them, soil, or microbial growth, they must be discarded and new buffers prepared. All pH buffers will support microbial growth, which will interfere with electrode function and change the buffer pH.

Measurements made using pH meters and electrodes are temperature-sensitive; that is, the pH reading obtained depends on the temperature of the solution. Some pH meters have a temperature probe (see Figure 9.6 [A]), such that a temperature correction is automatically made during the measurement. However, if this is not the case, the pH meter must be set to the proper temperature if accurate measurements are to be obtained.

Fouling of the reference electrode or the reference side of a combination electrode is a common problem in soil pH measurements. Fouling can be caused by salts, organic matter, and clay. Each electrode manufacturer will provide specific cleaning procedures that help to keep electrodes functioning properly; however, in many cases, no amount of cleaning is effective and the electrodes need to be replaced.

Depending on the extraction method to be used or developed, determination of pH using a standard method should be used. In different countries and geographic areas, different standard methods will be in common use; for example, in Ohio (United States), a 1:1 soil:water suspension is used, while in Zimbabwe, a 1:1 soil:0.02 M $CaCl_2$ is commonly used. The pHs determined by the standard method, used in a particular area, are used in many other procedures and are used to make recommendations and predictions about the environment. Changing the method will mean that the validity of all these relationships will need to be reestablished or new relationships determined. For these reasons, it is generally not advantageous to spend time and money developing a new methodology for determining soil pH unless there is a highly significant economic benefit in doing so [4–6].

9.3.2   Ion-Selective Electrodes

Ion measurements using ion-selective electrodes (ISEs)[2] (see Figure 9.6 [D]) are similar to pH measurements and typically are carried out using a pH meter capable of accepting ISEs. Each ion requires a specific electrode, some of which will be combination electrodes similar to combination pH electrodes, while others are only the sensing electrode, called a *half-cell*. For these half-cell electrodes, special reference electrodes (see Figure 9.6 [B]), which have a high flow of reference electrode solution, are used. There are two important differences between pH and selective ion measurements: (1) the latter may require that the solution be separated from the soil before measurement is made and (2) the reference electrode and meter must be specially designed for use with this type of electrode. ISEs and the reference electrode used with them are shown in Figure 9.6 (C and B, respectively). Table 9.1 gives common ISEs

[2] ISEs were originally and sometimes still are called ion-specific electrodes, and this term may still be encountered.

TABLE 9.1. Ion-Selective Electrodes and Their Characteristics

Electrode/Ion[a]	Type	Range Molar	Interferences[a]
Ammonia	Gas sensing/ combination	1.0 to 5×10^{-7}	Volatile amines, Hg^+
Bromide	Solid state	0.0 to 5×10^{-5}	$S^=$ must not be present, CN^-, I^-, NH_3, Cl^-, HO^-
Cadmium	Solid state/ combination	10^{-1} to 10^{-7}	
Calcium	Half-cell	1.0 to 5×10^{-7}	Zn^{2+}, Pb^{2+}, Fe^{2+}, Cu^{2+}
Carbon dioxide	Gas sensing	3×10^{-2} to 10^{-5}	Volatile organic acids
Lead	Solid state/ half-cell	0.1 to 10^{-6}	Cu^{2+}, Ag^+, Fe^{3+}, Cd^{2+}
Nitrate	Half-cell	1.0 to 7×10^{-6}	ClO_4^-, I^-, ClO_3^-, CN^-, Br^-, NO_2^-, HS^-, $CO_3^=$, HCO_3^-, Cl^-
Nitrogen oxide	Gas sensing/ combination	5×10^{-3} to 3.6×10^{-6}	CO_2, volatile weak acids
Oxygen	Gas sensing/ combination	0–14 ppm	–
Perchlorate	Half-cell	1.0 to 7×10^{-6}	I^-, NO_3^-, Br^-, ClO_4^-, ClO_3^-, CN^-, NO_2^-, HCO_3^-, $CO_3^=$, Cl^-
Potassium	Half-cell	1.0 to 10^{-6}	Cs^+, Na^+, NH_4^+, H^+
Redox/ORP	Combination	–	–
Sodium	Half-cell	Saturated to 10^{-6}	Ag^+, K^+, H^+, Li^+, Cs^+, Tl^+

[a]Not an exhaustive list of ISE electrodes or number of interferences. Also, manufacturer specifications must be consulted as different materials and construction may result in different ranges and interferences.

ORP = oxidation reduction potential.

useful in soil analysis. This is not an exhaustive list and new selective electrodes are being developed on a daily basis.

ISEs are standardized using standard solutions of the ion dissolved in water or in a solution designed to keep all samples at about the same ionic strength. These solutions can be purchased or prepared in the laboratory and typically cover several orders of magnitude, often between 1 and 10^{-6} or 10^{-7} M. Measurements are made at the various concentrations and a standard or calibration curve prepared (see Section 14.9.2). Usually, the meter can be programmed to read the concentration of the ion directly once a suitable curve is obtained. Raw data can also be entered into a spreadsheet, which can be programmed to calculate the amounts of ion present in any units desired.

As with pH measurements, a specific amount of water or ionic strength adjusting solution is added to soil, mixed, and allowed to stand. For some analyses, the resulting solution is filtered to separate the soil from the liquid.

The ISEs are then inserted into the suspension or the filtered solution to obtain an electrical signal, the magnitude of which is directly related to the concentration of the desired ion.

ISEs are subject to interferences from ions other than the one they are designed to measure. The Na^+ ISE is susceptible to interference from other single positive species (i.e., K^+, NH_4^+), and the same situation will hold for ISEs designed to measure negatively charged species (see Table 9.1). Generally, the electrode will be less sensitive to these interfering ions than to the ion it is designed to measure so that low levels of interfering ions may not make a significant difference in the measurements being made. In other cases, interfering ions can be precipitated or complexed to remove them from solution before measurements are made.

Interfering ions are problematic when analyzing environmental samples, particularly soil and soil extracts. These samples may contain unknown combinations and concentrations of ions. For these reasons, ISEs are most useful in two situations. The first is where the soil composition is well known and routine repetitive analysis is to be made. The second is where a preliminary screening is to be done and followed by detailed laboratory analysis. In the latter case, potential interfering ions will be determined and the validity of the original screening assessed [7,8].

9.3.3   Redox

Redox is an abbreviation for *red*uction and *ox*idation. It is based on the fact that one component cannot be oxidized without another being reduced. When a species is oxidized, it loses electrons. Conversely, reduction is the gain of electrons. Thus, oxidizing agents cause another species to lose electrons, and reducing species cause something to gain an electron. The most common oxidizing agent is oxygen. In soil, other electron-accepting agents such as carbon, ferric iron, and nitrate can also serve as oxidizing agents. Likewise, there are a number of reducing (electron donating) agents. Chief among these is hydrogen (H_2). Iron and metals can also act as reducing agents. In soil, reduced species are more soluble and generally more easily removed from the soil than are oxidized species. For this reason, the oxidation–reduction condition of a soil sample can be important in any extraction or analytical procedure.

In assessing the redox potential, it is also essential know the pH of the medium being analyzed. There is a direct relationship between pH and the potential measured, as shown in the equation in Figure 9.8. For each unit of increase in pH (decrease in H^+), there is a decrease in millivoltage. This simple equation does not hold in soil because it does not take into account the

$$E = -59 \, \text{mv} \times \text{pH}$$

Figure 9.8. Relationship between electrode potential and pH.

$$Eh = Eh^0 - \frac{RT}{nF} \ln \frac{(Red)}{(Ox)^n (H^+)^m} \tag{1}$$

$$Eh = Eh^0 - \frac{m}{n} 0.059 \, pH \tag{2}$$

$$\frac{m}{n} = \frac{H^+}{e^-} = 1 \tag{3}$$

Figure 9.9. Equations relating Eh, Eh^0, and pH, where Eh^0 is the standard electrode potential; R is the gas constant; T is the absolute temperature; n is the number of electrons; F is the Faraday constant; (Red), (Ox), and (H^+) the concentration of the reduced and oxidized species and the hydrogen ion, respectively.

complex electrical characteristics of soil and thus cannot be used in soil analysis. However, it does illustrate that any measurement or consideration of oxidation–reduction potential of a soil sample must also take its pH into account.

In soil, redox reactions occur between the limits of the oxidation and reduction of water as shown in Figure 9.4. The y-axis shows millivolts (mV) and the x-axis shows the pH. Chemists use these two terms to describe what happens in a redox reaction. Environmental and soil chemists refer to the measured voltage (expressed as millivolts) as Eh. Thus, these types of graphs are called Eh–pH diagrams. Eh is defined by the Nernst equation (Figure 9.9, equation 1), which can be simplified to equation (2) when the activities of (Red) and (Ox) are equal. Many redox couples have the relationship shown in Figure 9.9, equation (3), and thus their reactions as electron acceptors or donors are straightforward (however, in soil, significant variations from this simple relationship exist where $m \neq n$).

In addition to the simple Eh–pH graph shown in Figure 9.3, three-dimensional Eh–pH graphs can be produced. Known quantities of a pollutant can be added to a number of different soil suspensions and its degradation at different combinations of Eh and pH measured. In this way, the optimum conditions for the decomposition of the pollutant in question can be determined. An excellent example of this is the decomposition of pentachlorophenol and hexahydro-1,3,5-trinitro1,3,5-triazene in soil and water under various Eh–pH conditions as illustrated in the papers by Petrie et al. [9] and Sing et al. [10].

In soil analyses, knowledge of the Eh–pH can be used in three ways. It will provide information as to the form or species of pollutant present (see also Chapter 6). It can also be used to determine which extraction procedure is best suited to extract a component from a soil sample. Potential changes in species, movement in the environment, and conditions suitable for bioremediation or natural attenuation can also be derived from this type of measurement.

While not stated explicitly, in this discussion so far, it has been assumed that all the systems were well defined, at equilibrium, and at a constant 25°C. None of these conditions occur in soil in the environment. Soil is not a pure system and, often, all the components affecting redox reactions are not known, defined, or understood, and a host of different redox couples are likely to be present. Unless it is possible to take into account all couples present, it is not possible to describe the exact redox conditions in a soil without measuring it.

Even though very small soil samples may be well defined, large samples and field size areas are not. Soil is never at equilibrium, even though it may appear to be so over short periods of time. Reactions occur and microorganisms continue to function in soil samples after they are taken. In tropical conditions, soil temperatures will vary significantly even where there is little change in air temperature. Heating by sunlight and cooling by radiation will always occur. Variable shading by trees and other vegetation will add to soil temperature variability, and both rain and its subsequent evaporation will have a cooling effect on soil.

In spite of the limitations, Eh–pH data provide information about the condition of a soil in terms of it being in an oxidizing or reducing condition. Thus, it will indicate the prominent redox conditions of the species present (see also Chapter 6). It will also indicate what changes in Eh or pH may be desirable to effect the desired extraction or analysis for the compound or species of greatest concern [9–11].

9.3.4   Gas Electrode

Gas electrodes are similar to ISEs and usually work on the same basic principles. They look much like a standard ISE except that they have a gas-permeable, water-impermeable membrane at the tip. Gas present in the environment passes through the membrane, reacts with reagents in the interior, and produces a voltage. This voltage is related to the partial pressure of the gas being measured.

Electrodes are available for most common gases including oxygen, carbon dioxide, and ammonia. For oxygen and carbon dioxide, their natural concentration in air can be used in standardization, while other gases require samples of standard gas concentrations. Because of the importance of oxygen in biological processes, a number of different types of oxygen sensing electrodes have been developed. Some are compatible with pH meters, while others are not. It is thus important to make sure that the correct electrode is obtained for the instrument to be used.

There are three concerns in using gas sensing electrodes in soil: (1) some electrodes have membranes that have a limited shelf life and must be changed regularly; (2) the membrane is relatively delicate, so electrodes must be placed in soil carefully and cannot be subject to movement; and (3) many membranes must be kept moist to function properly, and thus they cannot be used in dry soils or situations where the soil may dry out during measurement [7].

9.4 VOLTAMMETRY

Voltammetry is the oxidation or reduction of a species at an electrode. In this case, the electrode is the source or the sink for the electrons being exchanged during the reaction. By measuring current (amperage) and potential (voltage) in a system, either stirred or unstirred, information about the species present and their concentrations can be obtained. In a typical experiment, only the oxidized form of a component might be present and the negative potential might gradually increase until a spike is observed. This is the potential at which the oxidized species is being reduced. Because each oxidized species has a different potential where it is reduced, this spike can be used to identify the species. Numerous analytical techniques are based on this electrochemical phenomenon. Some common examples are stripping voltammetry, cyclic voltammetry, and polarography.

The standard potentials of practically all oxidation and reduction reactions, especially those common in the environment and soil, are known or can easily be determined. Because of the specificity and relative ease of conducting voltammetric measurements, they might seem well suited to soil analysis. There is only one major flaw in the determination of soil constituents by voltammetric analysis and that is that in any soil or soil extract, there is a vast array of different oxidation–reduction reactions possible, and separating them is difficult. Also, it is not possible to start an investigation with the assumption or knowledge that all of the species of interest will be either oxidized or reduced.

In well-aerated soil, it is expected that all species will be in their highest oxidation states. However, this does not happen for reasons elucidated in previous chapters. In well-aerated soil, both ferrous and ferric iron can exist along with elemental iron.[3] Zinc, copper, and especially manganese can apparently exist in a mixture of oxidation states simultaneously in soil. Add to this a multitude of organic species that are also capable of oxidation–reduction reactions and the result is truly a complex voltammetric system [12,13].

9.5 ELECTRICAL CONDUCTIVITY

Salts are always present in soil (see Figure 9.2) and the EC of soil or water depends on the types and amounts of salts present. In humid regions, they are at low concentration and do not affect plant growth, while in semiarid and arid regions and near salt lakes or oceans, they may be at high concentration and may have detrimental effects on plant growth. The EC is simply measured by determining the amount of electricity passing through a cell of known dimensions and configuration when it contains a salt solution. In the case of soil, the

---

[3] The occurrence of elemental iron in soil is not common but is possible where there is contamination, for example, from an accident.

EC of a soil paste or solution extracted from soil is related to the soil's salt content.

9.5.1   Whole Soil Paste

Direct determination of the conductivity of a soil sample is carried out by making a paste and placing it in a special, standard cell containing two electrodes. The cell is made of a circular nonconductor with flat-strip electrodes on opposite sides of the cell. A soil paste in water is prepared and added to the cell. The top is leveled and the electrodes attached to a meter that measures its resistance, which is then related back to the salt content of the soil.

9.5.2   Water and Soil Extracts

EC is usually determined on solutions of salt in water. A soil sample is mixed with water until a paste, which is allowed to stand overnight, is obtained. The paste is then filtered and the EC of the solution measured. A conductivity cell for water is shown in Figure 9.6 (E). A sample is simply put in the cavity, or the cell is inserted into the extract, and a measurement made. Standardization is carried out by preparing standard solutions of salt, usually sodium chloride (NaCl), in distilled water.

The basic unit used to represent EC is siemens (S). For direct current, it is the reciprocal of the resistance, and for alternating current, it is the reciprocal of impedance (both in ohm). For soils, dS/m are the units used, where 0.1 S/m = 1 dS/m.[4]

Soils with pH < 8.5 and an EC of <4dS/m are considered normal. Soils in the same pH range but with EC > 4 are said to be *saline*. Soils high in sodium but with an EC < 4 are called *sodic soils*, while those with EC > 4 and high sodium are saline–sodic soils. Sodium is represented as either the ratio of exchangeable sodium to total exchangeable cations called the *exchangeable sodium percentage* or the *sodium absorption ratio*, which is the ratio of sodium ion to the square route of the calcium and magnesium ions [2,14–17].

9.6   TIME-DOMAIN REFLECTOMETRY

Time-domain reflectometry (TDR) involves the use of two or more substantial metal rods inserted into soil. The rods are parallel and are attached to a signal generator that sends an electrical input down the rods. The time it takes the signal to travel down the rods is dependent on the soil's apparent dielectric constant, which, in turn, is proportional to the amount of water in the soil. Upon reaching the end of the rods, the signal is dissipated and the amount of

---

[4] From *Encyclopaedia Britannica Deluxe Edition* 2004 CD-ROM.

dissipation is related to the amount of salt in the soil. This instrument can thus measure both the water content and the salt content of soil. Depending on the type of data needed and the experiment to be done, rods can either be moved from place to place or left in place for measurements made over a period of time [18].

9.7 POROUS BLOCK

The porous block is used to determine the water in soil by changes in the resistance between two electrodes encased in a porous material buried in soil. In a common porous block design, two electrodes are encased in gypsum and connected to wires. The blocks are buried in the soil and their moisture content comes to equilibrium with the soil water content. The resistance between the two electrodes decreases as the water content of the block increases. The change in resistance can then be related back to the soil water content. This basic idea is also used with "blocks" constructed of materials other than gypsum. The gypsum block is designed for fieldwork and finds it greatest use there (see Figure 5.12 [18]).

9.8 OTHER METHODS

There are a number of other types of measurement made in soil that involve electrodes that are not directly in contact with the soil. An example is the thermocouple psychrometer, which involves a Thomson thermocouple in a ceramic cell buried in soil. The thermocouple cools when a current is passed through it, causing water to condense on the thermocouple. When the electricity is turned off, the condensate evaporates at a rate inversely proportional to the relative humidity in the soil. A voltage generated by the cooling junction is measured and related to the soil moisture content. This moisture content is related to both the matrix and osmotic potentials of the soil being investigated.

There are many additional methods and variations on the methods discussed earlier (see Section 5.13). Most are designed for determination of soil water content in the field and are not generally used in laboratory analysis of soil components such as available plant nutrients or contaminants [18–20].

9.9 CONCLUSIONS

Soil has electrical characteristics associated with its components, salts, ions in solution, and the diffuse double layer. All, singly or in combination, can affect electrical measurements in soil. Electrodes inserted into soil are used to measure various soil characteristics, most often soil pH, salt, and water content. Fouling of electrodes by salts, organic matter, or inorganic components, including clay, is an important potential source of error in any soil measurement involving electrodes. Because of the potential errors, electrodes must be standardized frequently during procedures that involve multiple measurements over an extended period of time. Analytical procedures for the determination of soil characteristics using electrodes have been developed and are used in conjunction with other soil procedures and measurements. The development of a new method or procedure will require detailed investigation of the relationship of this new method to previously developed methods and to associated or dependent procedures or measurements.

PROBLEMS

9.1. Diagram a basic electrochemical cell. Diagram a similar cell using soil instead of water as the supporting medium.

9.2. What do the terms Eh and Eh^0 stand for? What types of electrodes are used for the determination of Eh in soil?

9.3. Describe the basic design of a pH electrode. What "kinds" of protons in soil can a pH electrode measure?

9.4. Diagram an ISE electrode. What characteristics of ISE electrodes make them hard to use for direct soil measurements?

9.5. In terms of Eh–pH, what limits the nonbiological range of Eh–pH values in soil?

9.6. Explain and give examples of why voltammetry is not generally useful in direct soil measurement of components present.

9.7. Describe the two common methods of determining salt content in soil using EC.

9.8. Soil water content can be measured in the field using a number of different types of electrodes and arrangements of electrodes. Describe two of these methods in some detail.

9.9. Describe some of the limiting or complicating factors involved in using ISE electrodes in making measurements on soil or soil extracts.

9.10. What types of metal electrodes are best suited to making electrical measurements in soils?

REFERENCES

1. Yang J, Lu F, Kostiuk LW, Kwok DY. Electrokinetic microchannel battery by means of electrokinetic and microfluidic phenomena. *J. Micromech. Microeng.* 2003; **13**: 963–970.

2. Carrique F, Arroyo FJ, Jiménez ML, Delgado ÁV. Influence of double-layer overlap on the electrophoretic mobility and DC conductivity of a concentrated suspension of spherical particles. *J. Phys. Chem. B* 2003; **107**: 3199–3206.

3. Patrick WH, Jr., Gambrell RP, Faulkner SP. Redox measurements of soils. In Bartels JM (ed.), *Methods of Soil Analysis: Part 3 Chemical Methods*. Madison, WI: Soil Science Society of America and American Society of Agronomy; 1996, pp. 1255–1274.

4. Fujitake N, Kusumoto A, Tsukamoto M, Kawahigashi M, Suzuki T, Otsuka H. Properties of soil humic substances in fractions obtained by sequential extraction with pyrophosphate solutions at different pHs I. Yield and particle size distribution. *Soil Sci. Plant Nutr.* 1998; **44**: 253–260.

5. Schlüter M, Rickert D. Effect of pH on the measurement of Biogenic Silica. *Mar. Chem.* 1998; **63**: 81–92.

6. Thomas GW. Soil pH and soil acidity. In Bartels JM (ed.), *Methods of Soil Analysis: Part 3 Chemical Methods*. Madison, WI: Soil Science Society of America and American Society of Agronomy; 1996, pp. 475–490.

7. Phene CJ. Oxygen electrode measurement. In Klute A (ed.), *Methods of Soil Analysis: Part 1 Physical and Mineralogical Methods*, 2nd ed. Madison, WI: Soil Science Society of America and American Society of Agronomy; 1986, pp. 1137–1160.

8. Willard HH, Merritt LL, Jr., Dean JA, Settle FA, Jr. *Instrumental Methods of Analysis*. Belmont, CA: Wadsworth Publishing Co.; 1988, p. 682.

9. Petrie RA, Grossl PR, Sims RC. Oxidation of pentachlorophenol in manganese oxide suspensions under controlled Eh and pH Environments. *Environ. Sci. Technol.* 2002; **36**: 3744–3748.

10. Singh J, Comfort SD, Shea PJ. Iron-mediated remediation of RDX-contaminated water and soil under controlled Eh/pH. *Environ. Sci. Technol.* 1999; **33**: 1488–1494.

11. Zhang TC, Pang H. Applications of microelectrode techniques to measure pH and oxidation–reduction potential in rhizosphere. *Soil Environ. Sci. Technol.* 1999; **33**: 1293–1299.

12. Abate G, Masini JC. Complexation of Cd(II) and Pb(II) with humic acids studies by anodic stripping voltammetry using differential equilibrium functions and discrete site models. *Org. Geochem.* 2002; **33**: 1171–1182.

13. Shuman LM. Differential pulse voltammetry. In Bartels JM (ed.), *Methods of Soil Analysis: Part 3 Chemical Methods*. Madison, WI: Soil Science Society of America and American Society of Agronomy; 1996, pp. 247–268.

14. Gardner DT, Miller RW. *Soils in Our Environment*, 10th ed. Upper Saddle River, NJ: Pearson Education Inc.; 2004, p. 261.

15. Janzen HH. Soluble salts. In Carter MR (ed.), *Soil Sampling and Methods of Analysis*. Ann Arbor, MI: Lewis Publishers; 1993, pp. 161–166.

16. Rhoades J. Salinity: electrical conductivity and total dissolved solids. In Bartels JM (ed.), *Methods of Soil Analysis: Part 3 Chemical Methods*. Madison, WI: Soil Science Society of America and American Society of Agronomy; 1996, pp. 417–436.

17. Lucian WZ, He L, Vanwormhoudt AM. Charge analysis of soils and anion exchange. In Bartels JM (ed.), *Methods of Soil Analysis: Part 3 Chemical Methods*. Madison, WI: Soil Science Society of America and American Society of Agronomy; 1996, pp. 1231–1254.

18. Campbell GS, Glendon GW. Water potential: miscellaneous methods. In Klute A (ed.), *Methods of Soil Analysis Part 1 Physical and Mineralogical Methods*, 2nd ed. Madison, WI: Soil Science Society of America, American Society of Agronomy; 1994, pp. 619–632.

19. Rawlins SL, Campbell GS. Water potential: thermocouple psychrometry. In Klute A (ed.), *Methods of Soil Analysis Part 1: Physical and Mineralogical Methods*, 2nd ed. Madison, WI: Soil Science Society of America and American Society of Agronomy; 1994, pp. 597–617.

20. Reeve RC. Water potential: piezometry. In Klute A (ed.), *Methods of Soil Analysis Part 1: Physical and Mineralogical Methods*, 2nd ed. Madison, WI: Soil Science Society of America and American Society of Agronomy; 1994, pp. 545–560.

BIBLIOGRAPHY

Atkins P, de Paula J. *Physical Chemistry*, 7th ed. New York: W.H. Freeman and Company; 2001, p. 252.

Bard AJ, Faulkner LR. *Electrochemical Methods Fundamentals and Applications*, 2nd ed. New York: John Wiley and Sons; 2000.

Bohn HL, McNeal BL, O'Connor GA. *Soil Chemistry*, 3rd ed. New York: John Wiley and Sons; 2001, p. 109.

Wang J. *Analytical Electrochemistry*, 2nd ed. New York: Wiley-VCH; 2006.

CHAPTER

10

TITRIMETRIC MEASUREMENTS

Titration is a general word used in many different disciplines. Any time a solution of known concentration is used to find the amount of an unknown component in another solution, it can be called a titration. Although this type of analysis is very old, it still finds widespread used in chemical analysis. Titrations are used in soil analysis to measure soil acidity, soil organic matter content, and various constituents isolated from soil, particularly ammonia.

Common chemical titrations include acid–base, oxidation–reduction, precipitation, and complexometric analysis. The basic concepts underlying all titration are illustrated by classic acid–base titrations. A known amount of acid is placed in a flask and an indicator added. The indicator is a compound whose color depends on the pH of its environment. A solution of base of precisely known concentration (referred to as the titrant) is then added to the acid until all of the acid has just been reacted, causing the pH of the solution to increase and the color of the indicator to change. The volume of the base required to get to this point in the titration is known as the end point of the titration. The concentration of the acid present in the original solution can be calculated from the volume of base needed to reach the end point and the known concentration of the base.

The same analysis can be conducted using a pH meter to detect the end point as shown in Figure 10.1, which illustrates a titration curve. A

Introduction to Soil Chemistry: Analysis and Instrumentation, Second Edition.
Alfred R. Conklin, Jr.
© 2014 John Wiley & Sons, Inc. Published 2014 by John Wiley & Sons, Inc.

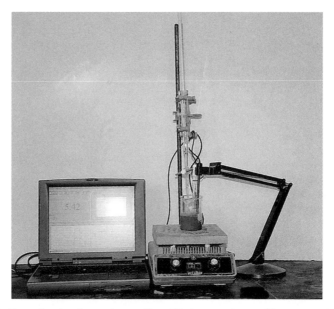

Figure 10.1. Setup for titrating and recording pH change using a pH meter card in a laptop computer.

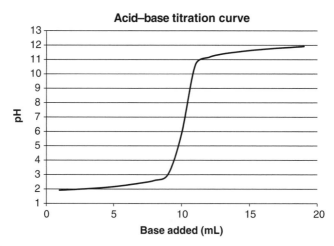

Figure 10.2. Curve obtained by titrating a standard acid with a standard base.

titration curve is a plot of the milliliters of acid added versus pH as in Figure 10.2.

The end point in a titration is a little different from the end of a reaction. What is desired is to know when all the acid is titrated. This happens when the titration curve, shown in Figure 10.1, reaches its maximum slope. This change occurs at a pH of 7 for a typical strong acid–base titration), but occurs at pH values above or below 7 for all other acid–base titrations (i.e., those involving weak acids or bases).

In most cases, a curve is not drawn and the end point is taken as the milliliters (mL) used just when the color change takes place. There should be a half-drop of titrant difference between the change from one color to the next. In many cases, the color change is very light but distinctive. If the titration curve is plotted, then the end point can be determined by inspection or by taking the first or second derivative of the data to find the point of maximum slope.

Although this explanation of titration has been simple, the same basic idea is applied to all forms of titration. In each case, there is a slight change in pH, millivoltage, or ion-selective electrode (ISE) reading with the added titrant, followed by a sharp change through the end point, followed again by a slight change.

Soil and soil suspensions are colored and hard to see through. Thus, it is hard to directly titrate them using colored indicators. There are typically only two cases where direct titrations of soil are carried out. The first is to determine the amount of amendment needed to bring the soil to a desired pH. The second is in the determination of soil organic matter where organic matter is oxidized with chromate and the unreacted chromate is titrated (actually called a *back titration*) to determine, by subtraction, the amount of dichromate reduced and thus the amount of organic matter present.

In other titrations, the component to be titrated is separated from soil and subsequently titrated. The simplest of these is the determination of soil ammonia. However, all forms of nitrogen in soil are important, so methods of converting other nitrogen-containing compounds to ammonia, distilling it, and determining its concentration by titration are important.

Other environmental analytical procedures using titration can be found in the United States Environmental Protection Agency's (USEPA) Web site compilation of methods, particularly the 9000 series methods [10].

10.1   SOIL TITRATION

Typically, acid soils are titrated with a sodium or calcium hydroxide [NaOH or $Ca(OH)_2$] solution and basic soils with hydrochloric acid (HCl), and pH changes are most commonly followed using a pH meter. Carbonates in basic soils release CO_2 during treatment with HCl, thus making the titration more difficult. For this reason, carbonates are often determined by other methods. It is important to keep in mind that basic solutions react with carbon dioxide in air and form insoluble carbonates. This means that either the basic titrant is standardized each day before use or the solution is protected from exposure to carbon dioxide in air. Specific descriptions of titrant preparation, primary standards, and the use of indicators and pH meters in titrations can be found in Harris [1] and in Skoog et al. [2].

The complex nature of soil makes hydrochloric acid the preferred acid titrant because the other common acids can be involved in reactions that are preferably avoided. Both sulfuric and phosphoric acids are di- and triprotic, respectfully, and each proton has a different pK_a. In addition, in high concentration, both are dehydrating agents, and in low concentration are hydrating agents and thus can be involved in reactions other than the desired titration reactions, potentially leading to erroneous results. Nitric acid, on the other hand, is monoprotic but is also an oxidizing reagent and can react with organic matter to produce nitro compounds and thus produce erroneous results. The use of sulfuric, phosphoric, and nitric acids may not be a problem with well-defined systems but can be problematic when applied to undefined systems, especially soils.

Oxidation–reduction reactions and titrations are often easy to carry out, and in many respects, oxidation–reduction titrations are the same as acid–base titrations. For instance, a standardized oxidizing solution, often permanganate, is added to a solution of an easily oxidized species of interest. Permanganate (i.e., the oxidized from of manganese) is dark purple in color and gets converted to a colorless compound when reduced, making the end point easy to determine. Other oxidizing reagents can be used and there are strongly colored molecules that can be used as indicators in the same way as in acid–base titrations. Also, a platinum electrode coupled to a reference electrode can be used to determine the end point using most pH meters.

Most oxidation reactions are between specific metal cations or metal oxyanions and cations. The problem that arises when applying oxidation–reduction reactions to soils is that all soils contain a complex mixture of oxidizable and reducible cations, anions, and organic matter, which means that it is impossible to determine which is being titrated. An exception to this is the oxidation of organic matter where an oxidation–reduction titration is routinely carried out. Organic matter determination will be discussed in Section 10.3.

Precipitation titrations are typified by the titration of chloride with silver or vice versa. In this case, interferences with the precipitation reaction may occur because of components in the soil, and the soil itself may interfere with detection of the end point. Thus, complexation reactions are rarely applied directly to soil; however, they can be applied to soil extracts. Common environmental titration methods described in the United States Environmental Protection Agency (USEPA) methods are summarized in Table 10.1 [1,2].

10.1.1 Back Titration

In a back titration, an excess amount of standardized reagent is reacted with an unknown amount of a component of interest. When the reaction is complete, the remaining unused reagent is titrated and the amount of component of interest is determined by difference. In the Kjeldahl procedures described next, ammonia is distilled into an acid of known concentration. When all of the ammonia has been distilled, the remaining unreacted acid is titrated. The

TABLE 10.1. Titrimetric Methods Used by the USEPA

Method	Species Titrated	Titrant
9014	Cyanide	Silver nitrate
9034	Sulfides	Iodine
9253	Chloride	Silver nitrate

USEPA, United States Environmental Protection Agency.

amount of ammonia distilled is calculated (and thus in the original material being investigated) from the difference between the amount of acid present at the start and the amount remaining at the end.

Back titrations are common in soil analysis, as they are used in both nitrogen and organic matter determinations. Back titrations are highly valuable analytical techniques and are applicable to other environmental analyses as well.

10.2   TITRATION OF SOIL pH

Titration of soil pH is an old method that is not widely used today. Basically, an acid soil suspension is prepared and titrated with a standardized base, often sodium hydroxide, although various basic calcium compounds such as calcium oxide (CaO) and calcium hydroxide [$Ca(OH)_2$] can also be used. Because of the dark color of many soils, they are often titrated using a pH meter as the indicator of the end point. A setup for the titration of soil is shown in Figure 10.1. Titration is slow in that it takes some time after the addition of titrant for some semblance of equilibrium to be reached. Once this happens, a reading can be made or simply another addition of titrant made.

A titration curve for an acid soil suspension to which 1 mL of a calcium hydroxide titrant is added and the change in pH followed for 2.3 minutes is shown in Figure 10.3. As can be seen, the pH initially increases and then falls back toward the original pH. The curve not only has a sawtooth pattern but is also curved in the reverse direction from a standard titration of an acid with a basic solution.

The initial pH rise as shown in Figure 10.3 and fallback is interpreted as the result of two reactions. The initial rapid pH increase is a result of neutralization of free protons, or H_3O^+, in the soil solution. The slower subsequent decrease in pH is a result of re-equilibration between H_3O^+ in solution and on exchange sites (Figure 10.4). In addition, there may be a release of weakly held protons from either or both inorganic and organic constituents in soil. It might be envisioned that there is an equilibrium between all three sources of protons and that the decrease is a return to reestablishing this equilibrium or is an example of soil buffering.

These three sources of protons are illustrated in Figure 10.4, in which $H_3O^+_s$ designates hydronium ions in solution, $H_3O^+_e$ designates hydronium ions on

Soil titration with calcium hydroxide

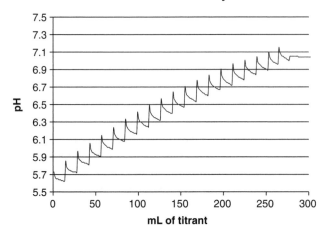

Figure 10.3. Stepwise titration of acid soil with calcium hydroxide.

$$H_3O^+{}_s \rightleftharpoons H_3O^+{}_e \rightleftharpoons H_b$$

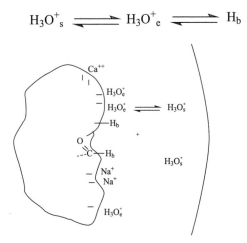

Figure 10.4. An equation showing the equilibrium between bonded protons, H_b, exchangeable protons, $H_3O^+{}_e$, and soluble protons, $H_3O^+{}_s$, is given above and illustrated below.

exchange sites, and H_b designates protons bonded to some soil constituent by either a covalent or polar covalent sigma bond. As discussed in the previous chapters, measuring soil pH using a salt solution results in a lower pH being found. Here, the cation provided by the salt replaces protons or hydronium ions on exchange sites, and thus they are in solution and can be directly titrated. When a base such as NaOH is added to soil, the Na^+ cation will exchange with protons or hydronium ions on exchange sites in a similar manner. In addition, every proton exposed to the soil solution will have a pK_a

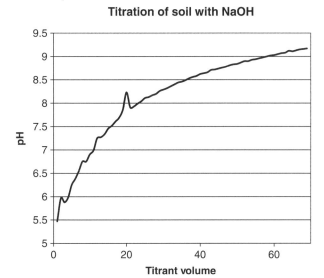

Figure 10.5. Titration of 50 g of soil suspended in 50 mL of distilled water with 0.1 M NaOH using a pH meter. The titrant was added slowly and continuously with stirring.

value and thus will be released or bonded depending on the pH of the solution. What is seen is that, as the solution is made more basic, protons from all these various sources are potentially released into solution.

If a slow continuous addition of base is made to the same soil used in Figure 10.3, a similar titration curve without the sawtooth pattern is seen. Figure 10.5 shows the titration curve obtained by the continuous slow addition of 0.1 M NaOH. Again, the curve is not a smooth line, and irregularities seen in this titration are seen in other titrations of this same soil. Note that no distinct titration end point is seen here as there is in Figure 10.2. However, it is possible to determine the amount of base needed to bring this soil to pH 6.5, which is a typical pH desired for crop production.

> **Caution:** Lowering or raising a soil's pH to effect remediation, especially on a large scale, is not feasible for four reasons: (1) changing the pH of soil to any great extent requires large amounts of acid or base because soil is highly buffered; (2) soil is destroyed at both very high and very low pHs; (3) a large amount of material that cannot be readily returned to the environment is produced; and (4) the material is no longer soil!

This also explains why the pH of any extracting solution is important. Depending on the pH of the extracting solution, the component(s) of interest may be in the form of an ion or polar or neutral molecules. This, in turn, determines if it will be solvated by the solvent chosen as the extractant. If a certain pH is needed for an extraction process, then titration of a soil sample can be carried out in order to determine how much base or acid would be needed for the process. This would be useful in cases where removal of a contaminant from a spill site or a field is required.

Because of the complex nature of soil and soil solutions, it is rarely possible to directly determine specific soil constituents by titrating soil or soil solutions using a pH meter, ion-specific electrode, or a platinum electrode (with appropriate reference electrode) [3].

10.3   ORGANIC MATTER

Soil organic matter can be divided into many fractions. The first distinction between fractions is the active fraction, which is the fraction undergoing active decomposition. The second distinction is the stable fraction, which is mostly humus. The most common method of determining soil organic matter does not differentiate between these two types. All organic matter is oxidized using a strong oxidizing agent, most often potassium or sodium dichromate in sulfuric acid. To effect complete oxidation, heating, by simply mixing the acidic and dichromate solutions, or using a hot plate, is required. When the reaction is complete, unreacted dichromate is back titrated in an oxidation–reduction titration, and the difference is used as the amount of organic matter present. This titration uses an indicator, the color of which is difficult to see because of the soil present. It is interesting to note that the indicator color change is much easier to see with natural rather than fluorescent lighting.

Other methods for determining the amount of soil organic matter are available [4]. However, they are not as commonly used as is chromate oxidation, which is commonly called the Watley–Black method. Usually, these other methods are both more time-consuming and less accurate than is the dichromate oxidation method. The dichromate oxidation of organic matter is the standard by which all other methods of determining soil organic matter must be compared [4,5].

Caution: Chromates, including potassium and sodium dichromate, are hazardous materials, as are sulfuric and phosphoric acid used in the oxidation of soil organic matter. Great care must be exercised in using these chemicals and in disposing of the waste generated.

10.4   AMMONIA

Ammonia is a gas that reacts with water to form ammonium ions as shown in Figure 10.6. The equilibrium lies to the right unless the solution is made basic, at which point the equilibrium shifts to the left and ammonia gas is released.

$$H_2O \ + \ NH_3(g) \ \xrightleftharpoons[^-OH]{H^+} \ NH_4OH$$

Figure 10.6. Reaction of ammonia gas with water to form ammonium.

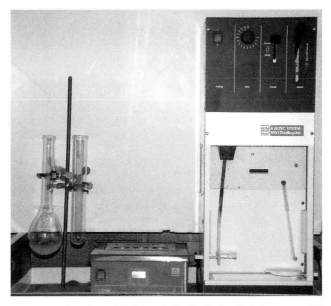

Figure 10.7. Setup for performing Kjeldahl analysis of soil. On the left are two different types of Kjeldahl flask; to the right of the flask is a heating block used to heat the test tube-shaped flask during digestion and on the right, a steam distillation unit.

This is the basis for a common method for the determination of ammonia in soil.[1] Soil is suspended in water and placed in a Kjeldahl flask. The suspension is made basic by the addition of a strong (5–50%) sodium hydroxide solution, and the flask is immediately attached to a steam distillation setup. Steam distillation of the suspension carries the released ammonia to an Erlenmeyer flask, catching the distillate in a standardized acid solution that is subsequently back titrated via acid–base titration. The amount of ammonia in soil can be calculated from the end point of the titration. This procedure is similar to a standard Kjeldahl determination and can be carried out using the same equipment, although no digestion is needed.

A steam distillation apparatus is shown on the right-hand side of Figure 10.7. This same apparatus can be used to determine ammonia in soil as described previously. A flow diagram for the determination of ammonia in soil using a Kjeldahl analysis is given in Figure 10.8 [6].

---

[1] A cation containing solution is added before ammonium determination to release exchangeable ammonium.

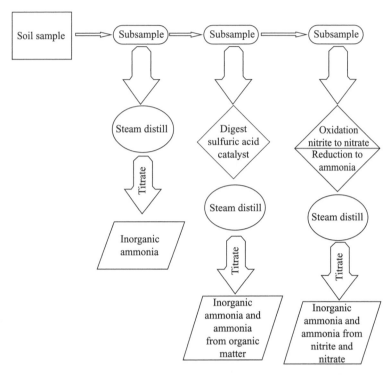

Figure 10.8. Flow diagram for the determination of ammonia, organic nitrogen, nitrite, and nitrate using Kjeldahl apparatus [6].

Caution: Extreme caution must be exercised in carrying out a Kjeldahl procedure. Digestion involves the use of concentrated sulfuric acid, and in the past, both selenium and mercury have been used as catalyst. Older equipment may be contaminated with these elements and should be treated with caution. During the digestion process, copious amounts of *choking, corrosive, and toxic* sulfur dioxide fumes are released. These fumes must be removed by evacuating them using an aspirator or some other appropriate method. Standard Kjeldahl equipment will come with components and directions for removing SO_2 safely. Ammonia distillation involves the use of highly concentrated caustic sodium hydroxide solutions, which must be handled with care. Sodium or potassium hydroxide spilled on skin is best removed with a dilute solution (1–2%) of acetic acid in water followed by washing with soap and water.

10.5 KJELDAHL: ORGANIC NITROGEN

Nitrogen in soil organic matter is mostly found in proteins and amino acids. Although the specific analysis for these important and interesting compounds can and is done (see Chapters 13 and 14), it is more often the case that the total inorganic and organic nitrogen in soil is determined. This is because inorganic nitrogen compounds are used by plants and are of environmental concern. Decomposition of organic nitrogen-containing compounds results in the release of ammonia into the soil solution, where it immediately reacts to form ammonium ions. Once in this form, it is readily oxidized by soil bacteria to nitrite and finally to nitrate. Because of the ready conversion of organic nitrogen into inorganic forms and the ready interconversion of inorganic nitrogen in soil, its total concentration, both inorganic and organic, is important.

In total soil nitrogen analysis, a soil sample is first digested in a Kjeldahl flask to convert all organic nitrogen into inorganic ammonia. Two Kjeldahl flasks are shown on the left-hand side in Figure 10.7. The flask with the bulb at the bottom is an older Kjeldahl flask, while the large test tube is a newer design. Digestion is accomplished using concentrated sulfuric acid and a catalyst. A salt such as potassium sulfate is added to increase the boiling point of sulfuric acid such that decomposition of organic matter occurs more readily. This mixture plus soil is heated until all organic matter has been destroyed. A heating block for heating the Kjeldahl flask is shown next to the distillation unit. After digestion, the solution is cooled and a concentrated basic solution, usually 50% NaOH, is added. The ammonia that is released is steam-distilled into a receiving flask containing a standard acid that reacts with the ammonia. Upon completion of the steam distillation, the unreacted acid is titrated and the amount of ammonia distilled is calculated by difference.

The Kjeldahl procedure has been used for many years to determine the nitrogen in human tissues and in both animal and human foodstuffs. For these materials, the procedure works well and is straightforward. This is not the case for soil. All soils naturally contain some ammonium, and when the steam distillation is conducted, this distills along with the ammonia produced by the decomposition of organic matter. This gives a measurement of the total ammonium in soil after digestion. It cannot distinguish between ammonium derived from organic matter and from the soil itself.

If a soil Kjeldahl organic nitrogen determination as described earlier has been carried out, it can be used along with simple ammonia steam distillation to measure the amount of nitrogen from each source, that is, inorganic ammonium and organic matter. However, this still does not provide a measurement of the total nitrogen in soil because it does not account for that present as either nitrite or nitrate. Figure 10.8 provides a flow diagram for determining all nitrogen in soil using the Kjeldahl procedure.

10.6 NITRITE AND NITRATE

Both nitrite and nitrate are highly mobile in soil and easy to extract. However, it is also possible to reduce each individually to ammonia, steam-distill the ammonia, capture it, and titrate it as described earlier for ammonia. If this procedure is followed, naturally occurring ammonia in soil must first be determined as described previously. After this step, a reducing agent is placed in the flask and nitrite and nitrate are reduced to ammonia. The soil is then rendered basic again and the ammonia steam-distilled and titrated. If both nitrate and nitrite are reduced at the same time, the combined amount of both is obtained. At this point, selective reduction of either nitrate or nitrite with subsequent distillation and titration will allow for the calculation of the amounts of all four forms of nitrogen in the soil sample.

The determination of nitrite in soil can be accomplished by reactions carried out before digestion or steam distillation to oxidize nitrite to nitrate and then to reduce nitrate to ammonia. Nitrite can also be reacted with an organic compound, typically salicylic acid, and subsequently digested in a typical Kjeldahl procedure. This procedure starts with an organic molecule that reacts with nitrite to form a nitro compound that is subsequently reduced to an amine. The amine is then subject to digestion just as with organic matter in soil. In this way, the total nitrogen content in soil can be determined. Also, when this procedure is combined with those described earlier, the amounts of all inorganic forms of nitrogen in soil (organic, ammonia, nitrite, and nitrate) can be determined. This is an extremely powerful method for elucidating the nitrogen status of soils. Figure 10.8 gives a flow diagram for determining nitrite and nitrate in soil using a Kjeldahl apparatus.

The procedures described earlier determine soil nitrogen that is often taken to be the total nitrogen in soil. Because there are very few other organic or inorganic nitrogen compounds commonly found in soil, there is little call for additional analytical procedures. However, there are some exceptions. It may be necessary to determine the gaseous oxides of nitrogen formed during denitrification. This is typically accomplished using gas chromatography (see Chapter 13). Also, using simple steam distillation, ammonia trapped in clay structures will not be determined. Such determination requires complete destruction of soil minerals and the subsequent analysis of the released ammonia.

Although Kjeldahl procedures are capable of providing information about all the common nitrogen components in soil, it is a time-, labor-, equipment- and reagent- intensive procedure. Typically, Kjeldahl equipment, namely, a digestion–steam distillation apparatus, can accommodate 6, 12, or more digestion tubes or flasks, and distillation and titration can be done quickly. However, digestion may take an hour or more, and thus it will take many hours to analyze any significant number of samples. For ^{15}N work, additional steps, time, and money may be required to convert ammonia to nitrogen gas and analysis of ^{15}N by mass spectrometry (see Chapter 14) [6].

10.7 CARBONATE DETERMINATION

Carbonates decompose under acidic conditions with the release of carbon dioxide as shown in Figure 10.9. To determine carbonate content, a soil

$$CaCO_3 + 2HCl \rightarrow CaCl_2 + CO_2 \uparrow + H_2O$$

$$NaHCO_3 + HCl \rightarrow NaCl + CO_2 \uparrow + H_2O$$

Figure 10.9. Reaction of carbonate and bicarbonate with acid.

sample can be placed in an Erlemneyer flask and a 0.1 M solution of hydro-chloric acid (HCl) added until no more carbon dioxide is released. The amount of HCl consumed is determined by back titration and used to calcu-late the amount of carbonate and bicarbonate present. The reactions shown are for calcium carbonate and sodium bicarbonate; however, all carbonates in the soil will be decomposed. Thus, this is a method for determining the total carbonate content, not just calcium carbonate. Additionally, inaccuracies can be caused by other components in soil that react with the HCl, such as bicarbonates.

Another approach in measuring carbonate is to measure the amount of carbon dioxide produced via reaction with HCl, either by measuring the volume of gas released or by reacting the carbon dioxide, in a separate flask, with base, and determining the amount of base remaining after the reaction.

Alternately, the weight lost when the carbonate is reacted with HCl can be determined. Heating carbonates results in their decomposition. Thus, soils containing carbonate can be heated to the appropriate temperature and the loss of weight measured. In this approach, the loss of organic matter and water of hydration of various components in soil must be accounted for in order to determine the weight of carbon dioxide [7].

10.8   HALOGEN ION DETERMINATION

Inorganic salts that contain halogens are usually soluble. They commonly occur as simple, single, negatively charged anions in soil. There are two common exceptions to this generalization. First, fluorine is commonly found bonded to phosphate in insoluble minerals called *apatites*, which are calcium phosphate fluorides.

Halogen anions are easily leached from soil with water and can be determined using silver nitrate as a titrant. The letter X is commonly used to represent halogens and thus may be interpreted to be any of them (i.e., F, Cl, Br, or I), but it is not generally used for At. Other possible counter ions, namely, nitrate and the cation associated with the halide, are ignored. Halogen salts can be determined by reacting them with silver ions to form a solid precipitate. The solid that is formed is weighed and the mass can be used to calculate the amount of halogen ions present in the original soil sample.

$$X^- + Ag^+ \rightarrow AgX \downarrow$$

$$Cl^- + Ag^+ \rightarrow AgCl \downarrow$$

Figure 10.10. Reaction of silver with halogen X^-.

Figure 10.10 shows the precipitation titration that can be done with or without an indicator. It should be noted, though, that silver ions are very reactive and act as oxidizing agents. For instance, silver ions can oxidize aldehydes including aldoses (sugars with an aldehyde functionality). For this reason, it is possible to obtain inaccurate or misleading results with titrating a soil extract with silver ions.

The most common halogen in soil is chloride, while both bromide and iodide occur but are uncommon [8]. The occurrence of either of these anions in soil would be cause for concern. Analysis for these other halides could be carried out using either capillary electrophoresis or high-performance liquid chromatography (HPLC) (see Chapter 13).

Organic compounds that contain halogens are not titrated but determined using chromatography and spectroscopy (see Chapters 13 and 14).

10.9 pH–STAT TITRATIONS

In another type of titration, the system is maintained at one fixed pH during a reaction. This type of titration, termed pH–stat,[2] has been applied to bioreactors where neither the starting material nor the product of the reaction is titrated, but rather, acidic or basic by-products or coproduced acid or base are measured. In biological reactions, these products may be CO_2, HCO_3^-, or $CO_3^=$. The system, maintained at a set temperature during the reaction of interest, is titrated with an automatic titrator set to maintain the specific pH. Bioreactors may be sealed such that no liquid or gas is lost or gained or they may be open and allow the exchange of gases.

The titrant used in pH–stat titration is usually a dilute (perhaps 0.01 M) acid or base. Two different sets of data can be obtained from a pH–stat titration. The amount of a substrate consumed or product formed can be determined by the total amount of titratant used. Because the titrant is added over a period of time, the rate of reaction can also be determined.

Even a small soil sample can be considered a bioreactor, and so the method of pH–stat is applicable to soil. Although this method is applicable to the study

---

[2] Stat stands for static and derives from the fact that the pH is held static during the progress of the reaction.

of processes occurring in soil, it is not particularly useful in routine analysis of soil components [9].

10.10 CONCLUSIONS

The titrimetric determination of soil constituents is most commonly applied to a limited number of soil analyses, namely, organic carbon, nitrogen compounds, carbonates, and chlorides. Determination of acid content by titration is generally not done because the titration curves are not amenable to typical titration analysis. Because of the color of soil and the fact that it is a suspension when stirred, it is often necessary to remove the constituent of interest before titration. In other cases, it is possible to do a direct titration using an appropriate indicator. However, even in these cases, detection of the end point is difficult.

Because of the complex nature of titration curves obtained using pH, ion-selective electrodes, or millivolt (Eh) measurements on whole soils, these methods, except for organic matter determination, are seldom used.

PROBLEMS

10.1. Explain why the end point of a titration is not at either end of a titration curve.

10.2. Explain why titration is not a generally useful method for discovering the acidity of soil.

10.3. Suggest areas in soil where there might be organic matter that is not determined by dichromate oxidation (refer to earlier chapters).

10.4. Make a flow diagram that shows how to determine all different forms of nitrogen found in soil.

10.5. Look up the titrimetric method for the determination of cyanide and describe it.

10.6. Give the equation for the reaction of carbonate with acid. Describe two ways titration might be used to determine carbonate.

10.7. In environmental analysis, ^{15}N can be used to determine where nitrogen moves in the environment. Explain how ^{15}N containing inorganic compounds might be isolated from soil and how it could be specifically determined. (*Note:* You may wish to consult Chapters 13–15 in answering this question.)

10.8. Describe pH–stat titration in detail.

10.9. Describe the two types of information that can be obtained about a reaction by using the pH–stat method of titration.

10.10. During most titrations, the solution or suspensions are mixed sometimes continuously. Considering this, why might it be a good idea to use an indicator during a complexometric titration?

208 TITRIMETRIC MEASUREMENTS

REFERENCES

1. Harris DC. *Quantitative Chemical Analysis*, 6th ed. New York: Freeman; 2006.
2. Skoog DA, West DM, Holler FJ, Couch SR, Holler J. *Analytical Chemistry: An introduction*, 7th ed. Pacific Grove, CA: Brooks Cole; 1999.
3. Thomas GW. Soil pH and soil acidity. In Bartels JM (ed.), *Methods of Soil Analysis Part 3: Chemical Methods*. Madison, WI: Soil Science Society of America and American Society of Agronomy; 1996, pp. 475–490.
4. Nelson DW, Sommers LE. Total carbon, organic carbon and organic matter. In Bartels JM (ed.), *Methods of Soil Analysis Part 3: Chemical Methods*. Madison, WI: Soil Science Society of America and American Society of Agronomy; 1996, pp. 961–1010.
5. Storer DA. A simple high sample volume ashing procedure for determination of soil organic matter. *Soil Sci. Plant Anal.* 1984; **15**: 759–772.
6. Bremner JM. Nitrogen—total. In Bartels JM (ed.), *Methods of Soil Analysis Part 3: Chemical Methods*. Madison, WI: Soil Science Society of America and American Society of Agronomy; 1996, pp. 1085–1122.
7. Loeppert RH, Suarez DL. Carbonate and gypsum. In Bartels JM (ed.), *Methods of Soil Analysis Part 3: Chemical Methods*. Madison, WI: Soil Science Society of America and American Society of Agronomy; 1996, pp. 437–474.
8. Frankerberger WT, Jr., Tabatabai MA, Adriano DC, Doner HE. Bromine, chlorine, fluorine. In Bartels JM (ed.), *Methods of Soil Analysis Part 3: Chemical Methods*. Madison, WI: Soil Science Society of America and American Society of Agronomy; 1996, pp. 833–967.
9. Ficara E, Rozzi A, Cortelezzi P. Theory of pH-stat titration. *Biotechnol. Bioeng.* 2003; **82**: 28–37.
10. U.S. Environmental Protection Agency. Test Methods SW-846 online. 2013. Available at: http://www.epa.gov/epawaste/hazard/testmethods/index.htm. Accessed June 3, 2013.

EXTRACTION OF INORGANICS

Inorganic components in soil are all the elements and compounds that do not contain carbon and hydrogen. All of the naturally occurring elements are found in soil, although some are present at extremely low levels. Extractions are used to remove the elements and compounds from soil for analysis. Hydrogen, oxygen, and silicon, when bonded to themselves and other elements, are not generally extracted because when bonded to inorganic elements, they are not chemically reactive in soil. Silicon, however, is an essential element for some crop plants including rice. Additionally, the carbonates (Na_2CO_3 and K_2CO_3), bicarbonates ($NaHCO_3$ and $KHCO_3$), and some hydrides are of interest and can be analyzed. The occurrence of carbonates and bicarbonates is easily determined by treating soil with acid. Effervescence, which is caused by the release of carbon dioxide as shown in Equation 11.1,

Introduction to Soil Chemistry: Analysis and Instrumentation, Second Edition.
Alfred R. Conklin, Jr.
© 2014 John Wiley & Sons, Inc. Published 2014 by John Wiley & Sons, Inc.

is indicative of the occurrence or presence of either bicarbonates or, more commonly, carbonates:

$$3HA + M_2CO_3 + MHCO_3 \rightarrow 2HOH + 3MA + 2CO_2. \qquad (11.1)$$

Other than hydrogen, oxygen, and silicon, all elements and inorganic compounds are extracted from soil to determine if they are sufficient for optimum plant production or to make sure they are not at toxic levels. For nonessential elements, only the latter determination is important. However, some essential elements, such as boron, can be at toxic at higher levels and thus be of concern. While most metals occur as cations, some, notably molybdenum, along with the nonmetals boron and phosphate, occur as oxyanions. Of the nonmetals, carbon, hydrogen, oxygen, sulfur, nitrogen, phosphorus, boron, and chlorine are essential for plants. Of the metals, potassium, calcium, magnesium, iron, manganese, zinc, copper, and molybdenum are essential. Other metals such as nickel and sodium are sometimes reported as being essential, but there is still some question about their function in biological systems.

It is often assumed that if something is in the soil, it will be in plants. This is incorrect. Plants do not take up all of the elements or molecules present in their immediate environment. However, there are some plants, called hyperaccumulators, that accumulate higher than normal levels of some toxic elements. These plants still do not take up all the elements in their environment and they are often small, so the total amount of toxic elements removed from soil is limited. In addition, not all species of an element are toxic and some are not biologically available and thus do not enter biological systems. For example, chromium as Cr(VI) is more toxic and more biologically available than is Cr(III) [1].

Inorganic components in soil are extracted with water, acidic solutions containing highly soluble ligands and chelates, and basic solutions. Acidic solutions are typically used for extraction of metals and metal ions in both exchangeable and nonexchangeable forms. Basic solutions are used much less commonly, although they are important for oxyanions, particularly phosphate.

11.1   EXTRACTION EQUIPMENT

In most cases, extraction is carried out using simple equipment such as Erlenmeyer flasks or test tubes. A sample, plus the appropriate extractant, is added to the container and shaken for a specified amount of time (see the Procedures in the next sections). A flat top or wrist shaker (see Figure 11.1) is commonly used for Erlenmeyer flasks. In some instances, an end-over-end shaker is required.

When shaking is complete, the suspension is gravity filtered through a standard fine filter paper into a sample container. If the filtrate is cloudy, it can be filtered through the same filter paper again. If this is not sufficient, the filter

Figure 11.1. Wrist shaker. Arms extend to the left and right having several clamps for flasks as shown. Top plate also has clamps for flasks.

paper plus sample can be placed on top of a fresh piece of filter paper and the suspension filtered again.

In some cases, centrifugation is used to separate the solid sample from the extractant. This works well, but care must be taken not to mix the sample while removing it from the centrifuge. Regardless of the method of shaking, all extracts must be absolutely clear—meaning free of suspended particles before analysis is carried out!

11.2 WATER AS A SOIL EXTRACTANT

Water is modified by soil when added as either rain or irrigation [2]. The modifiers are plants, plant roots, organic matter, organic matter decomposition products, carbon dioxide and other gases in the soil atmosphere, and dissolved inorganic compounds, commonly salts. Of particular importance is the change in pH that accompanies this modification of water. Thus, components obtained from soil by added extraction water will be significantly different from components extracted by water already in the soil. However, both types of water are used to investigate soil chemistry [2–4].

The extraction of soil with water at pH 7 would seem to be a good way to study the soil inorganic chemistry. The most common solvent that soils are in contact with is water in the form of rain. However, rain is not neutral but acidic. Rainwater pH ranges from 3.8 to 5.6, depending on the air in which it forms. Acid rain that contains H_2SO_4 and HNO_3 created by the reaction of gases in the atmosphere with water can have a pH as low as 2.0 [2].

Because rain is acidic, extracting soil with distilled or deionized (DI) water will not mimic the chemistry of rainwater percolating through soil. However, extraction of soil with distilled or DI water has the advantage of not adding anything from the water to the soil. Much soil chemistry research is carried out on water extracts of soil. There are two ways in which these extracts can be obtained. Soil can be extracted, as indicated earlier, with water added to soil or by removing naturally occurring water soil.

In both types of extraction, it is not certain that all the water or even a representative sample of the water is removed from soil. Water in small pores, cracks, or held at greater pressures that those applied to remove water will not be removed and their constituents will not be included in the analysis. However, both these methods find wide use in soil analysis.

11.2.1   Water and Reagent Purity

The quality of the water used is important when soil is to be extracted with water to measure inorganic constituents, although it is of even greater importance when water is used to extract organics (see Chapter 12 for more information on this topic). Thus, in all cases, the water used must be of the highest quality. It is important to make sure that DI water is stored in containers that will not add inorganics to it. Therefore, for the extraction of inorganics, plastic storage containers are always preferred.

During manufacture, many reagents are exposed to metal. It is recommended that the highest-purity acids, bases, salts, ligands, chelates, and other reagents be used. It is best if none of the extracting solutions contain the analyte of interest. If this is not possible, then care must be taken to make sure that the appropriate blanks are used during analysis.

11.2.2   Extraction with Added Water

This type of extraction can be carried out in two different ways. A soil sample can be brought into the laboratory and extracted with relatively large amounts of water to try to determine its inorganic composition. The water-to-soil ratio can be either on a mass-to-mass basis or a volume-to-mass ratio. A one-to-one ratio is commonly used, although other ratios have been used. After a designated extraction time, with or without shaking, water is filtered from the soil and analyzed. A typical water extraction of soil is given in Procedure 11.1.

Procedure 11.1. Typical Water Extractants[1]

A 2-g sample of sieved (#10 sieve <2 mm), air-dry soil is shaken with 240 mL of distilled water at 20°C for 1 hour. Shaking is accomplished on an end-over-end shaker at 28 rpm (rounds per minute), filtered, and analyzed.

TABLE 11.1. Common Anions and Anion Species in Soil

Anion	Species
Chloride	Cl^-
Bromide	Br^-
Nitrate	NO_3^-
Nitrite	NO_2^-
Perchlorate	ClO_4^-
Sulfate	SO_4^{2-}
Phosphate[a]	$M_2PO_4^-$, MPO_4^{2-}, PO_4^{3-}

[a]M represents a monovalent cation.

Water extraction is used to extract anions from soil except in the case of phosphate, which is extracted using a different extraction procedure (see Bray in Section 11.3). Common anions and anion species in soil are given in Table 11.1 [4].

Extraction with heated water is also possible but is not commonly used to extract inorganic constituents.

11.2.3   Extraction of Pore Water

A second approach is to obtain (extract) water as it naturally occurs in soil pores. Typically, a porous ceramic cup (other materials are available) is placed in an unsaturated soil, either in the field or laboratory. A vacuum is applied (slightly more negative pressure than the water in the soil pores; see also Chapter 5) to the ceramic cup via tubing to move water into a receiving container. This water can be analyzed for all its constituents. A reason for obtaining this type of soil water sample is to analyze it for one specific constituent such as a herbicide, insecticide, or pollutant. Water extracted in this way may also better represent the concentration of the analyte of interest to which plant roots are exposed.

Both kinds of water extraction can be used to compare different soils or soils under different management.

---

[1] Procedures 11.1–11.10 are examples only, have been abbreviated to illustrate the general concept, and are not to be taken as complete procedures. For complete procedures, see the specific reference for the section.

Several points need to be considered when applying this pressure method of sample collection. All of the extraction system parts must be explosion proof and the distance from the ceramic cup to the sample bottle must be as short as possible. The sample collection bottle should be at the same level as the ceramic cup. If it is higher, additional vacuum will need to be applied to move the sample water into the sample bottle. Sample storage, once the water is collected, is determined by the analyte of interest.

There is a large variety of equipment used in this type of analysis. Ceramic cups may be made of materials other than ceramic. Connecting tubing may be of a number of different types. Vacuum sources and control vary. Additional equipment may be necessary or useful in different sampling situations. Some initial investigation of available equipment and methodologies is essential before starting this type of sampling [3,4]. A typical extraction of soil pore water is given in Procedure 11.2:

Procedure 11.2. Typical Extracted Water

Prepare all equipment and a suitable ceramic cup as per manufacturer's instructions. Place the ceramic cup in soil and apply a vacuum slightly greater than the water to be removed; that is, if soil water at −35 kPa is to be removed, a vacuum of −55 kPa is applied to the tubing connecting the ceramic cup to the sample container. Continue to apply vacuum until a suitable amount of sample is collected (adapted from Reference 2).

Another way to extract pore water is to use a displacement method, which is usually applied in the laboratory. In this method, a solvent immiscible with water is applied to the soil in a column or centrifuge tube, and gravity or centrifugation is used to move the solvent through the soil, displacing pore water. The displaced water can then be separated from the immiscible liquid and analyzed.

All filtering and extraction systems must be checked for compatibility with the analyte of interest. Some or all of the analytes may be absorbed or otherwise lost during the extraction process, and this must be known beforehand. Carbon may be lost by high vacuum differentials between sampler and soil water. Long extraction times may lead to distortions in the results of the sample analysis [5].

11.2.4   Extraction with Dilute Salt Solutions

A variation on simple extraction is extraction with a dilute salt solution. In this case, either a dilute solution of calcium or potassium chloride is commonly used. The concept is that these dilute solutions mimic the root environment to some extent. In addition to extraction, these solutions are often used in the determination of soil pH [6]. A typical dilute salt solution extraction of soil is given in Procedure 11.3:

Procedure 11.3. Typical Water Extractants with Salts

Air-dry soil is mixed with 0.02 M calcium chloride solution (1:2 ratio, for instance, 10-g soil 20 mL 0.02 M $CaCl_2$ solution) and mixed for 1 hour. The pH of the suspension can be measured directly. In addition, the solution can be filtered for the determination of aluminum or magnesium by atomic absorption spectroscopy (AAS) or inductively coupled plasma (ICP) spectroscopy (adapted from Reference 5).

Simple extractions of soil, with or without salts, however, are not the most common methods of extracting soil to determine its composition or chemistry, particularly with regard to aluminum or magnesium [7].

11.2.5 Extraction with Neutral Cation Exchange Solutions

A common procedure is to extract only the exchangeable cations. This is typically done with a neutral ammonium acetate (NH_4OAc) solution. The ammonium exchanges with exchangeable cations in the soil. A solution containing the soluble and exchangeable cations in the sample is thus obtained. This solution can then be analyzed by either AAS or ICP spectroscopy [8]. A typical cation exchange extraction of soil is given in Procedure 11.4:

Procedure 11.4. Extraction with Cation Exchange Solution

A 2-g air-dry soil sample (weighed to three place accuracy) is placed in a 50-mL Erlenmeyer flask or a similar size centrifuge tube and shaken with 20 mL of a 1 M, pH 7, NH_4OAc solution, by shaking for 2 hours. Suspensions can be filtered or centrifuged to obtain a solution suitable for analysis by AAS or ICP (adapted from Reference 8).

Procedure 11.4 can also be used to determine the amount of all the individual cations present. The sum of these plus the pH can be used to determine the cation exchange capacity (CEC) of soil.

11.2.6 Cation and Anion Exchange

CEC of soils is reported as centimoles of charge per kilogram of soil, $cmol_c$/kg (meq/100 g in the older literature). This is, in reality, a measure of the sum of the effective negative charges on soil components and exchange sites as given in Equation 11.2a, where S represents soil negative sites. The reactions are often illustrated using equations that are similar to that given in Equation 11.2b. This is an equilibrium reaction; thus, it is thermodynamic and the equilibrium constant can be calculated using standard chemistry nomenclature as in Equation 11.2c. The quantities can be converted to activities, which are more appropriate, to produce a true thermodynamic equilibrium constant [9]:

$$CEC = \sum S^-. \tag{11.2a}$$

$$\boxed{Soil}{}^{-}\!\!\begin{smallmatrix}Na^{\cdot}\\Na^{\cdot}\end{smallmatrix} + Ca^{2+} \rightleftharpoons Ca^{2+}\!\!\begin{smallmatrix}-\\ \end{smallmatrix}\boxed{Soil} + \begin{smallmatrix}Na^{\cdot}\\Na^{\cdot}\end{smallmatrix} \tag{11.2b}$$

$$K_{eq} = \frac{\left(Ca^{2+}{}^{-}\boxed{Soil} \right)\left(Na^{\cdot} \right)^2}{\left(\boxed{Soil}{}^{-}\begin{smallmatrix}Na^{\cdot}\\Na^{\cdot}\end{smallmatrix} \right)^{\!2} \left(Ca^{2+} \right)} \tag{11.2c}$$

Under the same conditions and concentration, the larger the charge on a cation, the stronger it is attracted to the cation exchange sites on soil. Thus, a cation with higher charge will replace a cation with lower charge (see top of Figure 11.2). However, if a cation of lower charge but at a higher concentration is added, it can replace the higher charge ion. Both situations are shown in Equation 11.3. This is also illustrated at the bottom of Figure 11.2:

$$M^{3+} > M^{2+} > M^{+}; 2 \text{ molar } M^{+} > 1 \text{ molar } M^{2+}. \tag{11.3}$$

When different researchers study this selectivity, it is observed that H^+ does not easily fit into this scheme. This is probably due to its reactivity in terms of protonating various soil groups not associated with its CEC. Also, as can be determined by careful examination of Figure 11.2, some inaccuracy in determining CEC is possible, especially when exchange sites on soil components do not exactly match the charges on exchangeable cations.

The CEC of soil is important from two perspectives. First, it retards the leaching loss of important cations from soil. Each year, calcium and magnesium are leached out of the soil, causing soils in humid regions to become acidic. This increased acidity must be ameliorated by the application of limestone ($CaCO_3$). Second, exchangeable cations are available to plants, and this characteristic allows soil to serve as a store of important soil nutrients such as potassium, calcium, and magnesium.

In addition to using neutral ammonium acetate extraction, the CEC of a soil can be measured directly. A soil sample is leached with a solution containing a single cation, commonly, barium, at a relatively high concentration until all exchangeable cations in the sample have been removed (see top graphic in Figure 11.3). Complete removal can be determined by analyzing the leachate for cations other than the one being added. Once the soil exchange sites are saturated, the soil is leached with alcohol, usually 1-propanol, until no added cation is found in the leachate. Alternately, the counter ion (anion) added with the cation can be determined and its absence taken to indicate that all the nonexchangeable cation has been removed (see Figure 11.4). At this point, the cation can be removed by leaching the soil sample with a solution containing

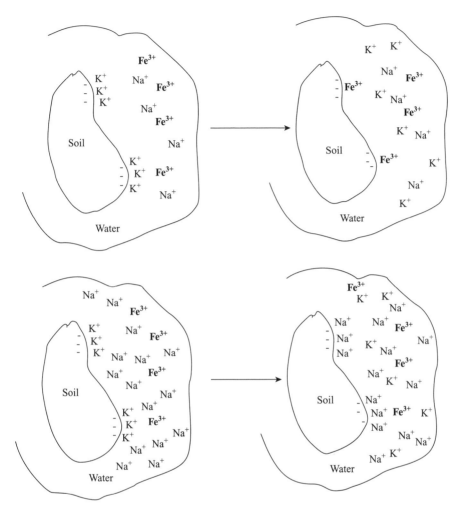

Figure 11.2. Top shows exchange of higher valent iron for lower valent potassium when sodium and iron are at the same concentrations. Bottom shows exchange of sodium for potassium when sodium is at a much higher concentration than iron or potassium. The common anions in soil, chloride, sulfate, nitrate, carbonate, bicarbonate, and the anions of small acids such as acetate, have been omitted for clarity.

another cation, often calcium, sodium, or ammonium, at high concentration (see Figure 11.5).

Anion exchange capacity can be investigated in a similar manner except that the soil is first leached with an anion [4].

When comparing the CEC of different soil samples, it is essential to do so with soil extracted at the same pH. Protons, H^+, will act as exchangeable cations and thus the measured CEC will be different at different pHs. This comes from exchange sites on clay, which are permanent, and those associated

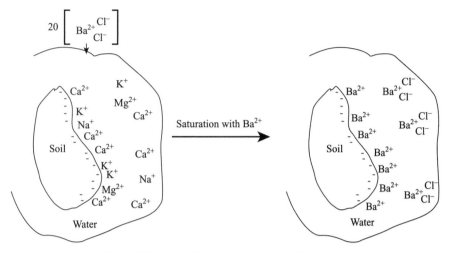

Figure 11.3. Saturation of exchange sites with barium.

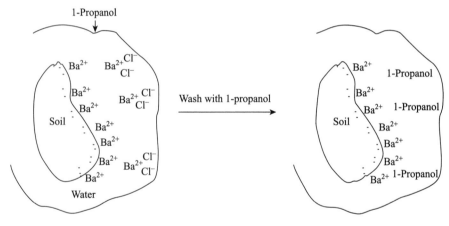

Figure 11.4. Washing with 1-propanol to remove nonexchangeable barium.

with organic matter, which are variable, depending on the pK_as of the components of the organic matter.

11.3 ACID EXTRACTANTS

A majority of soil extractions of inorganic constituents are carried out to determine the metal content of a soil sample. In these cases, it is common to use acid extracting solutions. Dilute hydrochloric and sulfuric acids are most commonly used. Nitric acid is an oxidizer, so undesired oxidation can occur during the extraction process using this reagent. Phosphate is a natural

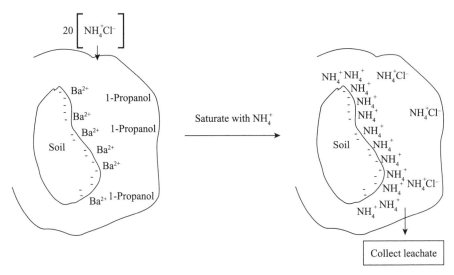

Figure 11.5. Leaching with excess ammonia to remove all barium from exchange sites for determination of CEC.

component of soil, so the use of phosphoric acid in soil extractions can lead to confusion. Also, insoluble metal phosphates that will prevent the determination of the metal will be produced. These latter two acids are not commonly used in metal extraction because of these interactions. Acetic acid has also been used as a soil extractant. It has the disadvantage of being a weaker acid than the mineral acids but has the advantage of being decomposed readily in soil.

As indicated previously, soils can contain carbonates and bicarbonates that release carbon dioxide when treated with acid. The release of CO_2 can be dramatic and can cause splattering and loss of sample. A preliminary step in any acid extraction is to check the soil sample to determine if it contains carbonate or bicarbonates. If it does, particular caution must be taken in extracting the sample. Also, because the acid extractant is neutralized by the carbonate or bicarbonate, this must be taken into account in carrying out the extraction.

It is common for acid extracting solutions to contain ligands, such as fluoride, that, when combined with metals, increase their solubility. Chelates such as ethylenediaminetetraacetic acid (EDTA), which combine with metals and either bring them into solution or keep them in solution, are also commonly used.

Two different objectives are common when extracting soil to determine its metal content. The first is to determine the amount of biologically active metal present. This is typically the goal for determining the levels of plant-required metals. In terms of total samples analyzed, this is the most common reason for extracting soil. Metals in soil that are essential to plants and that are most often extracted are potassium, calcium, and magnesium. The other metals, required in lower concentrations by plants, are primarily zinc and copper, although molybdenum is also important. Iron and manganese are usually sufficient for plant needs and are therefore less often extracted. Availability of these metals to plants is highly dependent on soil pH, so the likelihood of them being extracted is also highly dependent on the soil pH. Varying the pH of the extraction solution is carried out when it is important to determine if a contaminant is biologically available under some specific or unique soil conditions.

The second type of extraction involves removing metals from contaminated soil such that their concentration is below some defined level. Sometimes, this is called the "actionable level." Above this level, some action needs to be taken to remove the contaminating metal. Below this level, no action need be taken. This type of extraction is limited to contaminated soil such as would be found at Superfund sites. Metals commonly of concern are chromium and cadmium; other metals extracted are given in Table 11.2. At high levels, plant essential metals, mentioned earlier, are toxic and so may also be of concern under these conditions.

11.3.1   Biologically Available Nutrients

Biological extractions are carried out to determine if biologically important elements are at levels that are sufficient, yet not toxic, for plant needs. Acid soil extraction to determine the biologically available plant nutrients is the most common type of extraction of soil carried out. The objective is to extract a portion, not all, of a particular nutrient or metal that is correlated to the amount available to plants. The plants of primary interest are crop plants such

TABLE 11.2. Metals Extracted by USEPA Method 3050B[a]

FLAA/ICP-AES		GFAA/ICP-MS
Aluminum	Magnesium	Arsenic
Antimony	Molybdenum	Beryllium
Arsenic	Lead	Cadmium
Barium	Nickel	Cobalt
Beryllium	Potassium	Iron
Cadmium	Silver	Lead
Calcium	Sodium	Molybdenum
Cobalt	Thallium	Selenium
Copper	Vanadium	Thallium
Iron	Zinc	

[a]Methods for FLAA/ICP-AES and GFAA/ICP-MS require slightly different digestion procedures which must be followed [13].
FLAA, flame atomic absorption, is termed AAS in most instances in this book and in other places; ICP, inductively coupled plasma; AES, atomic emission spectroscopy; GFAA, graphite furnace atomic absorption; ICP-MS, ICP coupled to mass spectrometry.

as corn, soybeans, and wheat. However, the same extractions can be used for determining fertilizer needs for other crops, ornamentals, and plants used for remediation of contaminated or disturbed sites. For instance, these extractions are used to determine fertility needs of plants to be planted on reclaimed strip-mined land.

Extractants typically include soluble, exchangeable, and easily dissolved forms of the particular nutrient. Nutrients in the soil solution and on exchange sites are immediately available to plants. In addition, a small but significant amount of nutrients bound to or part of soil solid structure will become available to plants during the growing season, and so it is desirable to include them as part of the analysis. A typical acid nutrient extraction of soil is given in Procedure 11.5:

Procedure 11.5. Typical Nutrient Acid Extractants

To 1-g sieved air-dry soil add 20 mL acid extracting solution. Shake for 5 minutes and filter; if filtrate is not clear, repeat filtration. The filtrate can be analyzed by calorimetry or ICP (adapted from Reference 10).

Two typical acid extractants are the Bray (which has two forms, both of which are acidic) and the Mehlich-3. The Bray extractant is a dilute solution of hydrochloric acid and ammonium fluoride [11]. The Mehlich-3 extractant is a dilute solution of acetic and nitric acids and also contains ammonium nitrate and EDTA [11]. Both are designed to extract soluble, exchangeable, and easily dissolved nutrients, particularly phosphate. While the Bray extractant is designed to extract plant available phosphorus, the Mehlich-3 extractant also extracts potassium [10–12].

11.3.2   Environmental Extractions

Environmental extractions are carried out on contaminated soil to determine if the level of contamination is an environmental threat. Acid environmental extractants are most often used to determine the metal content of soil. A typical acid environmental extraction of soil is given in Procedure 11.6. This is an abbreviated and simplified version of the United States Environmental Protection Agency (USEPA) Method 3050B [13]. This method is not an absolute digestion of soil. It is intended to extract those metals shown in Table11.2 that might become available in the environment. It is a strong acid digestion, but metals entrained in silicon minerals are not dissolved. Metals in these minerals are not considered to become environmentally available. If complete digestion of soil is desired, then USEPA Method 3052 can be used [14]. However, it must be kept in mind that this type of digestion leaves results in a "nonsoil" mixture that must be disposed of carefully (see Figure 11.6).

Procedure 11.6. Typical Acid Environmental Extractants

Take 1 g (to nearest 0.001 g) air-dry, sieved soil and place in a 250-mL straight wall, no pore spout beaker. Add 10 mL of 1:1 diluted concentrated nitric acid; cover the beaker top with a ribbed watch glass. Heat the sample to 95°C for 15 minutes. The sample should not boil, but the condensate should condense on the watch glass and drip back into the beaker. Allow the sample to cool, then add 5 mL concentrated nitric acid and heat for 30 minutes. The sample should not boil, but the condensate should condense on the watch glass and drip back into the beaker. If brown fumes appear, repeat the step until there are no brown fumes. Maintaining the temperature (95°C), allow the sample to evaporate to 5 mL.

Cool the sample and add 2 mL of water and 3 mL of 30% hydrogen peroxide (H_2O_2). Warm the sample with the ribbed watch glass in place until effervescence slows. Allow the sample to cool and add 1 mL of water and again warm. Repeat this step until the sample does not change in appearance. Note: do not add more than 10 mL H_2O_2. Add to the ribbed watch glass and warm until the volume is 5 mL.

Add 10 mL of concentrated hydrochloric acid (HCl), cover with the ribbed watch glass, and heat (95°C) for 15 minutes.

Cool and filter the sample (Whatman No. 41 filter paper or equivalent) into a 100-mL volumetric flask and dilute to 100 mL with water. This sample can be analyzed by AAS or ICP (adapted from USEPA Method 3050B; see Reference 11).

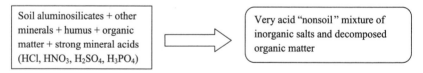

Soil aluminosilicates + other minerals + humus + organic matter + strong mineral acids (HCl, HNO_3, H_2SO_4, H_3PO_4) ⟹ Very acid "nonsoil" mixture of inorganic salts and decomposed organic matter

Figure 11.6. Results of extraction of soil with very strong mineral acids.

11.4 EXTRACTANTS FOR BASIC SOILS

Basic soils present a unique analytical challenge. Most of these soils contain calcium carbonate ($CaCO_3$) as the primary base. Basic soils also contain magnesium and, to a lesser extent, sodium carbonate. Although soils containing lithium and potassium carbonate are known, they are uncommon. These compounds produce a basic solution when dissolved in water. This means that adding either water as an extractant or water containing small amounts of salt is not effective because the soil already contains salts and solutions immediately become basic when added to these soils.

Adding acidic extractants to basic soils is ineffective for a number of reasons. First, the acid in acid extractants will be neutralized as illustrated in Equation 11.1 and Equation 11.4 and therefore will no longer be acid (in Equation 11.1, A is the anion of an acid and M is the metal of a base). Second, because the soil environment is basic, any components extracted by acid are not normally part of this environment and have no relevance to their availability. Acid will react with any carbonates present, as illustrated in Equation 11.4, and will release carbon dioxide. This gaseous release will cause splattering and a loss of sample.

$$HA + MOH \rightarrow HOH + MA \qquad (11.4)$$

For these reasons, basic soils are extracted using a basic solution. The most common is the Olson extractant, which is 0.5 M sodium bicarbonate ($NaHCO_3$).

11.4.1 Biologically Important Inorganics

Due to the effect of plant roots and the natural or normal decomposition of soil minerals, some phosphate and other nutrients will become available [7] during the growing season and thus must be accounted for. The soluble and exchangeable phosphate (i.e., the immediately available phosphate) and that portion of the soil phosphate that will become available to plants during the growing season are determined by the Olson extractant. A typical base extraction of soil nutrients is given in Procedure 11.7:

Procedure 11.7. Extraction of Basic Soils (Olson Extraction)

Add 1.5 g of air-dry soil to a 125-mL Erlenmeyer flask. Add 40 mL of 0.5N $NaHCO_3$, pH 8.5, and shake on a mechanical shaker for 30 minutes. Filter through a Whatman No. 2 filter paper. Refilter through the same filter paper if the filtrate is not clear. A 5 mL aliquot of this filtrate is taken for colorimetric determination of phosphate (adapted from Reference 10).

11.4.2 Environmental Extractions

There are only a few extractions other than the Olson extraction for basic soils [8,9]. A typical basic environmental extraction of soil is given in Procedure 11.8. Note that in this procedure, the components of interest are first extracted with basic solutions. Thus, the components present in the final analysis will only be those extracted by base. However, acid is used to adjust the pH before final analysis, although the solution is still basic (pH 7.5), at the end of the procedure [15].

Procedure 11.8. Basic Environmental Extraction

Take a 2.5-g (to 0.01-g) sample of field moist soil and add 50 mL of 0.28 M Na_2CO_3/0.5 M NaOH solution. Also add 400 mg of magnesium chloride ($MgCl_2$) and 0.5 mL of 1 M phosphate buffer to the suspension. Stir the samples for 5 minutes and then heat to 90–95°C for 60 minutes. Gradually, with continued stirring, cool and filter through a 0.45 μm membrane into a 250-mL flask, rinsing the digestion flask three times with water and adding these with filtration (0.45 μm membrane) to the 250-mL flask. With continued stirring, adjust the pH of the solution to 7.5 using 5 M nitric acid (HNO_3). (**Caution: carbon dioxide gas may be released during the neutralization and may cause loss of analyte.**) If cloudiness or a precipitate forms, filter through a 0.45 μm membrane again. Transfer the neutral solution to a 100-mL volumetric flask and dilute with water to the mark. This solution can be analyzed for Cr(IV) by the appropriate method (adapted from USEPA Method 3060A; see Reference 15).

11.5   MICROWAVE-ASSISTED EXTRACTION

Soil extraction is a time-consuming process in most analyses. Thus, to decrease the time needed for extraction, soils can be extracted with the assistance of microwaves. A microwave oven that is similar to a household microwave, except that it is specially designed for laboratory use, is used, along with special digestion vessels in the extraction. Extracting solutions are the same or very similar to those used without microwave assistance, although some variations may be necessary. Samples and extracting solutions are placed in a digestion container designed for microwave-assisted extraction and are placed in the microwave for the specified period of time.

During extraction, both the temperature and pressure in the extraction container increase, and thus care must be taken. The container must be able to withstand the temperatures and pressures and must be compatible with the extractant used. When the microwave extraction time is complete, the digestion container is removed and the digestate is handled and analyzed in the same way as a nonmicrowaved sample [15]. An example of a microwave extraction is given in Procedure 11.9:

Procedure 11.9. Microwave-Assisted Extraction

A 1-g air-dry, sieved (#10 sieve <2.00 mm) soil is added to a **polytetrafluoroethylene** (PTFE) digestion vessel. To this is added 10 mL of 70% nitric acid (HNO_3), and the mixture is microwaved at 15% power for 10 minutes. At this point, the extract can be filtered and analyzed or more extensive microwave digestion carried out as outlined in Sastre et al. (adapted from Reference 16).

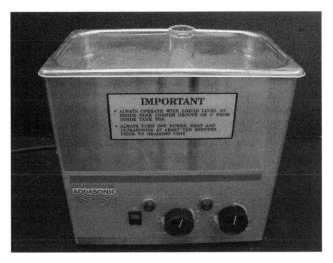

Figure 11.7. Sonic bath with flask containing soil and extractant.

Microwave extractions can be carried out by any of the above-mentioned extractions resulting in the extraction of the same components. The advantage is that microwave extraction is typically much faster.

11.6 ULTRASONIC EXTRACTION

Soil extraction using all types of aqueous solutions and ultrasonic agitation has been carried out. Simple apparatus such as extractant and aqueous solvent in an Erlenmeyer flask or test tube are used. An ultrasonic bath with Erlenmeyer flask is shown in Figure 11.7 (the use of an ultrasonic horn is shown in Figure 12.9). Ultrasonic extraction is typically carried out when the soil particles are not well separated, such as high clay soils, and thus the surfaces are not exposed to the extracting solutions.

11.7 SEQUENTIAL EXTRACTION

In some cases, it is desirable or necessary to extract components from soil that are not all removed by one extractant and thus require sequential extraction. Another use of sequential extraction would be to extract different ionic forms or species of the same or different elements or compounds. In this case, several different extractants may be used one after another. Once separated, the components can be analyzed by standard methods. A typical sequential extraction of soil is given in Procedure 11.10 [17–21]:

Procedure 11.10. Sequential Extraction of Soil

Step 1: To an appropriate centrifuge tube add 1-g (0.001-g) of air-dry, sieved soil and 40 mL of acetic acid. The mixture is shaken overnight, centrifuged, and the supernate separated from the residue for analysis.

Step 2: To the residue is added 40 mL of hydroxylammonium chloride, adjusted to the appropriate pH using nitric acid, and the extraction repeated as in step one.

Step 3: The residue from step two is treated twice with 8.8 M H_2O_2 and evaporated to near dryness. To this is added 50 mL of ammonium acetate adjusted to pH 2 using nitric acid. Extraction is repeated as in step 1.

Step 4: The residue from step three is microwave digested using 20 mL of aqua regia (adapted from Reference 7).

In Procedure 11.10, step 1 is designed to extract soluble species, carbonates, and species on exchange sites. Step 2 is designed to extract reducible iron and magnesium oxyhydroxides. Step 3 extracts oxidizable organic matter and sulfides, while step 4 extracts any metals remaining after the completion of the previous extractions. Sequential extraction methods have also been used to extract and quantify the amounts of various arsenic species, primarily as As(III) and As(IV) in soil [21].

11.8 ION EXCHANGE RESIN EXTRACTIONS

It is possible to extract or remove ionic species, both anions and cations, from soil using ion exchange resins. Both anion and cation exchange resins have been used as well as combinations of the two. Resins can be added to the soil and mixed, or they can be contained in a bag (Procedure 11.11), on a strip, or in capsules buried in soil. Mixing resins with soil allows for more intimate contact with soil and with the soil solution. However, one is faced with separation of the resin from soil at the end of some "extraction" time. Resins in bags, on strips, or as capsules can easily be removed from soil. However, the resins do not have as intimate contact with soil in this procedure. Good relationships between all these methods and standard extraction methods have been obtained and all approaches have found utility in determining the amounts of various ions in soil.

In some cases, resins have been used to try to determine only the plant or more generally the biological availability of an ionic species. Resins placed in soil have also been used to study ion speciation, soil microbiology, various phosphorus measurements, soil nutrient supply rate, nutrient transformations and movement, and micronutrient and metal toxicity [22–25].

Procedure 11.11. Ion Exchange: Resin Bag Method

Mixed bed, cation, and anion exchange resins (17 g) in a nylon bag are prepared. These are placed 6–8 cm deep in 300-g soil pots. After the assay period (5 months), bags are removed and extracted with 1N KCl (adapted from Reference 23).

It is common to concentrate organic components extracted from soil before analysis is conducted. Concentration of ionic species is not as common. However, the use of ion exchange resins to remove ionic species from soil is a well-established ion removal method. Although this method is not commonly discussed in terms of concentration of ions found in soil, it can lead to increased ion concentration and increased ability of analytical methods to measure trace amounts of ions in soil [26].

11.9 SURFACTANTS

Surfactants can be either found or synthesized to contain both hydrophobic and hydrophilic regions and can be cationic, anionic, or neutral. Cationic surfactants can be chosen to extract metals from soil. Of particular interest are those anionic surfactants that are naturally produced by plants and microorganisms because they can easily be broken down in soil once extraction has been accomplished. However, the use of surfactants for extracting metals has found limited use [27].

11.10 CONCLUSION

Inorganic species in soil are generally extracted with either water or an acid solution typically containing hydrochloric acid. Various other components that aid either in the solubilization, extraction, or stabilization of extracted inorganics, such as chelates, are also often added during the extraction process. Basic extractions are not as commonly used as are acid extractions with a few notable exceptions. The use of ion exchange resins to extract ions from soil is well established.

PROBLEMS

11.1. Explain why the extraction of soil with distilled water with pH 7 will not duplicate the loss of inorganic components from soil due to percolation of rainwater though soil.

11.2. Describe how the pH of water is influenced by various soil components. Include all soil components in your explanation.

11.3. From common observations of the effects of acids on metals, explain why acids would be used to extract metals from soil.

11.4. There are a number of exceptions to the observations alluded to in Problem 11.3. Explain two of these.

11.5. What base is commonly used to extract soil and what component, in what form, is this base used to extract?

11.6. What might be expected to be accomplished by a sequential extraction of soil?

11.7. Describe the different objectives in extracting a soil sample for biologically active fractions and for environmental concerns.

11.8. What do FLAA, ICP, AES, GFAA, and MS stand for and in which chapters might you find out more about these?

11.9. What is the danger associated with extracting basic soils with acid? (Hint: something is rapidly released.)

11.10. In some instances, organic matter is removed by oxidation before extraction of metals. Would you expect there to be more metal or less in the extract when this is done?

REFERENCES

1. Shanker AK, Cervantes C, Loza-Tavera H, Avudainayagam H. Chromium toxicity in plants. *Environ. Int.* 2005; **31**: 739–753.

2. Willey JD, Bennett RI, Williams JM, Denne RK, Kornegay CR, Perlotto MS, Moore BM. Effect of storm type on rainwater composition in southeastern North Carolina. *Environ. Sci. Technol.* 1988; **22**: 41–46.

3. Sornsrivichai P, Syers JK, Tillman RW, Cornforth IS. An evaluation of water extraction as a soil-testing procedure of phosphorus. Glasshouse assessment of plant-available phosphate. *Fert. Res.* 1988; **15**: 211–223.

4. Westergaard B, Hansen HBC, Borggaard OK. Determination of anions in soil solutions by capillary zone electrophoresis. *Analyst* 1998; **123**: 721–742.

5. Gollany HT, Bloom PR, Schumacher TE. Rhizosphere soil-water collection by immiscible displacement-centrifugation technique. *Plant Soil* 1997; **188**: 59–64.

6. Hendershot WH, Lalande H, Duquette M. Soil reaction and exchangeable acidity. In Carter MR, Gregorich EG (eds.), *Soil Sampling and Methods of Analysis*, 2nd ed. Boca Raton, FL: CRC Press; 2008, pp. 173–178.

7. Hoyt PB, Nyborg N. Use of dilute calcium chloride for the extraction of plant available aluminum and manganese from acid soil. *Can. J. Soil Sci.* 1972; **52**: 163–167.

8. Simard RR. Ammonium acetate extractable elements. In Carter MR (ed.), *Soil Sampling and Methods of Analysis*. Boca Raton, FL: CRC Press; 1993, pp. 39–42.

9. Summer ME, Miller WP. Cation exchange capacity and exchange coefficients. In Bartels JM (ed.), *Methods of Soil Analysis. Part 3. Chemical Methods*. Madison, WI: Soil Science Society of America; 1996, pp. 1201–1229.

10. Bray RH, Kurtz LT. Determination of total, organic, and available phosphorus in soils. *Soil Sci.* 1945; **59**: 39–46.

11. Sawyer JE. Differentiating and understanding the Mehlich 3, Bray and Olsen soil phosphorus tests. Available at: http://www.agronext.iastate.edu/soilfertility/presentations/mbotest.pdf. Accessed June 3, 2013.

12. Mehlich A. Mehlich 3 soil test extractant—a modification of Mehlich 2 extractant. *Com. Soil Sci. Plant Anal.* 1984; **15**: 1409–1416.

13. Method 3050B. Acid Digestion of Sediments, Sludges, and Soils. Available at: http://www.epa.gov/epawaste/hazard/testmethods/sw846. 2012. Accessed June 3, 2013.

14. Method 3052. Microwave Assisted Acid Digestion of Siliceous and Organically Based Matrices. http://www.epa.gov/epawaste/hazard/testmethods/sw846. 2012. Available at: http://www.epa.gov/epawaste/hazard/testmethods/sw846. Accessed June 3, 2013.

15. Method 3060A. Alkaline Digestion for Hexavalent Chromium. Available at: http://www.epa.gov/epawaste/hazard/testmethods/sw846. Accessed June 2, 2013.

16. Sastre J, Sahuquillo A, Vidal M, Rauret G. Determination of Cd, Cu, Pb and Zn in environmental samples: microwave-assisted total digestion versus aqua regia and nitric acid extraction. *Anal. Chim. Acta* 2002; **462**: 59–72.

17. Davidson CM, Duncan AL, Littlejohn D, Garden LM. A critical evaluation of the three-stage BCR sequential extraction procedure to assess the potential mobility and toxicity of heavy metals in industrially-contaminated land. *Anal. Chim. Acta* 1998; **363**: 45–55.

18. Mossop KF, Davidson CM. Comparison of original and modified BCR sequential extraction procedures for the fractionation of copper, iron, lead, manganese and zinc in soils and sediments. *Anal. Chim. Acta* 2003; **478**: 111–118.

19. Tessier A, Campbell PGC, Bisson M. Sequential extraction procedure for the speciation of particulate trace metals. *Anal. Chem.* 1979; **51**: 844–851.

20. Gleyzes C, Tellier S, Astruc M. Fractionation studies of trace elements in contaminated soils and sediments: a review of sequential extraction procedures. *TrAC Trend Anal. Chem.* 2002; **21**: 451–467.

21. Wenzel WW, Kirchbaumer N, Prohaska T, Stingeder G, Lombi E, Adriano DC. Arsenic fractionation in soils using an improved sequential extraction procedure. *Anal. Chem. Acta* 2001; **436**: 309–323.

22. Qian P, Schoenau JJ. Practical applications of ion exchange resins in agricultural and environmental soil research. *Can. J. Soil Sci.* 2002; **82**: 9–21.

23. Binkley D, Matson P. Ion exchange resin bag method for assessing forest soil nitrogen availability. *Soil Sci. Soc. Am. J.* 1983; **47**: 1050–1052.

24. Gibson DJ. Spatial and temporal heterogeneity in soil nutrient supply measured using in situ ion-exchange resin bags. *Plant Soil* 1986; **96**: 445–450.

25. Sibbesen E. A simple ion-exchange resin procedure for extracting plant available elements from soil. *Plant Soil* 1977; **46**: 665–670.

26. Weng L, Temminghoff EJM, van Riemsdijk WH. Determination of the free ion concentration of trace metals in soil solution using a soil column Donnan membrane technique. *Eur. J. Soil Sci.* 2001; **52**: 629–637.

27. Mulligan CN, Youg RN, Gibbs BF. Surfactant-enhanced remediation of contaminated soil: a review. *Eng. Geol.* 2001; **60**: 371–382.

BIBLIOGRAPHY

Kim C, Lee Y, Ong SK. Factors affecting EDTA extraction of lead from lead-contaminated soils. *Chemosphere* 2003; **51**: 845–853.

Wilson N. *Soil Water and Groundwater Sampling*. Boca Raton, FL: CRC Press; 1995.

CHAPTER

12

EXTRACTION OF ORGANICS

Organic soil, organic matter, and organic contaminants, except those capable of hydrogen bonding and having low molecular weight, have limited or no solubility in water (see Table 12.1). Water is therefore not usually a solvent of first choice when extracting organic compounds from soil. Organic solvents such as hexane and dichloromethane are more commonly used for this purpose. Supercritical carbon dioxide and organic solvents under supercritical conditions[1] are also commonly used to extract organic compounds from soil. In addition, extraction of organic compounds usually involves the use of specialized glassware or high-pressure apparatus.

However, for some compounds, hot water and pressurized hot water have been used successfully to extract contaminating organic compounds from soil. This includes both low soluble and insoluble compounds [1] (see Section 12.3.3).

Humus, which is extracted using a basic aqueous solution, is an exception to this general rule. Fractionation of humus (see Figure 12.1) is also mostly

---

[1]Supercritical conditions occur when an extractant is kept under conditions where it would otherwise be a gas, such as temperatures above its boiling point. It is kept as a liquid by being kept under high pressure.

Introduction to Soil Chemistry: Analysis and Instrumentation, Second Edition.
Alfred R. Conklin, Jr.
© 2014 John Wiley & Sons, Inc. Published 2014 by John Wiley & Sons, Inc.

TABLE 12.1. Common Functional Groups Found in Organic Compounds and Their Solubility in Water

Compound Type	Functional Group	Salt Formation	Solubility in Water
Alkanes	none	NA	Insoluble
Alkenes	$R_2C=CR_2$	NA	Insoluble
Alkynes	$RC\equiv CR$	NA	Insoluble
Ethers	$R_3C-O-CR_3$	NA	Insoluble
Alcohols	R_3C-OH	NA	Five carbons or less soluble, more than five carbons insoluble
Aldehydes	RCHO	NA	Low molecular weight soluble, high molecular weight insoluble
Ketones	R_2CO	NA	Low molecular weight soluble, high molecular weight insoluble
Acids	RCOOH	Yes	Low molecular weight soluble, high molecular weight insoluble
Acid salts	RCOOM[a]	Yes	Generally soluble
Amines	RNH_2, R_2NH, R_3N	Yes	Low molecular weight soluble, high molecular weight insoluble
Amine salts	$RNH_2 \cdot HCl$, $R_2NH \cdot HCl$, $R_3N \cdot HCl$	Yes	Generally soluble
Esters	RCOOR	NA	Insoluble
Anhydrides	RCOOOCR	NA	Generally insoluble, very reactive
Acid halides	RCOX[b]	NA	Generally insoluble, very reactive
Amides	$RCONH_2$, RCONHR, $RCONR_2$	Yes	Low molecular weight soluble, high molecular weight insoluble
Amide salts	$RCONH_2 \cdot HCl$, $RCONHR \cdot HCl$, $RCONR_2 \cdot HCl$	Yes	Generally soluble

[a]M is a general representation for metal.
[b]X always represents a halogen, generally, F, Cl, Br, or I.

carried out using aqueous solutions that are either acidic or basic. Humus is extremely important in soil chemistry and has been studied extensively. A number of chemical structures have been proposed as being part of its molecular composition; however, no definitive structure for the whole molecule has been elucidated. Nevertheless, many of the humus moieties derived from

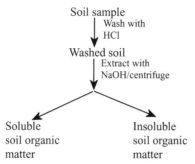

Figure 12.1. Flow diagram for extraction of soil organic matter and isolation of humus.

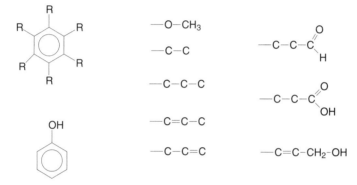

Figure 12.2. Various organic groups found in soil organic matter. Left top: benzene ring substituted with various groups represented by R. Lower left: benzene ring with –OH group. Benzene rings with –OH groups are called phenols. At the middle top is a methoxy group. Below this are various arrangements of two and three carbon groups with and without double bonds. Right: three carbon groups with, top to bottom, aldehyde, acid, and alcohol functional groups. In those cases where carbons have less than the required four bonds, the bonds are to hydrogen atoms.

ligands, carbohydrates, cellulose, proteins, lipids, and so on, have been isolated and their structure determined (see Figure 12.2).

In most cases, including natural and contaminating organic matter, it is desirable to extract all the organic analyte of interest from soil. Plants do not, in general, take up significant amounts of organic compounds from soil and organic compounds are not required as nutrients. However, it is essential to differentiate between unavailable organic compounds and those that are biologically available, or slowly available, to microorganisms, animals, and humans.

Biologically available organic contaminants may find their way into the food chain and be toxic, but they are also more easily decomposed and are thus removed from the environment. Slowly decomposed biologically available organic compounds, such as pesticides, may have long lifetimes in soil and thus pose a hazard to animals and humans. Biologically unavailable compounds, such as tars, produce undesirable characteristics in soil, such as water repellency, and are therefore undesirable. All types of compounds must be removed, their concentration determined, and methods of removal or remediation determined.

The biological availability of contaminating and naturally occurring organic compounds can be estimated by finding their solubility in water. The more soluble the compound, the more available it is for decomposition. While this is true for most organic compounds, there are some that are soluble but also recalcitrant to decomposition. This is the result of complex, sometimes multicyclic, structures that inhibit decomposition, such as those of polysaccharides and lignins [2].

The availability of contaminating organic compounds is also often described by their partition coefficient between water and a solvent, most often octanol. The more soluble a compound is in water, the more available for decomposition. By describing the solubility ratio, the relative rates of organic compound decomposition in soil can be estimated [3].

Another approach is to determine the effect of contaminating organic compounds on soil fauna such as worms. Although this can be informative, it is often impossible to differentiate between the effects of active uptake and passive uptake or simple exposure [4].

It is important to keep in mind that any extraction of organic matter from soil will include both naturally occurring organic matter and organic contaminants. Separating the two at some later stage of analysis is thus an essential analytical step. For example, extraction of soil with hexane or dichloromethane will extract both 1,1,1-trichloro-2,2-di(4-dicholorphenyl)ethane (DDT), a contaminant, and octadecanoic acid, a natural fatty acid. Also, the herbicide 2,4-dichlorophenoxy acetic acid, a contaminant, and indole-3-acetic acid, a natural plant hormone, are both extracted by water (see Figure 12.3). These

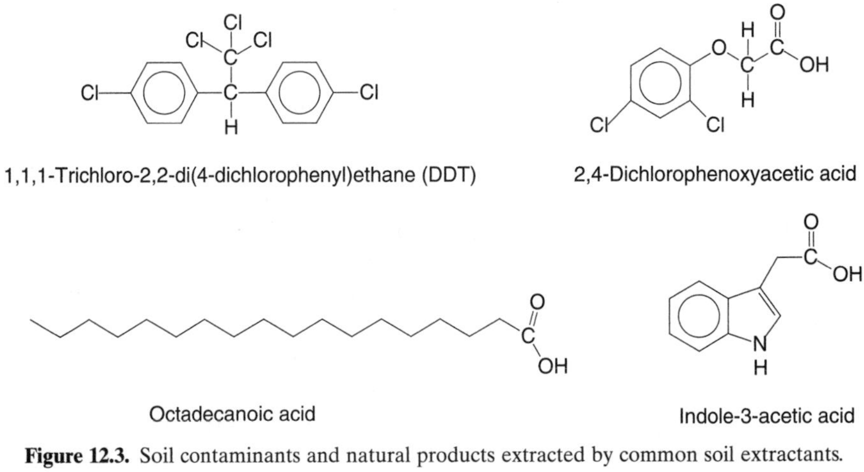

1,1,1-Trichloro-2,2-di(4-dichlorophenyl)ethane (DDT)   2,4-Dichlorophenoxyacetic acid

Octadecanoic acid   Indole-3-acetic acid

Figure 12.3. Soil contaminants and natural products extracted by common soil extractants.

are but two examples of the extraction of both naturally occurring organic compounds and contaminants by the same extractants.

12.1 SAMPLING HANDLING BEFORE EXTRACTION

It is extremely important that all sampling equipment and containers be clean and compatible with the sample to be contained before they are used. This must include caps and cap liners. In many cases, containers will be new and used containers never used. In some cases, containers will be certified as being contaminant free. These types of containers are always to be preferred for sample storage.

In many cases, new samplers will not be used to obtain each sample from a contaminated site, which will often be analyzed for very small amounts of contaminants. Even if this is not the case, it is important that the sampler be cleaned thoroughly between samples. As with sample containers, samplers must be compatible with the sample and with the analysis to be done. Metal samplers may add small but measurable amounts of metals to the sample, causing interference with the analytical procedure to be preformed.

In addition, many organic compounds are volatile or have a relatively high vapor pressure even when they are normally solids at room temperature. Thus, it is extremely important to handle soil samples carefully during sampling and between sampling and actual analysis. This means placing samples in sealed containers immediately after removing them from the soil and keeping the samples sealed and under cool conditions. In some cases, it will be important to slow or prevent microbial activity, often by freezing, between sampling and analysis. It is also important to keep samples under conditions as close to those occurring in the soil during sampling as possible, including moisture content, temperature pressure, pH, and Eh. Samples must also be protected from contamination by organics in the environments encountered during transport and storage.

12.2 EXTRACTION EQUIPMENT

As with sampling, laboratory extraction equipment must be clean. In addition, extraction equipment must be compatible with the analyte of interest. It also must not add any of the analyte or an interfering analyte to the sample during extraction. This is particularly important because of the low levels of analyte being determined and because most laboratories will not have new, disposable equipment.

It is best to wash extraction equipment using the pure extracting solution to be used in the primary extraction. For example, if an aqueous acid solution is to be used, then this same solution should be used in cleaning equipment. This is also true when an organic extractant is to be used. When water is to be used at any stage of the extraction process, it is extremely important to take special care to make sure the water is clean (see Section 12.3, Soil Organic Matter Extraction Solvents).

Equipment used for extracting organic compounds varies from the very simple to that requiring high pressure and moderately high temperatures (e.g., a simple capped Erlenmeyer flask attached to a pressurized system capable of keeping a liquid solvent liquid when it is heated above its boiling point), as it extracts organic matter. In this latter case, the most common solvent is liquid carbon dioxide. Liquid carbon dioxide has the advantage of being a clean solvent to begin with and a gas at STP.[2] Thus, once the pressure is released, the carbon dioxide changes into its gaseous state, leaving any extracted organic material behind. This step can be carried out at low temperature, further decreasing the chances of loss or organic compounds.[3]

The same or similar conditions can be used to keep any common solvent, such as hexane, above its boiling point during the extraction process. Using solvents under these conditions requires that both the temperature and the pressure be carefully decreased before isolation of the extracted organic material is carried out.

12.2.1 Simple Equipment

It is possible to place a soil sample in an Erlenmeyer flask or other suitable container, add the desired extractant, stopper and shake for a specified period of time, filter, and analyze the extract for the analyte of interest. For an example of this type of extraction, see Procedure 12.1.[4] The extract obtained from the simple extraction of 10 g of soil with water using a magnetic stirrer is shown in the right-hand flask in Figure 12.4 (see also Section 12.2.2).

In this simple extraction, the composition of the stopper or liner of the cap must be compatible with the solvent being used. If this type of extraction is

---

Procedure 12.1. Simple Extraction

A 1-g sample of sieved (#10 sieve, <2.00 mm) air-dry soil is placed in a 20-mL glass centrifuge tube and 10 mL of dichloromethane (CH_2Cl_2) added. The mixture is shaken for between 30 minutes and 1 hour. Extracts are centrifuged to separate the extract from soil before analysis (adapted from Reference 5).

---

[2]STP, standard temperature and pressure, 0°C (centigrade) and 760 mmHg (mercury).

[3]Carbon dioxide is a gas at atmospheric pressure above –40°C.

[4]All procedures, that is, Procedures 12.1–12.7, are examples only, have been abbreviated to illustrate the general concept, and are not to be taken as complete procedures. For complete procedures, see the specific reference for the section.

Figure 12.4. Solutions obtained by Soxhlet and magnetic stirring of soil with either water or hexane.

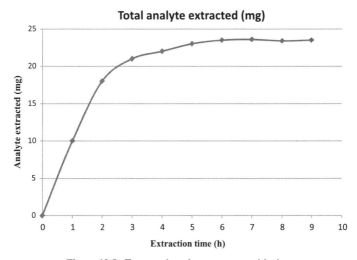

Figure 12.5. Extracted analyte recovery with time.

standard, then the time of shaking or mixing will be specified. If this is a new extraction, then the extraction can be carried out for various lengths of time. The amount of analyte extracted versus time can be plotted (see Figure 12.5). The time necessary to obtain maximum extraction can be determined from this graph. In the example shown, virtually all the analyte has been extracted after 5 hours. This then would be the time to use in all extractions of this analyte and this type of extraction. Note that from 5 to 9 hours, the graph is not flat but varies a little. This is normal as explained in Chapter 1 and illustrated in Figure 1.1 [5]. Simple extractions are used when the analyte is in high concentration or is not associated with any soil components.

The extraction procedure described in Procedure 12.1 can be used to extract hydrocarbons, gasoline, diesel, and various oils from soil when these contaminants are present in relatively high concentration.

12.2.2 Soxhlet Extraction Equipment

Soxhlet extraction usually involves low boiling hydrophobic solvents, although water can be used. A sieved, air-dry soil sample is placed in a thimble that is placed in the Soxhlet extraction body. Solvent vapor from a round bottom flask situated below the extraction body is directed around it and into a condenser where it condenses and drops into the thimble containing the soil sample. When the liquid gets to a predetermined level, an automatic siphon siphons the liquid back into the boiling flask. Figure 12.6 shows a Soxhlet extraction apparatus and Figure 12.7 is a diagram showing the various parts of the Soxhlet extraction apparatus. The solvent is continuously boiled such that vapor continuously condenses into the sample. Higher boiling extracted material remains in the round bottom flask from which it is isolated once the extraction is complete.

The time of extraction for a standard Soxhlet extraction will be specified when following an established procedure. If a new extraction procedure is

Figure 12.6. Soxhlet setup for extraction of solid samples.

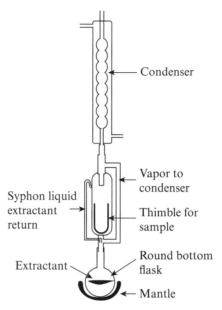

Condenser

Vapor to
condenser

Syphon liquid
extractant
return

Thimble for
sample

Round bottom
flask

Extractant

Mantle

Figure 12.7. A diagram of a Soxhlet extraction setup.

being developed, then samples can be taken from the round bottom flask at various time intervals and analyzed to determine the concentration of analyte present. When these data are graphed, a plot similar to Figure 12.5 is obtained. The optimum time for extraction can be found from such a plot [6–8]. Soxhlet extractions have been used to extract hydrocarbons for determination of the total hydrocarbon content.

Soxhlet extraction, as described in Procedure 12.2 is usually carried out when relatively low levels of contamination are present. It has been used extensively to extract hydrocarbons and similar compounds from soil.

Procedure 12.2. Soxhlet Extraction

Three- to ten-gram soil samples are sieved (#10 sieve, <2.00 mm) and mixed with sodium sulfate to dry the soil and are paced in an extraction thimble. The thimble is placed in the extraction body of a Soxhlet extraction setup and is extracted with either dichloromethane or dichloromethane/ acetone mixtures for 18 hours (adapted from Reference 6).

Soxhlet extraction can be used with organic solvents or water and often results in a solution that can be analyzed directly. Figure 12.4 shows the results of extracting 10 g of soil using Soxhlet extraction. The two flasks on the left were extracted for 8 hours using a Soxhlet extraction procedure. The leftmost flask is a hexane extract that is absolutely colorless, while the center flask is a water extract. The right-hand flask is the solution obtained when 150 mL of distilled water is mixed with soil and stirred with a magnetic stirrer for 8 hours followed by filtration. The aqueous Soxhlet extraction has a deep yellow color, while the magnetic stirred extract is cloudy even after being filtered twice. All three solutions absorb ultraviolet light as a result of components extracted from the soil [8].

Soxhlet is an older method that is subject to inaccuracies; however, it is still a recognized method particularly for investigating contaminated soil. It is a method accepted by the United States Environmental Protection Agency (USEPA).

12.2.3   Supercritical Extraction

Supercritical extraction is an important methodology for extracting organics from soil. As the name implies, the solvent is kept in its liquid state but above its boiling point by keeping it under pressure. Figure 12.8 shows a simplified supercritical carbon dioxide extraction setup. It is possible and common to have pumps, in addition to the one shown, to add chemical modifiers to the solvent, to increase extraction efficiency, or to extract a specific organic component [6,9–12].

An example of a supercritical CO_2 extraction is given in Procedure 12.3. Note that methanol is added as a modifier and that extraction is rapid, taking only 15 minutes. Supercritical extractions are used when the analyte is difficult to extract, as when it is strongly associated with soil solid, and when it is particularly desirable to have an extract free of solvent contamination that might interfere with subsequent analysis.

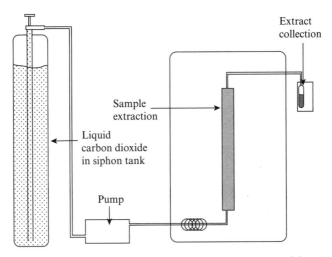

Figure 12.8. Supercritical carbon dioxide extraction setup. If the siphon tank is not used, a pump would be needed to liquefy the CO_2.

Procedure 12.3. Supercritical Extraction

A 1-g sieved, air-dry soil sample is placed in an extraction cell with methanol as a modifier. The sample is extracted at a CO_2 flow rate of 1.5 mL/min with supercritical carbon dioxide for 15 minutes and analytes trapped in an octadecyl siloxane microextraction disk for subsequent analysis (adapted and condensed from Reference 9).

Supercritical extractions such as that illustrated in Procedure 12.3 have been used to extract phenols, organochlorine, organophosphate compounds, and amines from soil [9,10,12].

12.2.4 Microwave Extraction

Microwave and microwave-assisted extraction require both a special extraction vessel and a microwave oven designed for the extraction process. When organic solvents are being used, it is important that the extraction vessel be compatible with the solvent or solvents being used because they will be subject to high pressure and temperature. Both specially designed Teflon and quartz vessels are available and are generally inert to common extracting solvents. However, in all cases, it is essential that compatibility be tested before extraction of samples [7,9,13–15].

Microwave extractions such as that illustrated in Procedure 12.4 have been used to extract chlorinated biphenyls, phenols, sulfonylurea herbicides, and triazines from soil [7,9,13,15].

Procedure 12.4. Microwave Extraction

A 1-g sieved, air-dry soil sample is placed in a polytetrafluoroethylene (PTFE)-lined extraction vessel, with a mixture of hexane and acetone, and microwaved at full power for 10 minutes (adapted and condensed from Reference 14).

In Procedure 12.4, sample and solvent are added to the extraction vessel at room temperature and are heated to 130°C during the extraction process [9]. The extraction time is short when compared to other extractions procedures, being only 10 minutes. Microwave extraction is used when a rapid analysis is desired and the analyte of interest is not affected by high temperature and pressure.

12.2.5 Sonication: Ultrasound

Sonication is commonly used in situations where simple extraction procedures are being used. In these cases, the suspension of soil in solvent is sonicated

Figure 12.9. Ultrasonic "horn" for disruption and extraction of soil samples. The horn tip is small enough in diameter and long enough to reach the bottom of a test tube.

either in an ultrasonic bath or by using a sonic probe (see Figure 12.9). The sonication procedure separates the soil and its components into small fractions that allow for a more efficient and complete extraction [16,17]. However, in some cases, sonication does not lead to increased extraction of contaminants, and thus, in developing new extraction procedures, the efficiency of sonication for the soil, and organics extracted, must be checked.

Sonication is most commonly used with aqueous solutions. This can include both strong and mild acids. However, it can also be used with organic solvents, although this is less common [8,9,16,17]. An example of the use of sonication with dichloromethane, an organic solvent, is given in Procedure 12.5. Sonication is particularly called for when the analyte of interest is strongly associated with soil solids and may be trapped in components associated with each other.

Procedure 12.5. Sonication-Assisted Extraction

A 2-g air-dry soil sample is placed in a suitable container and sonicated three times for a specified length of time with 10 mL of dichloromethane. Extracts are then analyzed (adapted and condensed from Reference 9).

Procedure 12.5 illustrates the use of sonication in extraction procedures. It has been used in the extraction of phenols, amines, and polycyclic aromatic hydrocarbons. It has also been used to extract hydrocarbons from clay soils [8,9,17].

12.3 SOIL ORGANIC MATTER EXTRACTION SOLVENTS

Caution: Common organic extracting solvents are both volatile and flammable. Ignition can be caused by hot surfaces without the need of sparks or flames. Some solvents (particularly diethyl ether) can also form peroxides that are explosive when concentrated, especially when heating is involved. Some solvents may also be toxic, carcinogenic, and/or teratogenic. Always consult the Material Safety Data Sheets (MSDS) before using any solvent.

Organic solvents at STP and under supercritical conditions are the most common extractants for soil organic matter. Supercritical CO_2 and, to a lesser extent, N_2O have also been used to extract both native and organic contamination from soil. Humus is extracted using aqueous solutions, but otherwise, water is rarely used to extract organic compounds from soil. A list of common soil organic matter extractants is given in Table 12.2.

12.3.1 Organic Solvents

Specially purified solvents are available from chemical supply houses for many extraction procedures. In situations where this is not the case, then the highest-purity solvent and other reagents should be used and should be checked to make sure they are compatible in all respects with the analysis to be carried out. This means that solvents do not add analytes of interest nor do they add impurities that could interfere with the analysis.

For organic compounds, particularly organic pollutants, extraction is most often carried out using common organic solvents. Hexane and trichloromethane (chloroform) are two of the more common solvents, as are dichloromethane (methylene chloride) and propanone (acetone). Halogenated solvents, which can cause interference in subsequent analyses, have been used extensively to extract hydrocarbons from soil. An example of an organic solvent extractant procedure is given in Procedure 12.2 [18]. Table 12.2 gives the characteristics of common organic extraction solvents. All of these solvents have

TABLE 12.2. Soil Extractant Characteristics

Solvent	Imperial Formula	Water Soluble	Boiling Point (°C)
Hexane	C_6H_{14}	No	68–69
Dichloromethane	CH_2Cl_2	No	39.6
Tricholormethane	$CHCl_3$	No	61.2
Water	H_2O	Yes	100
Acetone	C_3H_6O	Yes	56–57
Diethyether	$(C_2H_5)_2O$	No	34.6
Carbon dioxide	CO_2	No	−57
Nitrous oxide	N_2O	No	−88.45

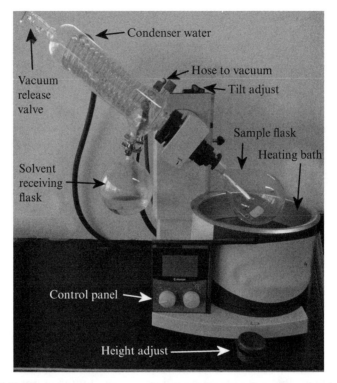

Figure 12.10. Flash evaporator for removing solvent from samples under reduced pressure.

low boiling points, making them easy to remove from the analyte of interest once extraction is complete. These solvents can also be used as supercritical extractants by placing them under pressure and heating [19].

Solvents can be evaporated using a number of different methods. They may be removed by gentle heating, passing a compressed gas over the surface, or flash evaporation. A flash evaporator is used by placing the extract in a round bottom flask and heating while applying a vacuum. Because the system is under vacuum, the solvent boils at a lower temperature and is removed quickly (see Figure 12.10). Both of these situations decrease the likelihood of extracted components decomposing during the evaporation process.

12.3.2   Inorganic Extractants

Carbon dioxide is the most common inorganic extractant used for the extraction of organic compounds in soil. Under pressure, it remains in the liquid state and can be used to extract organic compounds from soil. When the pressure is released, the carbon dioxide becomes a gas and is thus removed from the extracted components. An additional benefit is that liquid carbon dioxide is converted to gas at relatively low temperatures, thus limiting the loss of volatile organic compounds. A diagram of a simple supercritical carbon dioxide extraction setup is shown in Figure 12.8.

Various modifying agents such as methanol, as shown in Procedure 12.3, and ethylenediaminetetraacetic acid (EDTA, a chelate) can be added to liquid CO_2 to either make the extraction more effective or more specific. In either case, these additives must be removed before further analysis of extracted organics is carried out.

Other supercritical extractions using inorganic gases have also been carried out, most notably, those using nitrous oxide, N_2O. These have not, however, been used as extensively as supercritical CO_2 extraction [9–12].

12.3.3   Water

Organic compounds are sometimes extracted with both room temperature and hot water [1,20]. For example, low-molecular-weight acids can be extracted by room temperature water. Low water solubility organic compounds can be extracted by hot water. In other cases, water is used in the cleanup of extracted material. Many laboratories do not purchase specially purified water but rather purify it before use. Caution must be used when water is used in any step during the isolation of organic components from soil. Water will absorb materials from any container with which it comes in contact. If placed in a glass container, some silicon will be dissolved. If placed in a plastic container, some plastic components will be dissolved. Many laboratories use ion exchangers and reverse osmosis in purifying water. In this case, the water is exposed to various kinds of plastics during the purification process. If an analysis of soil contaminated with chemicals used in manufacturing plastics is to be carried out, then water purified using plastic components may interfere with and confound the analytic results.

In addition to absorbing components of containers, water will absorb materials from the atmosphere to which it is exposed. The simplest and most common example is carbon dioxide and this results in water becoming acidic through the formation of carbonic acid. However, any component that is present in the atmosphere that is in contact with water will be absorbed by it. This type of interference can be overcome by having a suitable sorbant, in a drying tube (see Figure 12.11), between the water and the atmosphere to which the water is exposed. Another approach is to store water in a container where the air has been replaced by a pure gas such as nitrogen, helium, or argon.

Very pure water can be produced by first passing it through an ion exchanger to remove cations and anions. This is followed by reverse osmosis, where everything except H_2O is removed. The final step is to follow this with simple distillation in an all-glass distillation setup to remove components picked up from plastics, for example, bisphenol A. Although room temperature and hot water can be used to extract low-molecular-weight organic polar compounds and low-solubility organic matter, it is not generally the solvent of first choice in extracting organic compound from soil.

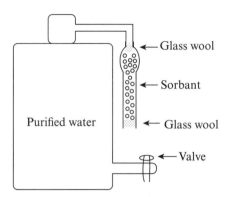

Figure 12.11. Purified water container protected from atmosphere by sorbant in the drying tube. Agents, such as activated carbon, can also be used to protect water from compounds in air.

Although water and aqueous solutions are not generally used to extract organic compounds from soil, it is the most important extractant for extracting humus and its constituents. Typically, a sodium hydroxide solution (0.5 M) is used in the initial extraction of soil after an initial washing with dilute hydrochloric acid as shown in Figure 12.1. This solution (NaOH) solubilizes the base soluble organic components and leaves the insoluble material that is called humin. The extracted or solubilized organic matter can further be fractionated by acidifying the extract to a pH of 1.5 using hydrochloric acid as shown in Procedure 12.6. Fractionation yields humic acid, which is soluble in the acidified extractant, and fulvic acid, which is insoluble. In a like manner, the organic matter that is insoluble in the initial extraction can be further fractionated [1].

Procedure 12.6. Extraction of Soil Organic Matter (Humus) Using Aqueous Solutions

A soil sample is placed in a centrifuge tube, mixed with 0.5 M HCl for 1 hour, and centrifuged. The HCl solution is removed and the soil washed with water to remove the acid. The soil is then extracted with 0.5 M NaOH for 18 hours with shaking, followed by centrifuging to remove dissolved organic matter from soil (adapted and condensed from Reference 21).

Newer and more complex humus extractions have been developed. These typically involve more steps such as both physical separation on the basis of density and particle size (related to the size of soil inorganic components), and chemical separation based on extractions and washings with hydrofluoric acid (HF), hydrochloric acid (HCl), and sodium hydroxide (NaOH). The products of such separations are then subjected to spectroscopic analysis and interpretation [22,23].

12.3.4 Surfactants

Another approach has been to use surfactants in water for extracting organic matter. Both synthetic and natural surfactants, called saponins, have been used; however, they have been more commonly used in the removal of organic matter in field contamination situations. In using surfactants, it is important that they do not produce undesirable effects in the soil such as toxicity to plants and animals. Natural surfactants are produced by both plants and microorganisms and have the advantage of being naturally broken down by soil organisms.

Surfactants can be either found or synthesized to contain both hydrophobic and hydrophilic regions and can be cationic, anionic, or neutral. Thus, surfactants can be chosen to extract a specific type of soil organic matter, whether natural or a contaminant.

Common surfactants such as Tween, Triton,[5] and sodium dihexyl sulfosuccinate, among many others, have been used to extract organic compounds from soil. In the field, they have been particularly useful in the remediation of soils contaminated with halogenated organic compounds, oils, and other hydrocarbon compounds [24].

12.4 CLEANUP

In many cases, extraction of organic compounds requires that the extract be cleaned up before analysis. This may be as simple as centrifugation, but it may also involve concentration or evaporation of the solvent. Another approach is to absorb the analyte of interest on an absorbant and subsequently to extract it into another solvent. This may be necessary to ensure that the extract is compatible with the analytical procedure to be used.

A cleanup column from a commercial source can be used or can be constructed easily from a Pasteur pipette (see Figure 12.12). A small piece of cotton is placed in the bottom of the pipette. A suitable absorbant, such as a solid coated with or bonded to a high-molecular-weight hydrocarbon such as octadecanol (this is often simply called a C-18 extraction column), can be loaded into the pipette and covered with another small amount of a cotton ball. A list of common extract cleanup absorbants and their characteristics is given in Table 12.3.

Samples can also be dried by constructing a column similar to the cleanup columns shown in Figure 12.12 containing one of the drying agents given in Table 12.4. Passing the sample through the column will dry it and there will be no need to filter the drying agent from the sample.

An example of a solid-phase extraction (SPE) used as a cleanup procedure is given in Procedure 12.7. The column is conditioned by passing water and the

[5]Both Tween and Triton are brand names.

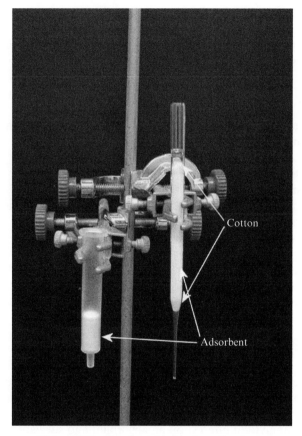

Figure 12.12. Absorbant for cleaning soil extracts.

TABLE 12.3. Some Common Extract Cleanup Adsorbents

Method Based on Sorbent Used	Advantages	Disadvantages
Alumina	Different pH and activity ranges available for different cleanup needs	May decompose or irreversibly adsorb some compounds
Florisil	Cleanup of chlorinated hydrocarbons, nitrogen, and aromatic compounds	Has basic properties and may not be compatible with acids
Silica gel	Separation on the basis of differing polarity	Components with the same polarity will not be separated
Gel permeation	Components separated on the basis of size	Different components of the same size will not be separated

TABLE 12.4. Common Drying Agents and their Characteristics

Drying Agent	Characteristics Capacity/Speed/Intensity[a]	Restrictions
Calcium chloride	High/medium/high	Reacts with many functional groups
Calcium sulfate	Low/fast/high	Applicable in most situations
Magnesium sulfate	High/fast/moderate	Applicable in most situations
Potassium carbonate	Medium/medium/ moderate	Reacts with acids and phenols
Sodium sulfate	High/low/low	Hydrate decomposed on heating
Molecular sieve 5 Å	High/high/high	Suitable for most solvents

[a]Capacity = amount of water absorbed; speed = rate of water removal; intensity = level of water remaining after drying, that is, high means that the smallest amount of free water remains after drying.

pure extractant through it. The extractant containing the analyte is then passed through the column that retains the analyte. The column is washed to remove impurities contained in the extract. The column is then washed with fresh clean solvent to remove the analyte. Depending on the conditions, the analyte may or may not be concentrated by this procedure [25].

Procedure 12.7 is an example of a basic approach that has also been used to clean up polycyclic aromatic hydrocarbons, polychlorinated biphenyls, and N-, P-, and Cl-containing pesticides before further analysis [25].

Procedure 12.7. Extract Cleanup

A diol SPE column is conditioned by passing 10 mL of methanol through the column followed by 10 mL of purified water. Soil extract (10 mL) is passed through the column and the column washed with an additional 10 mL of water. The column is then eluted twice with 4 mL of a 3:2 v/v solution of acetonitrile:0.1 M aqueous ammonium acetate solution. The eluate is then analyzed (adapted from Reference 25).

As a last step, extractants, extracts, and/or solvents may need to be dried before analysis. Drying may be accomplished as described previously, or a drying agent, such as calcium chloride, may be added directly to the solution. Although calcium chloride is a common drying agent, there are others with varying drying characteristics and reactivities in terms of organic functionalities. A list of common drying agents and their characteristics is given in Table 12.4.

Although it is not common, sometimes a drying agent is added to soil before extraction. In these cases, the same drying agents are used and are added and mixed with soil before starting the extraction process.

12.5 CONCLUSION

Extraction of organics from soil typically requires or normally involves equipment and solvents significantly different from those involved in the extraction of inorganic compounds. The equipment involves Soxhlet setups, supercritical solvent extractions, sonication, and microwave digestion. While some of these may also be used in the extraction of inorganics, they are much more commonly used with organics. The solvents used are also different. The most commonly used solvents are hexane, dichloromethane, acetone, and methanol. Although acids or bases are not commonly used, they may be used in conjunction with these solvents. Both room temperature and hot water can also be used to extract organic materials from soil.

All of the solvents that are liquid at STP can be used in the extraction of organics in conjunction with any extraction equipment. In addition to using these same solvents, supercritical extraction using CO_2 or N_2O may also be carried out. Both CO_2 and N_2O will change to gas when the pressure is removed and thus will leave the sample free of solvent.

PROBLEMS

12.1. Considering that the choice of extractant depends on the size and polarity of the analyte of interest, what extractant would be used to extract all the hydrocarbons from a fuel-contaminated soil? Would the type of fuel spilled make a difference in the type of extractant used? If so, how?

12.2. What would be the best eluent for a soil sample contaminated with a mixture of low-molecular-weight alcohols?

12.3. What types of precautions must be taken when extracting analytes that have a high volatility?

12.4. Describe the basic efficiencies of simple extraction, supercritical extraction, and Soxhlet extraction.

12.5. In addition to agitation, what does sonication do to the sample that might increase extraction efficiency?

12.6. Describe the most important difference between hexane and dichloromethane as extractants, and acetone and methanol as extractants.

12.7. EDTA is called a chelate. Describe specifically what it does. You will need to explore other chapters to answer this question.

12.8. Sodium sulfate is often used to dry either soil or extractant or both. Why would drying be important?

12.9. Explain why cleanup of supercritical extraction products is usually not necessary.

12.10. In some cases, components extracted with supercritical carbon dioxide are immediately dissolved in a solvent after extraction. Under what conditions might this be advisable?

REFERENCES

1. van Bavel B, Hartonen K, Rappe C, Riekkola ML. Pressurised hot water/steam extraction of polychlorinated dibenzofurans and naphthalenes from industrial soil. *Analyst* 1999; **124**: 1351–1354.

2. Southgate DTA. Determination of carbohydrates in foods. 2.—Unavailable carbohydrates. *J. Sci. Food Agric.* 1969; **20**: 331–335.

3. Kenaga EE, Goring CAI. Relationship between water solubility, soil sorption, octanol-water partitioning, and concentration of chemicals in biota. In Eaton J, Parrish P, Hendricks A (eds.), *Aquatic Toxicology proceedings of the Third Annual Symposium on Aquatic Toxicology.* Philadelphia, PA: American Society for Testing and Materials STP 707; 1980, pp. 78–115.

4. Neuhauser EF, Loehr RC, Malecki MR, Milligan DL, Durkin PR. The toxicity of selected organic chemicals to the earthworm *Eisenia fetida. J. Environ. Qual.* 1985; **14**: 383–388.

5. Schwab AP, Su J, Wetzel JSS, Pekarek S, Banks MK. Extraction of petroleum hydrocarbons from soil by mechanical shaking. *Environ. Sci. Technol.* 1999; **33**: 1940–1945.

6. Hawthorne SB, Miller DJ. Direct comparison of Soxhlet and low-temperature and high-temperature supercritical CO_2 extraction efficiencies of organics from environmental solids. *Anal. Chem.* 1994; **66**: 4005–4012.

7. Zuloaga O, Etxebarria N, Fernández LA, Madariaga J. Comparison of accelerated solvent extraction with microwave-assisted extraction and Soxhlet for the extraction of chlorinated biphenyls in soil samples. *Trends Anal. Chem.* 1998; **17**: 642–647.

8. Buddhadasa S, Barone S, Bigger S, Orbell JD. Extraction of hydrocarbons from clay soils by sonication and Soxhlet techniques. *J. Jpn. Pet. Inst.* 2001; **44**: 378–383.

9. Llompart MP, Lorenzo RA, Cela R, Li K, Bélanger JMR, Paré JRJ. Evaluation of supercritical fluid extraction, microwave-assisted extraction and sonication in the determination of some phenolic compounds from various soil matrices. *J. Chromatogr. A* 1997; **774**: 243–251.

10. Snyder JL, Grob RL, McNally ME, Oostdyk TS. The effect of instrumental parameters and soil matrix on the recovery of organochlorine and organophosphate pesticides from soils using supercritical fluid extraction. *J. Chromatogr. Sci.* 1993; **31**: 183–191.

11. Brady BO, Kao CPC, Dooley KM, Knopf FC, Gambrell RP. Supercritical extraction of toxic organics from soils. *Ind. Eng. Chem. Res.* 1987; **26**: 261–263.

12. Ashraf-Khorassani M, Taylor LT, Zimmerman P. Nitrous oxide versus carbon dioxide for supercritical fluid extraction and chromatography of amines. *Anal. Chem.* 1990; **62**: 1177–1180.

13. Font N, Hernandez F, Hogendoorn EA, Baumann RA, van Zoonen P. Microwave-assisted solvent extraction and reversed-phase liquid chromatography-UV detection for screening soils for sulfonylurea herbicides. *J. Chromatogr. A* 1998; **798**: 179–186.

14. Lopez-Avila V, Young R, Beckert WF. Microwave-assisted extraction of organic-compounds from standard reference soils and sediments. *Anal. Chem.* 1994; **66**: 1097–1106.

15. Shen G, Lee HK. Determination of triazines in soil by microwave-assisted extraction followed by solid-phase microextraction and gas chromatography-mass spectrometry. *J. Chromatogr. A* 2003; **985**: 167–174.

16. Cambardella CA, Elliott ET. Methods for physical separation and characterization of soil organic matter fractions. *Geoderma* 1993; **56**: 449–457.

17. Sánchez-Brunete C, Miguel E, Tadeo J. Rapid method for the determination of polycyclic aromatic hydrocarbons in agricultural soils by sonication-assisted extraction in small columns. *J. Sep. Sci.* 2006; **29**: 2166–2172.

18. Schlüsener MP, Spiteller M. Bester K. Determination of antibiotics from soil by pressurized liquid extraction and liquid chromatography–tandem mass spectrometry. *J. Chromatogr. A* 2003; **1003**: 21–28.

19. Alzaga R, Bayona JM, Barceló D. Use of supercritical fluid extraction for pirimicarb determination. *J. Agric. Food Chem.* 1995; **43**: 395–400.

20. Hawthorne SB, Grabanski CB, Hageman KJ, Miller DJ. Simple method for estimating polychlorinated biphenyl concentrations on soils and sediments using subcritical water extraction coupled with solid-phase microextraction. *J. Chromatogr. A* 1998; **814**: 151–160.

21. Anderson DW, Schoenau JJ. Soil humus fractions. In Carter M (ed.), *Soil Sampling and Methods of Analysis*. Boca Raton, FL: CRC press; 1993, pp. 391–396.

22. Cao X, Olk DC, Chappell M, Cambardella CA, Miller LF, Mao J. Solid-state NMR analysis of soil organic matter fractions from integrated physical-chemical extractions. *Soil Sci. Soc. Am. J.* 2011; **75**: 1374–1384.

23. Anderson DW, Schoenau JJ. Soil humus fractions. In Carter MR, Gregorich EG (eds.), *Soil Sampling and Methods of Analysis*, 2nd ed. Madison, WI: CRC Press; 2008, pp. 675–680.

24. Mulligan CN, Youg RN, Gibbs BF. Surfactant-enhanced remediation of contaminated soil: a review. *Eng. Geol.* 2001; **60**: 371–380.

25. Dąbrowska H, Dąbrowski Ł, Biziuk M, Gaca J, Namieśnik J. Solid-phase extraction cleanup of soil and sediment extracts for the determination of various types of pollutants in a single run. *J. Chromatogr. A* 2003; **1003**: 29–42.

BIBLIOGRAPHY

Cohen DR, Shen XC, Dunlop AC, Futherford NF. A comparison of selective extraction soil geochemistry biochemistry in the Cobar area, New South Wales. *J. Geochem. Explor.* 1998; **61**: 173–189.

Dean JR. *Extraction Methods for Environmental Analysis*. New York: Wiley; 1998.

Environmental Protection Agency Test Methods SW-846 On-Line. Test Methods for Evaluating Solid Wastes Physical/Chemical Methods. 2012. http://www.epa.gov/epawaste/hazard/testmethods/sw846. Accessed June 3, 2013.

Gopalan AS, Wai CM, Jacobs HK, eds. *Supercritical Carbon Dioxide: Separations and Processes*. Washington, DC: American Chemical Society; 2003.

Holmberg K, Jönsson B, Kronberg B, Lindman B. *Surfactants and Polymers in Aqueous Solution*. New York: John Wiley and Sons, Ltd; 2002.

Martens D, Gfrerer M, Wenzl T, Zhang A, Gawlik BM, Schramm K-W, Lankmayr E, Kettrup A. Comparison of different extraction techniques for the determination of polychlorinated organic compounds in sediment. *Anal. Bioanal. Chem.* 2002; **372**: 562–568.

Miege C, Dugay J, Hennion MC. Optimization, validation and comparison of various extraction techniques for the trace determination of polycyclic aromatic hydrocarbons in sewage sludges by liquid chromatography coupled to diode-array and fluorescence detection. *J. Chromatogr. A* 2003; **995**: 87–97.

CHAPTER

13

CHROMATOGRAPHY

Chromatography is a powerful, essential tool for the analysis of soils. Of the many forms of chromatography, gas and high-performance liquid chromatography are most commonly used in the analysis of soil extracts. Chromatographic and spectroscopic methods are almost exclusively referred to using

Introduction to Soil Chemistry: Analysis and Instrumentation, Second Edition.
Alfred R. Conklin, Jr.
© 2014 John Wiley & Sons, Inc. Published 2014 by John Wiley & Sons, Inc.

TABLE 13.1. Fundamental Types of Chromatography

Name of Chromatography	Stationary Phase	Mobile Phase	Types of Compounds Amenable to Separation
Gas (also GLC and GSC)	High-boiling liquid	Gas	Compounds to be separated must be in gaseous phase.
HPLC	Solid	Solvent, aqueous, aqueous solution, or mixture of solvents	Compounds or ions must be soluble in mobile phase and have appreciable attraction to the solid phase.
TLC	Solid	Solvent, aqueous, aqueous solution, or mixture of solvents	Compounds or ions must be soluble in mobile phase and have appreciable attraction to the solid phase.
Electrophoresis	Solid or semisolid in a buffer solution	Electricity	Compounds or ions must be ionized to be separated.

GLC, gas-liquid chromatography; GSC, gas-solid chromatography—no high-boiling liquid, only solid; HPLC, high-performance liquid chromatography; TLC, thin-layer chromatography.

their acronyms: GC—gas chromatography, LC—liquid chromatography, HPLC—high-performance liquid chromatography,[1] TLC—thin-layer chromatography, CE—capillary electrophoresis (a more complete list is given on p. xiv). Also, because there are subdivisions of each type of chromatography, the abbreviations will be lengthened, such as GLC—gas-liquid chromatography (liquid because there is a liquid coating the stationary phase).

Chromatographic methods are also often used as part of systems that are called "hyphenated methods," (see Chapter 15) where the output of the chromatographic section is used as the input for an identification method such as mass spectrometry. These hyphenated methods are also most often referred to by their acronyms, for example, GC-MS—gas chromatography-mass spectrometry and HPLC-MS—high-performance liquid chromatography-mass spectrometry. Note that although ultraviolet-visible (UV-Vis) is hyphenated, it is not a hyphenated method in that it does not consist of two different methods of analysis. Hyphenated methods will be discussed fully in Chapter 15.

Table 13.1 summarizes the common chromatographic methods and the characteristics that differentiate them from other methods. In this list, HPLC

[1] High-performance liquid chromatography is also called *high-precision* and *high-pressure liquid chromatography*.

and GC are most commonly used in analyzing soil extracts, particularly when they are used in combination with mass spectrometric methods.

13.1 FUNDAMENTALS OF CHROMATOGRAPHY

All chromatographic methods function on the same principle, which is the partitioning of components in a mixture between two phases: (1) a stationary phase, which may be a solid, liquid, or gel, and (2) a mobile phase, which may be gas, liquid, solution, or a varying mixture of solvents. When a mixture is introduced into a chromatographic system, its components are alternately absorbed and desorbed, that is, partitioned between, the stationary and mobile phases. Partitioning is caused by different polarities of the stationary and mobile phases and the compounds being separated. Compounds in the mixture have different affinities for the phases and they will move at different rates in the chromatographic system and thus be separated.

This partitioning–separation process is diagrammed in Figure 13.1 and Figure 13.2. In Figure 13.1, a mixture is introduced into a chromatographic column. The components are absorbed into the stationary phase, and the component that is more soluble in the mobile phase moves into this phase and is swept along the column, where it is absorbed again. The less soluble component then moves into the mobile phase and is again reabsorbed by the stationary phase. In Figure 13.2, a mixture is introduced into a chromatographic column; and as its components move down the column, they are separated into discrete "packets" of the same compound. Upon exiting the column, they are detected.

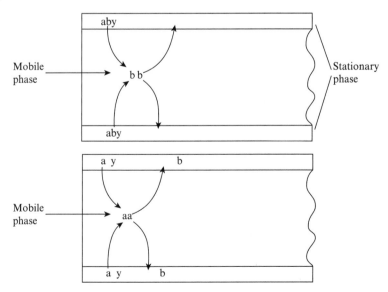

Figure 13.1. Partitioning of the components of a mixture in a chromatographic column.

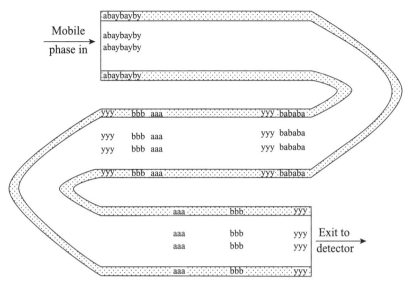

Figure 13.2. Separation of a complex mixture by chromatography. The top shows the column just after application of mixture. As the sample progresses through the column, separation of mixture occurs as a result of the chromatography. The first component exits the column and is detected, followed by the other two components.

13.2 GAS CHROMATOGRAPHY

GC is a powerful, rapid method for separating mixtures of gases and compounds with boiling points below $\sim 400°C$.[2] The sample, when introduced into a gas chromatograph, must either already be a gas or be immediately turned into a gas on injection. A complete chromatogram is usually obtained in less than 20 minutes, although some analyses can take much longer.

13.2.1 Sample Introduction

For GC, a 1–0.1-μL syringe, shown in Figure 13.3, is frequently used for the introduction of the sample into the injection port. The needle must be of the correct type and length for the gas chromatograph being used, or poor results will be obtained. Injection is effected through a septum made of special rubber. Teflon-backed septa are also available and must be in good condition for accurate chromatograms to be obtained. Injection ports without septa are also available. When a new septum has been placed in the injection port, piercing it with a syringe needle will be harder, and care must be taken not to bend the needle or syringe plunger. When the needle is fully inserted, the plunger is depressed and a recording of the detector output is started. This is the "0 time"

---

[2] The temperature value 400°C is not an absolute upper limit of GC, but it is near the upper limit of most GC columns.

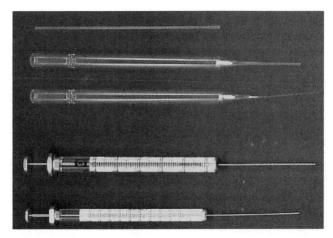

Figure 13.3. Sample application tools from top to bottom, glass capillary, Pasteur pipette with tip drawn out, syringe for HPLC, and syringe for GC.

used in calculating the retention times of emerging peaks (R_t is described later). Needle insertion and depression of the plunger must be done in one continuous motion to obtain a good chromatogram. GCs can also be fitted with automatic samplers that inject the sample automatically.

13.2.2 Mobile Phases

Common GC mobile phases (see Table 13.2) are hydrogen, argon, helium, nitrogen, and air. Helium and nitrogen are the most commonly used. Because gas chromatographic detectors are extremely sensitive and it is desirable to keep the noise level as low as possible, it is always advisable to use very high-purity gas as the mobile phase.

13.2.3 Stationary Phases

Gas chromatographic stationary phases are either solids or liquids coated or bonded to solids. Common stationary phases are listed in Table 13.2. For GC, the solids are activated carbon of various sorts, polymers, molecular sieves, or similar materials, and the chromatography may be referred to as *gas-solid chromatography* (GSC). This type of stationary phase and chromatography is commonly used for compounds that are gases at room temperature or are low-boiling compounds, and thus this will be used to investigate gases in the soil atmosphere.

There are two types of GC columns that employ a liquid phase. Packed columns, usually 3 or 6mm (~1/8 or ¼ in.) in diameter and 2m (~6ft) long, are filled with an inactive solid coated with a high-boiling grease or oil or a similar compound bonded to the solid. Also, compounds having various

TABLE 13.2. Common Stationary and Mobile Phases Used in Chromatography

Chromatographic Method	Stationary Phases	Mobile Phases
Gas chromatography	Solid or inert solid covered by high-boiling liquid[a]	Gas usually helium, nitrogen, argon, or hydrogen
Liquid chromatography	200–400 mesh silica and alumina	Organic solvents
High-performance liquid chromatography	Finer porous silica and alumina particles are used, increasing efficiency but requiring high pressure to push eluent through column[b]	Aqueous solutions and solvents used in liquid chromatography but at high pressure
Thin-layer chromatography (applies also to paper chromatography)	Thin layer, commonly 250 μm thick, of silica gel, alumina, cellulose,[b] or a sheet of chromatographic paper	Aqueous and solvent solutions
Electrophoresis	Solid or semisolid absorbant saturated with buffer	Current

[a]Liquid can be varied such that the stationary phase has varying degrees of polarity.

[b]Surface modifications commonly involve phases that change the surface polarity such as attaching a long-chain hydrocarbon, usually 18 carbons in length. This would be called a reverse-phase C18 impregnated stationary phase.

functionalities and polarities can be bonded to the stationary phase to produce packings with specific separation capabilities.

The second type is a capillary column with an inside diameter of 1 mm or less and a length ranging from 10 to 60 m, although some are much longer. The inside of the capillary is commonly coated with the same stationary-phase materials as used in packed columns. In either case, the layer must have a sufficiently high boiling point or must be attached such that it will not be lost when the column is heated during use. Most columns are usable to 250–300°C and in some cases even higher. A capillary gas chromatographic column (A) and a packed column (B) are shown in Figure 13.4.

Some GC packings or stationary phases are sensitive to either or both oxygen and water. Because soil always contains both, it is important that care be taken with the analysis of soil air or soil extracts to make certain that they do not contain any components that will degrade the chromatographic column, which may cost between $500 and $1000.

13.2.4 Detection

There are four main types of detectors used in GC: thermal conductivity detector (TCD), also called a hot wire detector, flame ionization detector (FID), electron capture detector (ECD), and quadruple mass spectrometer (MS)

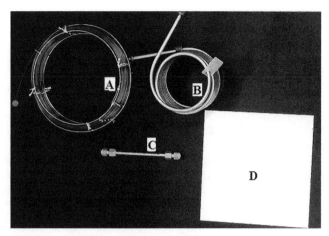

Figure 13.4. Columns used in chromatography and a thin-layer sheet. Columns A and B are a capillary and packed GC columns, respectively. Column C is for HPLC. D is a thin-layer sheet with plastic backing such that it can be cut into smaller pieces as needed.

TABLE 13.3. Common Detection Methods for Chromatography

Gas	Thermal conductivity (TC or TCD)
	Flame ionization (FID)
	Electron capture (ECD)
	Mass spectrometry (MS or GC-MS)
Liquid chromatography	Visual inspection of eluting fractions
	Ultraviolet analysis of eluted fractions
High-performance liquid	Ultraviolet or visible (UV-Vis)
chromatography	Refractive index
	Conductivity
Thin-layer	Fluorescence-indicator-impregnated
	Visualization reagents
	Charring with acid and heat
Electrophoresis	Staining
	UV-Vis
	Conductivity

(Table 13.3). A TCD consist of coils of high-resistance wire in a detector block where the carrier gas from the gas chromatograph exits the column and flows over the wire. The coils of wire are arranged in a Wheatstone bridge (see Chapter 1) with two arms receiving gas exiting the column and another two receiving the pure carrier gas. The heat capacity of the carrier gas changes when a compound in the carrier gas exits the column and flows over the coiled wire; this changes the resistance characteristics of the coil, which is recorded. The TCD is a universal detector, but it is also the least sensitive of the common detectors.

The FID has a small hydrogen flame into which the carrier gas exits. There is a voltage across the flame, which is nonconducting. When an organic compound exits the column, it is burned and produces ions, which conduct electricity. The electrical signal thus produced is recorded. FID is a highly sensitive detector although not the most sensitive. It will respond only to organic compounds or compounds that will burn, and exiting compounds are thus destroyed in the process.

The ECD is a very sensitive detector commonly used in analyzing soil extracts containing electronegative elements, particularly halogenated pesticides. It is usually constructed with a radioactive, β-radiation, ionizing source that ionizes compounds exiting the gas chromatographic column. The ions are measured electrically in a manner similar to an FID. However, it is typically 10 or more times more sensitive than an FID. Because of its high sensitivity, it can be overloaded if halogenated compounds other than the analyte of interest, such as halogenated solvents, are present in the sample [1].

The quadruple MS detector, or mass filter as it is sometimes called, is extremely sensitive and allows the identification of compounds exiting the gas chromatograph. The compounds are ionized as they enter the MS (shown in Figure 13.5), where they are fragmented and analyzed. Other sensitive but also more selective detectors are available and may be called for in certain analyses.

13.2.5 Gas Chromatography Applied to Soil Analysis

Soil extracts are most commonly characterized by GC and GC-MS. The composition of soil air can also be easily analyzed using GC. A gas chromatogram

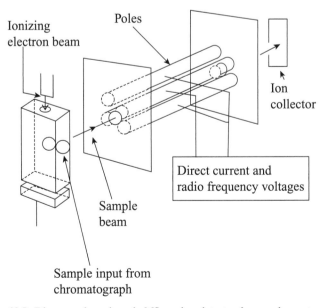

Figure 13.5. Diagram of quadrupole MS used as detector for gas chromatography.

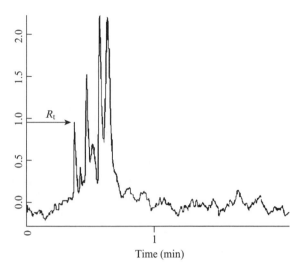

Figure 13.6. Gas chromatograph of soil air taken in Figure 13.7. R_t is the time it takes the first peak to come out. R_t values can be found for all peaks if needed.

of the soil air in a container with wheat growing in it is shown in Figure 13.6. A sample of the soil air for this analysis was obtained by simply inserting a small Phillips head screwdriver into the soil to make an access hole for a syringe needle. The syringe needle was then inserted into the hole and the plunger withdrawn to move soil air into the syringe (see Figure 13.7). This syringe had a valve in the needle hub such that the sample was sealed in the syringe between taking the sample and injecting it into the chromatograph. The GC used for this analysis was equipped with an FID, so each peak represents an organic compound. If standards with the same retention times are injected into the gas chromatograph under the same exact conditions, then each compound could be tentatively identified (other compounds could exist that also have the same retention time under the same conditions). A more definitive approach is to use a GC-MS, which would give a true identification of these components (see Figure 13.5).

It is common to extract a soil with an organic extractant using any of the many extraction procedures described in Chapter 12, and then to inject the extract into a GC or GC-MS. Although such extraction procedures can and are used to study natural organic components in soil, the greatest use of GC and GC-MS in soil analysis falls into two cases: (1) to assess the fate of a herbicide or insecticide once they have been applied to a field or in the case of a spill of these materials and (2) to determine the condition of soil contaminated by an industrial organic material such as from a spill or a by-product of a manufacturing process. There are no plant nutrients, other than carbon dioxide and water, that are volatile, so GC is not used to investigate soil fertility.

Figure 13.7. Gastight syringe used to sample soil air; the chromatograph for the sample taken is shown in Figure 13.6.

A number of techniques have been developed to directly volatilize soil components and contaminants and introduce them into the carrier gas of a gas chromatograph. Such procedures avoid the time and cost of extraction, cleanup, and concentration and also avoid the introduction of contaminants during the extraction process. However, these methods are not universally applicable and caution must be used when applying new or untested analyses or analytical procedures [2–5].

13.2.6   Gas Chromatography–Mass Spectrometry

GC-MS consists of a gas chromatograph where the chromatographic column exits into an MS (see Figure 13.5) that acts as a detector for the GC. As the GC carrier gas exits the column, it enters the MS where organic components are ionized, fragmented, and scanned, and the masses of the components and fragments are measured and recorded. The fragmentation pattern is used to identify the compound using a computer to match the pattern found with standard fractionation patterns for known compounds. The computer then produces a list of compounds that come close to matching the unknown along with a percentage (%) that indicates how well the patterns match. Caution must be exercised with these types of matching pattern computer programs as they can easily misidentify compounds, especially if the program is set up for the wrong analysis or the assumed composition of the unknown [6].

13.3   HIGH-PERFORMANCE LIQUID CHROMATOGRAPHY

In HPLC, the mobile phase is a liquid in which the sample must be soluble, and detection is most often accomplished by ultraviolet (UV) absorption. It is generally a slower process than GC; however, its advantage is that the compounds to be separated are not limited by their boiling point, although low-boiling compounds are almost never separated by HPLC. Solid mixtures, as long as they are soluble in the mobile phase, can be analyzed by HPLC, whereas the solids that are typically encountered in soil analysis are not usually volatile enough to analyze via GC.

13.3.1   Sample Introduction

For HPLC, the injector is a valve. In the charge position, a 50-μL syringe is used to fill the sample loop that holds a specific volume of sample solution. The valve is switched to the run position, and the eluent carries the sample out of the sample loop and into the column (auto samplers are also available for HPLC). A recording of the detector output is automatically started at the time of injection and produces a chromatogram of the separated components.

13.3.2   Mobile Phases

The mobile phase in HPLC is called the eluent and is a liquid or a mixture of liquids. Common eluents are water, aqueous solutions, acetonitrile, and methanol. Almost any other common solvent compatible with the column packing and the detector may be used. In some cases, the HPLC instrument will be capable of making a mixture of eluents or changing the mixture of eluents during chromatography. If this is done, care must be taken to make sure that the eluent mixture is compatible with the detector.

13.3.3   Stationary Phases

The most common packing material in HPLC columns (Figure 13.4, C) is a solid with an organic group attached to it. For instance, the solid may have a hydrocarbon chain containing 18 carbons attached to it, making it hydrophobic. This type of column is called a C18 column or a reverse-phase column. Columns can be made with varying polarities and functionalities and thus can be used to carry out a wide variety of separations.

13.3.4   Detection

Five different detectors are common in HPLC: (1) UV, (2) refractive index (RI), (3) conductivity, (4) inductively coupled plasma atomic emission, and (5) mass spectrometry. A UV detector passes a specific wavelength of UV light through the sample as it exits the chromatographic column. The absorbance of the compound is then recorded. The source of UV light may be a deuterium lamp with a filter to remove all but the desired wavelength of light. Alternately, it may be designed like a spectrometer such that the analytical wavelength of light being used can be changed. In the most sophisticated cases, the whole spectrum of the compound is taken as it exits the column. In this case, the UV spectrum may be used to identify the compound.

RI and conductivity detectors are much simpler but cannot identify compounds eluting from the chromatographic column. The RI of the eluent changes as its composition changes. Thus, as compounds elute from the column, the RI changes, and this is recorded to obtain a chromatogram. The conductivity detector is used when water is used as the eluent and the materials being separated are ionic. When no ions are present, the conductivity of water is very low; when ions emerge from the column, the conductivity of the water increases and is recorded, producing the chromatogram. Chromatographic procedures that involve changing the eluent during the analysis will not generally be suitable for use with these detectors. Retention times can be used for "identification" of eluting compounds in the same way as with GC [7,8].

The quadrupole MS detector is used in the same way as it is with GC, with the same sensitivity and ability to identify compounds. In use, most or all the solvent is removed before the sample enters the MS (see Chapter 14).

13.3.5 High-Performance Liquid Chromatography Applied to Soil

In soil analysis, HPLC is used much like GC in that soil is extracted and the extract, after suitable cleanup and concentration, is analyzed. One major difference between them is that HPLC does not require the components to be in the gaseous phase. They must, however, be soluble in an eluent that is compatible with the column and detector being used. A second difference is that both a syringe and an injector are used to move the sample into the eluent and onto the column. Detection is commonly by UV absorption, although RI, conductivity, and mass spectrometry are also commonly used. Conductivity or other electrical detection methods are used when analysis of ionic species in soil is carried out [3,7,8].

13.4 THIN-LAYER CHROMATOGRAPHY

TLC is carried out on a thin layer of adsorbent on a glass or plastic support (other supports have been used). It has sometimes been referred to as *planer* chromatography since the separation occurs in a plane. Paper chromatography, which is carried out using a piece of paper, usually filter paper, is very similar to TLC and will not be covered here.

13.4.1   Sample Application

In TLC, a 10-μL or larger syringe is often used to place sample spots on the thin layer before development. Alternately, a glass capillary tube or Pasteur pipette may be heated in a burner and pulled to obtain a fine capillary suitable for spotting (see Figure 13.3).

13.4.2   Mobile Phases

Development of the thin-layer chromatogram is accomplished by placing a small amount of eluent in the bottom of a suitable container, placing the spotted thin layer in the container, sealing it, and allowing the eluent to ascend the layer through capillary action.

13.4.3   Stationary Phases

A thin layer of adsorbent is applied to a support that may be a sheet of glass, metal, or plastic (Figure 13.4, D). Adsorbents are typically alumina, silica gel, or cellulose and may be mixed with gypsum to aid in adhering to the support. They may also include a fluorescent indicator that aids in visualization once the plate is developed. These adsorbents may also have hydrocarbons attached to them such that reverse-phase TLC can be carried out.

Caution: Spraying thin-layer plates with visualization reagents should always be done in a functioning hood. These reagents may be caustic and toxic, as is the case with sulfuric acid.

13.4.4   Detection

Thin-layer plates, once developed and dried, are sprayed with a visualization reagent that allows the detection of the separated components. The visualization reagent produces a colored spot where each of the components of the mixture is on the plate. For example, a 0.1% solution of ninhydrin (1,2,3-indantrione monohydrate) in acetone can be sprayed on a plate that has separated amino acids on it. The amino acids will show up as blue to brown spots on the plate. Sulfuric acid (note caution) can be sprayed on silica or alumina plates and heated to show the position of organic compounds as charred spots (this reagent cannot be used on cellulose plates because the whole plate will turn black). Reducing compounds can be visualized using a solution of ammonical silver nitrate, which will produce black spots where the silver is reduced by the reducing compound. This reagent is particularly sensitive for reducing sugars. There are a whole host of different visualizing reagents that can be found in the book by Hellmut et al. [9].

A method for detecting compounds that have UV absorbance is the use of thin-layer plates that contain a fluorescence dye or indicator. Plates are developed and then placed under UV light. Compounds on the plate may show up as bright or dark spots depending on their interaction with the UV light and the fluorescence produced by the fluorescence indicator [9]. It has been reported that flame ionization detection can also be used; however, this is not a standard method [10].

13.4.5   Soil Thin Layer

A variation on TLC is the use of soil thin layers to investigate the movement and degradation of organic compounds in soil. A soil is sieved using a number 200 or smaller sieve, and this soil used to produce a suspension of soil in distilled water that is then spread on a glass sheet to produce a soil thin layer. An organic component, often a herbicide, is spotted on the dried sheet, which is developed using deionized water. The movement and degradation of the organic compound under these conditions can then be related to its expected movement and degradation in the field. However, organic matter can confuse the results, so caution in interpreting R_f values is essential [11].

13.5   ELECTROPHORESIS

As with other chromatographic methods, there are a number of electrophoretic methods, including paper and gel electrophoreses and CE. Electrophoresis uses an electric current to move ionic species, either simple ions, amino acids, or complex proteins, through a medium (i.e., a gel) or a capillary (i.e., CE). During this process, typically, the ionic species move at different rates and are thus separated.

13.5.1   Sample Application

In paper or gel electrophoresis, the sample may be applied with a syringe or a micropipette similar to the application of samples to thin-layer plates. In some cases, there may be "wells" in the gel that accept the solution containing the species to be separated. In CE, samples may be applied using electromigration, hydrostatic, or pneumatic injection. In all cases, the ions to be separated must be soluble in and compatible with the stationary phases and buffers used.

13.5.2   Movement of Species

For electrophoresis, the paper or gel is saturated with the required buffer at the desired pH. The ends of the paper or gel are placed in a buffer reservoir that contains the buffer with which the paper or gel is saturated and that also have electrodes connecting one end to the positive DC terminal and the other to the negative terminal of a power source. It is the electrical current that causes the movement of ionic species through the medium.

In CE, a high voltage is used to produce electroosmotic flow. Both electricity and the buffer flow through the capillary, with the buffer flowing toward the cathode. Both carry the sample through the capillary and because of the flow of the buffer, both charged and uncharged species are separated.

13.5.3   Stationary Phases

Electrophoresis can be carried out using paper or a gel as the supporting medium. Typically, it can only be carried out in media compatible with water because buffers or salt solutions are required to carry the electric current required for separation. CE is carried out in a fused-silica capillary filled with buffer.

13.5.4   Detection

Once an electrophoretic separation has been accomplished, the paper or gel is sprayed or dipped in a visualizing solution similar to that used in the visualization of components on a thin-layer plate. Detection methods similar to those used in HPLC are used in CE. The type of analytes (e.g., inorganic ions, organic ions, and ionic biomolecules) will determine which detection method is best.

13.5.5   Electrophoresis Applied to Soil

Gel electrophoresis has been applied to soil DNA and RNA extracts using procedures similar to those used in DNA testing for forensic analysis. CE has also been applied to the analysis of ionic species extracted from soil. While these processes show promise for the elucidation of valuable information about soil, neither is used for common, routine soil analysis [12–14].

13.6   IDENTIFICATION OF COMPOUNDS SEPARATED BY CHROMATOGRAPHIC PROCEDURES

There are three ways that compounds are "identified" once separated. The simplest is that for which chromatography is named. If a mixture of colored compounds is separated, then when each compound elutes from a column, or where they are on a thin layer or gel is found by searching for the color. Most species, however, are not colored. Therefore, after separating colorless components, they must be identified by other means. This can be done in two ways. The first and simplest is by R_f or R_t, which are the distance, relative to some fixed point, that compounds move during a chromatographic procedure (R_f) or between the time they enter the chromatographic column and the time they exit (R_t) (see Figure 13.8, Figure 13.9, and Table 13.4 for further details on the use of R_f and R_t). Equations for calculating R_f and R_t are given in Figure 13.8 while Figure 13.9 and Table 13.4 shows how the calculations are used to determine identity of the component separated.

R_f is applied to TLC, while R_t is applied to GC and HPLC. For identification purposes, the R_f values of pure samples of all the compounds expected to be found in the unknown are determined, and these R_f values are used to "identify" these same compounds in the unknown mixture. The distance from the site where the spots were placed on the plate to the top of the plate where the

$$R_f = \frac{\text{Distance spot moved}}{\text{Distance eluent moved}}$$

$$R_t = \text{Time from injection to top of peak}$$

Figure 13.8. Equation for the calculation of R_f and R_t.

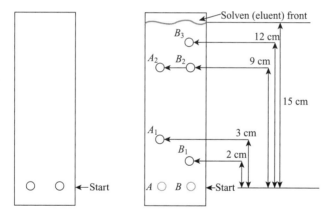

Figure 13.9. Thin layer at start (left) and after development (right). Right chromatogram shows the distance the eluent and spots moved during the development of the chromatogram.

TABLE 13.4. The R_f values of Spots on Thin-Layer Sheets Shown in Figure 13.9[a]

Spot	Calculation	R_f
A_1	3/15	0.20
A_2	9/15	0.60
B_1	2/15	0.13
B_2	9/15	0.60
B_3	12/15	0.80

[a]The distance the solvent front moved is 15 cm; therefore, this is the denominator for all calculations.

eluent stopped is recorded. This number is then divided into the distance the spot moved from the beginning, as shown in Figure 13.9.

The distances should always be measured from the starting point (spot), which is where the original mixtures were spotted and both from and to the midpoint of spots. The determination of the R_f of five spots on a thin-layer chromatogram is illustrated in Figure 13.9. The R_f values of the spots are given in Table 13.4. The materials in spot A_2 and B_2 are expected to be the same because they have the same R_f and the chromatogram was run under exactly the same conditions at the same time.

R_t is used in the same manner as is R_f except that it is the retention time of components in the GC or HPLC column (see Figure 13.6). The time is measured from the time of injection to the time the top of the peak elutes from the column. Another approach is to include a compound that is not retained by the solid or liquid phase in the injection and its peak used as the starting time (0 time) for measuring the time for each peak to exit the column. Two compounds exiting the column at the same R_t will be assumed to be the same compound. Often, R_t values are used for identification, although only a spectroscopic method can truly be an identification method because two different molecules can have identical R_t values.

Another method is to chromatograph the unknown and to determine the R_f values of the components present. When a compound is "identified" by its R_f, then a pure sample of that compound can be added (this is called "spiking") to the unknown and a sample of this mixture analyzed. If only one larger peak is obtained at the specific R_f, both the unknown and the known are the same. When mixtures of similar composition are to be analyzed repetitively by a chromatographic method, it is common to prepare a list of the compounds that are expected to be present along with either their R_f or R_t values. In this way, at least tentative assignments regarding the identity of components in the unknown mixture can be made, even if further testing is needed to confirm these identities.

13.7 QUANTIFICATION

GC and HPLC allow for ready quantification of the components exiting the column in that the area under the peak in the chromatogram is proportional to the amount of component present. However, to make a quantitative analysis, it is essential to have a calibration curve (see Chapter 14) for each of the components of interest. This means making solutions of differing but *known* concentrations, injecting them, and finding the relationship between peak area and amount of material present in a manner similar to that described in Section 14.9, for colorimetric analysis. In many cases, the software that controls the chromatograph can be set up to automatically do this analysis (see also Chapter 15).

For TLC, quantitative analysis is more difficult, although it can be accomplished in some instances. If such an analysis is undertaken, great care must be used to ensure that there is an excellent relationship between the spot characteristic measured and the amount of material present.

13.8   CONCLUSION

Chromatography in its various forms is extremely important in the isolation and identification of complex mixtures found in the environment and in soil. It is applied to components that are isolated from soil by extraction. The extraction solution is important in that it must be compatible with the chromatographic method and analyte detection method. Samples for GC must be volatile because they must be in the gaseous state for analysis. Samples for HPLC and TLC must be soluble to some minimal extent in the eluent being used. Detection of components after separation is either by optical, spectrophotometric, or electrical methods. In the case of TLC plates, separated components may be found by spraying the plates with a reagent that reacts with the component(s) present to produce a spot where the component has eluted. By chromatographing known compounds, a list of R_f values can be prepared and used to "identify" the components in an unknown mixture. In the case of GC and HPLC, the same thing can be accomplished by preparing a list of R_t values for compounds and this used to assign the identity of components in an unknown mixture. Identification by R_f and R_t values, however, is generally not sufficient for absolute identification. True identification occurs when HPLC and GC are coupled to a spectrometric and spectroscopic method such as MS, infrared spectrometry (IR), or UV-Vis (see Chapter 14).

PROBLEMS

13.1. Explain how all chromatographic methods are similar.

13.2. Explain how the basic chromatographic methods are different.

13.3. What physical characteristics must components have to be separated by GC?

13.4. What do the terms R_f and R_t refer to? Explain how they are used in determining what components are likely to be present in a mixture.

13.5. Describe detection methods used in GC, HPLC, and TLC.

13.6. What general characteristic must a component have to be separated by electrophoresis?

13.7. How are soil thin-layer plates used in environmental investigations?

13.8. Explain why it is advantageous to have separated compounds exiting a GC or HPLC column analyzed by MS (consult Chapter 14).

13.9. What is the difference between the mobile phase and the stationary phase in chromatography?

13.10. The chromatograms from two different injections have the following components (peaks) with the indicated retention times: injection 1: peak 1, $R_t = 0.55$; peak 2, $R_t = 1.25$; peak 3, $R_t = 2.44$; peak 4, $R_t = 5.65$; injection 2: peak 1, $R_t = 0.22$; peak 2, $R_t = 1.00$; peak 3 $R_t = 5.65$; peak 4, $R_t = 6.74$. Which of the peaks in the two chromatograms is likely to be the same compound?

REFERENCES

1. Walsh ME. Determination of nitroaromatic, nitramine, and nitrate ester explosives in soil by gas chromatography and electron capture detector. *Talanta* 2001; **54**: 427–438.

2. Wei M-C, Jen J-F. Determination of chlorophenols in soil samples by microwave-assisted extraction coupled to headspace solid-phase microextraction and gas chromatography-electro-capture detection. *J. Chromatogr. A* 2003; **1012**: 111–118.

3. Schwab W, Dambach P, Buhl HJ. Microbial degradation of heptenophos in the soil environment by biological Baeyer-Villiger oxidation. *J. Agric. Food Chem.* 1994; **42**: 1578–1583.

4. Shen G, Lee HK. Determination of triazines in soil by microwave-assisted extraction followed by solid-phase microextraction and gas chromatography-mass spectrometry. *J. Chromatogr. A* 2003; **985**: 167–174.

5. Miège C, Dugay J, Hennion MC. Optimization, validation and comparison of various extraction techniques for the trace determination of polycyclic aromatic hydrocarbons in sewage sludges by liquid chromatography coupled to diode-array and fluorexcence detection. *J. Chromatogr. A* 2003; **995**: 87–97.

6. Vryzas Z, Papadopoulou-Mourkidou E. Determination of triazine and chloroacetanilide herbicides in soils by microwave-assisted extraction (MAE) coupled to gas chromatographic analysis with either GC-NPD or GC-MS. *J. Agric. Food Chem.* 2002; **50**: 5026–5033.

7. Kerven GL, Ostatek BZ, Edwards DG, Asher CJ, Oweczkin J. Chromatographic techniques for the separation of Al and associated organic ligands present in soil solution. *Plant Soil* 1995; **171**: 29–34.

8. Krzyszowska AJ, Vance GF. Solid-phase extraction of dicamba and picloram from water and soil samples for HPLC analysis. *J. Agric. Food Chem.* 1994; **42**: 1693–1696.

9 Hellmut J, Funk W, Fischer W, Wimmer H. *Thin Layer Chromatography Reagents and Detection Methods*, Vol. 1A. New York: VCH; 1990.

10. Padley FB. The use of flame-ionization detector to detect components separated by thin-layer chromatography. *J. Chromatogr.* 1969; **39**: 37–46.

11. Sanchez-Cámazano M, Sánchez-Martín MJ, Poveda E, Iglesias-Jiménez E. Study of the effect of exogenous organic matter on the mobility of pesticides in soil using soil thin-layer chromatography. *J. Chromatogr. A* 1996; **754**: 279–284.

12. Gottlein A, Matzner E. Microscale heterogeneity of acidity related stress-parameters in the soil solution of a forested cambic podzol. *Plant Soil* 1997; **192**: 95–105.

13. Heil D, Sposito G. Organic matter role in illitic soil colloids flocculation: II. Surface charge. *Soil Sci. Soc. Am. J.* 1993; **57**: 1246–1253.

14. Griffiths RI, Whiteley AS, O'Donnell AG, Bailey MJ. Rapid method for coextraction of DNA and RNA from natural environments for analysis of ribosomal DNA- and rRNA-based microbial community composition. *Appl. Environ. Microbiol.* 2000; **66**: 5488–5491.

BIBLIOGRAPHY

Ardrey RE. *Liquid Chromatography-Mass Spectrometry: An Introduction*. New York: Wiley; 2003.

Grob RL, Barry EF. *Modern Practice of Gas Chromatography*, 4th ed. New York: Wiley-Interscience; 2004.

Niessen WMA, ed. *Current Practice of Gas Chromatography-Mass Spectroscopy*. New York: Marcel Dekker; 2001.

Snyder LR, Kirkland JJ, Glajch JL. *Practical HPLC Method Development*, 2nd ed. New York: Wiley-Interscience; 1997.

Tabatabai MA, Frankenberger WT, Jr. Liquid chromatography. In Bartels JM (ed.), *Methods of Soil Analysis Part 3: Chemical Methods*. Madison, WI: Soil Science Society of America and American Society of Agronomy; 1996, pp. 225–245.

U. S. Environmental Protection Agency Test Methods. SW-846 On-Line. 2012. http://www.epa.gov/epawaste/hazard/testmethods/sw846. Accessed June 3, 2013.

CHAPTER

14

SPECTROSCOPY AND SPECTROMETRY

The electromagnetic spectrum is a continuum of wavelengths, λ, which can also be expressed as frequency (proportional to the reciprocal of wavelength). Within this spectrum, visible light represents a very small part: the region of wavelengths around 10^{-5} cm. All regions of the electromagnetic spectrum have been used to analyze environmental constituents, including those in soil. In some cases, the radiation is passed through the material being investigated and absorbed frequencies related to the components it contains. In other cases, radiation is reflected or refracted from the sample and information gathered as a result of changes in the radiation occurring during these processes.

Introduction to Soil Chemistry: Analysis and Instrumentation, Second Edition.
Alfred R. Conklin, Jr.
© 2014 John Wiley & Sons, Inc. Published 2014 by John Wiley & Sons, Inc.

How electromagnetic radiation is used to investigate the characteristics of a sample depends on how it interacts with matter. X-rays are diffracted by atoms, and so they are used to elucidate the arrangement of atoms in a material, commonly crystals. Absorption of ultraviolet (UV) and visible light results in the movement of electrons from one orbital to another. Electrons in double or triple bonds, nonbonding electron pairs, and electrons in **d** orbitals are most susceptible to this type of interaction. Infrared (IR) radiation interacts with matter by changing its vibrational and rotational modes. When atoms are placed in a strong magnetic field, the spins of their electrons and protons interact with radio frequency (RF) wavelengths. The result of this interaction, namely, nuclear magnetic resonance (NMR) spectroscopy and electron paramagnetic resonance (EPR) spectroscopy, is used to determine the environment of electrons and nuclei.

The movement of electrons between orbitals in atoms can also be used to gain information about the elements. Energy from an outside source, such as heat, flame, furnace, an electrical arc, plasma, UV or visible light, X-rays or gamma rays, can move electrons from one orbital to another. If light is used, the absorbed wavelength can be used to determine the type and amount of element present. The other possibility is that the excited electron will fall back to its original position, emitting a wavelength of light specific for that particular transition. Measuring this wavelength and the amount of light produced provides information about the kind and amount of element present. The method of exciting the electrons varies with different instrumentation, but the basic process is the same. This technique, although technically applicable to all elements, is most sensitive and therefore most commonly used in the detection and quantification of metals.

All analytical methods that use some part of the electromagnetic spectrum have evolved into many highly specialized ways of extracting information. The interaction of X-rays with matter represents an excellent example of this diversity. In addition to straightforward X-ray absorption, diffraction, and fluorescence, there is a whole host of other techniques that are either directly X-ray-related or come about as a secondary result of X-ray interaction with matter, such as X-ray photoemission spectroscopy (XPS), surface-extended X-ray absorption fine structure (SEXAFS) spectroscopy, Auger electron spectroscopy (AES), and time-resolved X-ray diffraction techniques, to name only a few [1,2].

No attempt will be made to thoroughly investigate all of these specialized techniques. Only the main, common, or routine methods and instrumentation used in soil analysis will be discussed.

14.1 SPECTRAL OVERLAP

Spectra can be obtained as either absorption spectra, in which the material of interest absorbs unique wavelengths of radiation, or emission spectra, where the material of interest emits unique wavelengths of radiation. All compounds and elements absorb and emit numerous wavelengths or bands of electromagnetic radiation. If two compounds or elements are placed in a beam of electromagnetic radiation, all the adsorptions of both compounds will be observed. Or if the same mixture is excited, all bands emitted by both compounds present will be in the emission spectrum. This leads to the possibility of three types of interferences: (1) two different compounds absorbing in the same place in the spectrum, (2) two different compounds emitting in the same place in the spectrum, (3) one compound emitting while another is absorbing in the same place in the spectrum. This latter case, however, is usually only observed in atomic and X-ray spectroscopy. One or more of these interferences can be present in an analytical procedure, especially when applied to soil and soil extracts.

In situations where all the components of a mixture are known and all their characteristics fully understood, correction for any or all of these types of interferences can be made. However, in soil and soil extracts, it is generally impossible to identify all the components and fully understand all their characteristics. Thus, it is especially important to rule out, compensate for, or eliminate all possible interferences when carrying out an analysis of soil. This is generally accomplished by isolation of the component of interest from most, if not all, the soil matrix components with which it is associated (see Chapters 7 and 11–13) [3,4].

14.2 NOISE

Every measurement has noise—random changes in the results; that is, if an instrument is left to make measurement without any sample, the baseline will not be a straight line but will be a random recording of instrument output or noise. Figure 14.1 shows the noise in the baseline of a gas chromatograph at maximum sensitivity. When an absorption or peak is vastly larger than the noise, there is little question of its authenticity. When it is not much larger than the noise, there is question of its authenticity: is it real or is it noise?

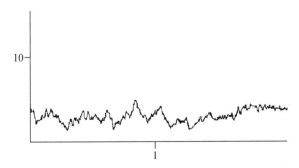

Figure 14.1. Baseline noise from a gas chromatograph.

There are two ways to approach this issue, and both should be investigated in any questionable measurement. First, it is often assumed that any absorption or other measurement that is three or four times larger than the noise is real. This is a good start. However, there are other more scientific approaches to this problem. If repeated measurements on different subsamples or aliquots produce the same absorption or measurement, with minor variations, then it is probably a real result and not noise. On the other hand, if, during repeated measurements, absorption features occur in *exactly* the same location and have *exactly* the same characteristics such as shape and area under the peak, then they are probably not due to the sample because some variation in measurement always occurs. This type of problem is usually a result of instrument malfunction and this must be investigated.

14.3   THE VISIBLE REGION

The visible region of the spectrum (between 400 nm [violet] and 900 nm [red]), is used extensively in soil analysis in the colorimetric determination of components extracted from soil. Once extraction is complete and the extract has been filtered, it is analyzed for the components of interest by treating it with a reagent to produce a colored product. The amount of color is directly related to the amount of component present.

An excellent example of this type of analysis involves the determination of phosphate in soil extracts. Soil is extracted with an appropriate extractant and added to a solution of acid molybdate, with which the phosphate reacts to produce a purple- or blue-colored solution of phosphomolybdate. Standard phosphate solutions are prepared, reacted with acid molybdate, and the intensity of the phosphomolybdate color produced is measured. A standard curve (also called a calibration curve) is prepared (see Section 14.10) from which the intensity of the color is directly related to the concentration of phosphate in the extract.

For this type of analysis to be accurate, three characteristics must exist:

1. The soil extract must not contain any components that absorb light at the same wavelength as the phosphomolybdate. This includes suspended material that will refract light. Refracted light does not get to the detector and so is recorded as being absorbed by the sample, thus giving an inaccurate result.
2. The color-producing reagent must not react with any other commonly occurring component in the extract or form a similarly colored product.
3. The soil must not contain any compound that inhibits or interferes with the production of the colored compound.

As described in previous chapters, soil contains many different inorganic and organic elements, ions, and compounds, plus both inorganic and organic colloids. Thus, it cannot be assumed that the soil being investigated does not contain any of the types of interferences mentioned earlier. Some soils high in organic matter may have dark-colored components that are released into the extracting solution and cause interference with the spectrophotometric analysis. Thus, all extracts must be analyzed to make sure they are free of interferences. For example, this can be accomplished for phosphate by adding a known concentration of phosphate to the sample, extracting, and determining the phosphate. If all the added phosphate can be accounted for, then no interference is present. If interference is present, the method of standard additions can be used in carrying out the analysis [5].

14.4   ULTRAVIOLET REGION

Information about a soil extract can be obtained from the UV and visible regions of the spectrum. These are wavelengths from 190 to 400 and 400 to 900nm, respectively (wider ranges of wavelengths are available on some instruments). The two regions of the spectrum are used very differently in soil analysis, although they are commonly found together in ultraviolet-visible (UV-Vis) spectrometers. The UV region can be used to obtain spectra of soil extracts; however, it is most commonly used as a detector for high-performance liquid chromatography (HPLC), while the visible region is used for colorimetric analysis discussed further.

Common solvents such as water, acetonitrile, and heptanes are transparent over both the UV and visible regions. Because these represent both hydrophilic and hydrophobic solvents, a very broad range of compounds can be easily analyzed by UV.

Adsorption in both regions involves promotion of electrons in double or triple bonds or nonbonding pairs of electrons in elements, such as those on oxygen or nitrogen, and d orbital electrons in metals, to higher energy levels. Conjugated double bonds and nonbonding electron pairs conjugated with double bonds absorb at longer wavelengths than do isolated double bonds or nonbonding electron pairs. More highly conjugated systems have longer wavelengths of the maximum absorption (referred to as *lambda max*, λ_{max}) (see also Figure 8.3). This means that all aromatic and other conjugated systems have strong UV adsorption bands. Unconjugated aldehydes, ketones, amides, acids, esters, and similar compounds also have UV absorptions, although they are not as strong.

Inorganic compound adsorptions, particularly those that are colored, involve the promotion of electrons in the d orbitals. Absorption can occur in both isolated compounds and inorganic moieties with organic ligands.

Any extract of soil containing large amounts of organic matter may have multiple absorption bands in the UV-Vis region of the spectrum. A UV-Vis spectrum of a simple aqueous, low-organic-matter soil extract is shown in Figure 14.2. As seen in this spectrum, even a simple soil extract can have

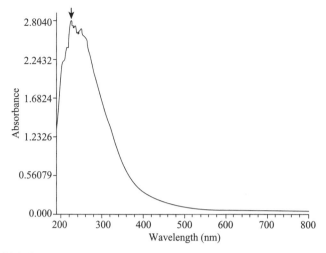

Figure 14.2. UV-Vis spectrum of simple deionized water extract of soil. Absorption maximum is indicated by an arrow. Note that the absorbance is above 1, and therefore the solution should be diluted before attempting to interpret the spectrum.

significant absorptions that can mask other potential absorptions of interest. Absorptions can also result in interferences with either or both a UV or visible determination of a component or the derivative of a component.

An interesting application of this region of the spectrum is the determination of nitrate and nitrite in soil extracts. Nitrate is very soluble in water and can be extracted using a simple water extraction procedure. Nitrate absorbs at 210 nm and nitrite at 355 nm, and both can be quantified using these absorption maxima (λ_{max}). Other materials extracted from soil, as noted previously, however, can obscure this region, and thus caution must be exercised to determine if there are any interfering components in such extracts [6–10].

14.4.1   Ultraviolet Sample Preparation

After extraction, samples to be analyzed are dissolved in a solvent, commonly acetonitrile, heptane, hexadecane, or water. A scan of the pure solvent in the sample cell, typically made of quartz (Figure 14.3 [A]), is first obtained. The dissolved sample is placed in the sample cell, which is placed in the sample compartment of the instrument, and a spectrum is obtained.

Many compounds have large molar absorptivities,[1] and thus only small amounts are needed to obtain a spectrum. Most UV-Vis spectrophotometers give the most accurate absorbance values when the absorbance measurements are between 0 and 1 absorbance unit. Often, a sample will have an absorbance

---

[1] Absorptivity is defined as $A = \varepsilon bC$ where A = absorbance, ε = molar absorptivity (L/mol/cm), b = path length of radiation through sample (cm), and C = molar concentration.

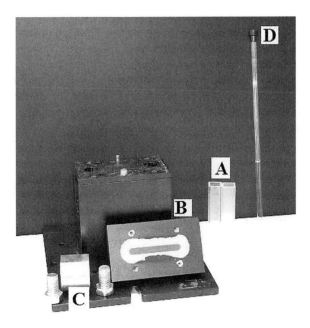

Figure 14.3. Sample holders for various spectrophotometric measurements. (A) UV-Vis cells; (B) infrared ATR plate and stand-plate behind which B is attached when in use; (C) simple KBr pellet maker using the two bolts and the center dye; (D) a sample tube for NMR spectroscopy.

above 1, as is the case in Figure 14.2. In these cases, it is usually useful to dilute the sample by a factor of 10 to bring the absorption below 1 before using the absorbance data.

14.5   INFRARED SPECTROSCOPY

The next section of the electromagnetic spectrum most often used for analysis is the mid-infrared (MIR) region. The absorption of IR light causes changes in the vibration, bending, and rotation of bonds in molecules, resulting in the absorption of radiation. Changes from one mode to another, higher-energy mode are caused by the absorption of IR light. Longer wavelengths of light are used in IR compared to UV-Vis spectroscopy, and spectra, for purely historical reasons, are usually reported in reciprocal centimeters, which is a measure of frequency rather than wavelength. In the older literature, this region is from 4000 to $250\,\mathrm{cm}^{-1}$ (2.5–$50\,\mu m$).

Today, the IR region is divided into three distinct regions[2]: the near infrared (NIR), 900–2000 nm; the MIR, 4000–$250\,\mathrm{cm}^{-1}$; and the far infrared (FIR), less

---

[2] These are typical instrument ranges. Ranges covered by an instrument depends on the manufacturer.

than $250\,cm^{-1}$. Instrumentation for both the NIR and FIR regions is not as readily available as for the MIR region, and so these regions are not as commonly used. However, there has been some substantial development of NIR for the analysis of components in soil, food, and feed such as oils and protein [11,12].

Generally, water is not used as a solvent in IR spectroscopy because water-soluble halogen salts of alkaline and alkaline earth metals, such as sodium chloride and potassium bromide, which are transparent in the IR, are used in spectrometer optical systems and accessories. Large amounts of water will dissolve these parts of the optical system, making it inoperable. Small amounts of water, such as that in high-humidity air, will lead to optical components being fogged and will make them less transparent or opaque. These same salts are used for sample cells, sample preparation, and analysis, and thus samples containing water will be deleterious to or destroy cells. Water has many strong, broad adsorption features in the MIR, and these make analysis of some components mixed with water difficult or impossible. For these reasons, water is generally avoided in IR spectroscopy. However, there are techniques for sampling aqueous systems.

IR spectroscopy is most frequently used for the identification of pure organic compounds. Some organic functional groups, particularly alcohol, acid, carbonyl, double bonds, triple bonds, and amines, have unique, strong, and easily identifiable absorption bands in the IR spectrum. Both methyl and methylene groups are insensitive to their environments and absorb strongly in narrow frequency ranges, making them easily identifiable and useful for quantification (see hydrocarbon analysis as discussed further). However, because these groups are almost always present in organic compounds, their absorption is seldom useful in identifying specific compounds. Carbonyl groups also have a strong, sharp absorption that is usually the strongest in the spectrum, which makes their absorption useful for identification (see Figure 14.4 and Figure 14.5).

A correlation chart giving the important and unique absorption frequencies of various organic functionalities can be found in many books on IR spectroscopy. These charts are useful, but the user needs to be familiar with the shape and size of typical functional group absorption bands before using them with any degree of accuracy. A broad absorption between 2500 and $3650\,cm^{-1}$ shows the presence of $-OH$. However, the absorption of alcohol $-OH$ is very different from that of acid, as can be seen in Figure 14.4. Another example is the carbonyl absorption, which occurs in the $1700\,cm^{-1}$ region of the spectrum. It is important to know not only that these are sharp absorptions but also that they are the strongest absorptions in the spectrum of aldehydes, ketones, and acids as seen in Figure 14.5. If there is an absorption in this region, but it is not the strongest absorption in the entire spectrum or it is not sharp, then it is not from a typical aldehyde, ketone, or acid but may be from a double bond, although this is not common. Table 14.1 gives the wave number ranges of some important functional group absorptions for the MIR region of the spectrum.

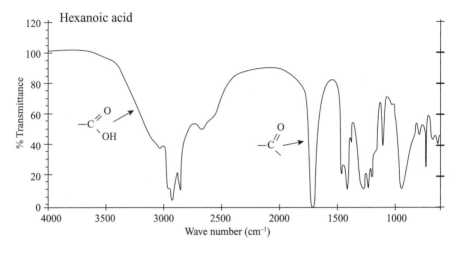

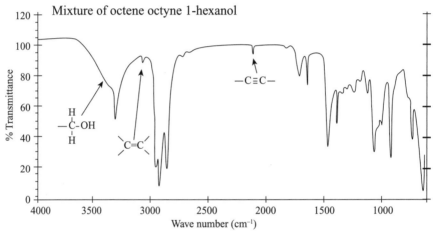

Figure 14.4. Infrared spectra showing common absorption bands. The −OH of acids and alcohols are hydrogen-bonded, leading to broadening. The −OH of hexanoic acid, which is unusually sharp, absorbs in the 3500–2500 cm^{-1} region. Bottom spectrum shows absorptions due to functional groups in the compounds making up the mixture.

IR is not typically used in common soil analytical procedures, although it has been used to investigate various soil components, the most common being humus and its various subcomponents (see Figure 14.6, top spectrum). It has also been used to identify soil clays, both crystalline and amorphous. Many IR spectra can be accessed on the Web via government laboratory sites.

There is one case in environmental work where the methyl and methylene absorptions are useful and used. This is in the United States Environmental Protection Agency (USEPA) method for the determination of total recoverable petroleum hydrocarbons (TRPHs) in a soil or other extract. Here, a supercritical carbon dioxide extraction of a hydrocarbon-contaminated soil is made (EPA Method 3560). The extracted hydrocarbons are dissolved in

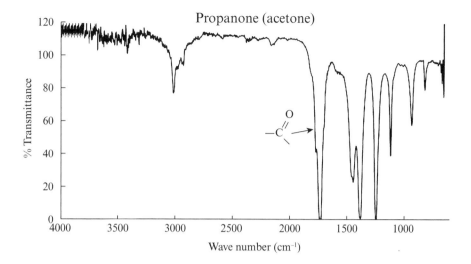

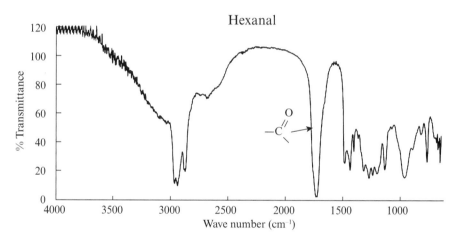

Figure 14.5. Infrared spectra of a ketone, upper spectrum, and aldehyde. Both have strong =O absorption **band**s. Hexanal has a chain of $-CH_2$ groups and $-CH_3$ group, and thus the strong absorption just less than $3000 cm^{-1}$.

tetrachloroethylene, which is transparent in the region of the MIR spectrum where methyl and methylene groups absorb. The intensity of the IR $-CH_3$ and $-CH_2-$ absorption bands in the $2800-3000 cm^{-1}$ region is related to the amount of hydrocarbons in the soil extract through standard curves, as described further [11–17].

14.5.1   Infrared Sample Preparation

Because IR absorption is caused by bond vibrations, and all solvents are molecules that, by definition, have atoms bonded to one another, all compounds have absorption features in the MIR region of the spectrum. Thus, most

TABLE 14.1. Important Diagnostic Absorptions of Major Organic Functionalities in Mid-Infrared, ^{1}H NMR, and ^{13}C Spectra

Functionality	Mid-Infrared (cm^{-1})	^{1}H (ppm)	^{13}C (ppm)
$-CH_3$	3000–2900 and 2900–2800 S-N	0–1 S	0–40 S
$-CH_2-$	2950–2850 and 2875–2800 S-N	1–2 S	10–50 S
H-C=C-H (alkene)	3100–3000 W-N	5–7 W-N	100–170 S
$-C\equiv C-H$	2050–2175 W-N	2.5–3.1 W-N	60–90 S
$\rangle C-OH$	3500–3100 B-S	2.0–6.0 S	50–90 S
$-C\langle^O_H$	2700–2900 (H) W-N; 1650–1750 (C=O) S-N	9.5–9.6 W-SS	160–215 S
$-C\langle^O_\backslash$	1650–1750 (C=O) S-N	NA	160–215 S
$-C\langle^O_{OH}$	3300–2900 (−OH) B-M; 1675–1775 (C=O) S-N	10–13 W	160–185 S
$\rangle C-NH-$	3200–3500 W-N	1.0–5.0 W-N	150–180 S
(benzene)-H	3030 W-N	6.0–9.5 S-N-SS	100–170 S

NA, not applicable; S, strong; N, narrow; B, broad; W, weak; M, moderate; SS, split.

common organic solvents contain groups that absorb throughout the region. Carbon tetrachloride, carbon disulfide, and mineral oil (often called *nujol*) are exceptions to some extent because they have limited numbers of features in the IR region. Thus, if solvents must be used in recording an IR spectrum, these solvents would seem to be preferable. However, it should be noted that mineral oil, for example, has strong absorbances due to $-CH_3$, $-CH_2-$ in its molecules. This obscures regions that may be of analytical importance (see petroleum hydrocarbon analysis as discussed previously). Additionally, carbon tetrachloride and carbon disulfide are generally avoided, if possible, because of their toxicity, difficulty in handling, and detrimental environmental effects.

Methods are available to remove the interferences due to solvents, such as subtraction and compensation. These may be accomplished manually or electronically depending on the instrumentation and how it is configured. However, the preference is always to obtain a spectrum without the interference of a solvent!

Perhaps the most widely and commonly used method for liquid sampling, used with Fourier transform infrared (FTIR) spectrophotometers, is

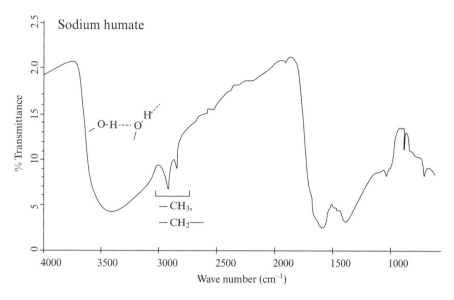

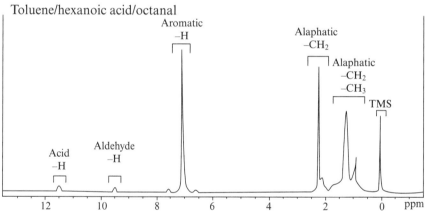

Figure 14.6. Infrared spectra of a KBr pellet of 3% sodium humate (Aldrich) and an NMR spectrum of a mixture of toluene, hexanoic acid, and octanal. The functionalities responsible for the absorption features are labeled.

attenuated total reflection (ATR). An ATR cell is shown in Figure 14.3. A liquid sample is placed in the trough, the bottom of which is a crystal that is situated such that the incident IR beam is totally reflected inside the crystal and exits into the instrument. When reflected, the beam slightly passes into the sample in contact with the crystal, allowing the sample to absorb energy, producing a spectrum. Another common method is to place a drop of pure liquid between two salt windows or in a sealed cell, where the space between the windows is predetermined and held constant. Sealed cells can be used where quantitation of the component is desired or where a volatile solvent or solution of a compound is being analyzed.

Solid samples or solid extracts can be mixed and ground with potassium bromide (KBr), pressed to form a transparent pellet, and a spectrum obtained from the pellet (see C in Figure 14.3). There are gas cells for obtaining spectra of gases and many other methods for obtaining spectra from liquid and solid samples that are not as frequently used as these [13–17].

Raman spectroscopy complements MIR and is relatively free of interference from water. However, it has not been used extensively to study soil chemistry.

14.6   NUCLEAR MAGNETIC RESONANCE

Although some useful and interesting data have been obtained by extracting various components from soil and analyzing them by NMR spectroscopy, it is not generally useful in the *in situ* analysis of soil or soil components. Most soils contain enough iron to potentially change the characteristics of the magnetic field, thus interfering with the analysis. Components extracted from soil, such as humus and phosphate, can be analyzed by NMR.

NMR is an extremely powerful method for observing the environment of an atom of interest. The element most commonly studied is hydrogen bonded to another element, usually carbon, and is referred to as NMR spectroscopy (sometimes called HNMR, ^{1}H NMR, P NMR, or proton NMR). The second most commonly studied element is carbon, specifically ^{13}C, bonded to other carbon atoms and to hydrogen atoms. Other elements commonly measured include fluorine, nitrogen, and phosphorus.

Details of NMR spectroscopy will not be dealt with here but can be found in references sited in the bibliography. NMR is most often carried out on liquid samples or solutions of pure compounds in deuterated solvents, although the NMR of solids can also be obtained.

Different organic functional groups (i.e., methyl, methylene, phenyl, and the hydrogen atoms adjacent to the carbonyl carbon in aldehydes and organic acid groups) absorb at different frequencies and thus can be easily identified. Similarly, different ^{13}C environments result in different absorption characteristics. For instance, carbon atoms in aromatic compounds absorb different frequencies than do those in carbonyl groups.

There are also a number of powerful NMR experiments that yield detailed information about the structure of pure organic molecules. However, as with other regions of the spectrum (see Figure 14.4), chemical mixtures produce spectra that are combinations of all of the absorption features of all of the components in the mixture. The ^{1}H NMR spectrum of a mixture of toluene, hexanoic acid, and octanal is shown in Figure 14.6. This spectrum shows the unique absorptions of each of these functional groups and also illustrates that mixtures of compounds give spectra containing the absorption bands of all the components. If an impure sample had such a spectrum, it would not be known if it was one molecule containing all the indicated functionalities or a mixture of several molecules containing these functionalities.

Table 14.1 gives some important diagnostic adsorptions for ^{1}H and ^{13}C NMR spectroscopy.

One place where there is potentially more use for NMR in soil and environmental analysis is in speciation of soil components. Using a broadband NMR instrument, many different elements can be analyzed, and it can be used to differentiate between species such as PO_4^{-3}, MPO_4^{-2}, and $M_2PO_4^{-1}$ (where M represents a metal). This potential, however, has not been exploited to any great extent [18–21].

14.6.1   Nuclear Magnetic Resonance Sample Preparation

Because most common solvents, including water, contain protons, and most NMR analyses involve the measurement of protons, a solvent without protons is generally used in NMR spectroscopy. Commonly, solvents in which the hydrogen atoms are replaced with deuterium (i.e., solvents that have been deuterated) are used, the most common being deuterochloroform. In addition, an internal standard, most commonly tetramethylsilane (TMS), is added to the sample in the NMR sample tube (see Figure 14.3, D) and all absorption features are recorded relative to the absorption due to TMS.

14.7   MASS SPECTROMETRY

Mass spectrometry, as the name implies, is a method by which the mass of a molecule, that is, its molecular weight, is determined. Spectrometry is different from spectroscopy in that, unlike spectroscopy, the sample is not interacting with light or electromagnetic radiation. Instead, a "spectrum" of the molecular weights of compounds present in a sample is measured. In mass spectrometry, a sample is introduced into the instrument, ionized, and fragmented during the ionization process. The sample fragments are accelerated down a path, which can be of several types, separating fragments on the basis of their mass and charge. After they are separated, the ions are detected and recorded. Because the method depends on the ions moving unimpeded through the path, the inside of the instrument must be under high vacuum at all times.

The ionization–fragmentation process can be accomplished by bombarding the sample with electrons or with a chemical species. This process removes an electron from the compound and fragments it, producing a positive species. Accelerating plates at high negative voltage attract the particles, which pass through a hole or slit in the plate and move down the path. The path may be a straight tube at the end of which is a detector. The time it takes an ion to travel the length of the straight tube will depend on its mass and charge. Such an arrangement is called a *time-of-flight* (TOF) mass spectrometer.

In a magnetic sector mass spectrometer, the path of the ions is through a curved tube with a magnet at the curved portion. The path of the charged species will bend in the magnetic field, depending on the strength of the magnetic field and the mass and charge of the ions. After passing the magnet, the charged species will impinge on a detector and be recorded. By changing the strength of the magnetic field, the masses of the ions reaching the detector are also changed. In this way, the presence of ions of different masses can be detected and recorded.

In a quadrupole mass spectrometer, the ions pass into a path between four rods that are attached to an electric circuit that applies a range of frequencies to the rods. Ions resonate in the quadrupole until a certain frequency, which depends on their mass and charge, is reached and then the ions exit the quadrupole and are measured. A diagram of a quadrupole mass spectrometer is given in Section 13.2.3, Figure 13.5.

Other types of mass spectrometers are available, but those mentioned earlier are the most commonly used in analyses of soil and soil extracts.

As with all spectroscopic methods discussed previously, this method is best suited to measurement and elucidation of the characteristics of pure compounds. For this reason, MS is often used as a detector for gas chromatographs. The GC separates the mixture into pure compounds and the MS then analyzes each pure chemical as it exits the column. The most common MS for this application is the quadrupole mass spectrometer. For this reason, it is discussed in Chapters 14 and 15.

Using mass spectrometry, it is possible to determine the molecular weight of the compound being analyzed. It is also possible to distinguish between isotopes of elements. Thus, ^{14}N and ^{15}N can be separated and quantified using mass spectrometry.

Extensive work on the nitrogen cycle and fate of nitrogen compounds in the environment has been carried out using ^{15}N mass spectrometry. Fertilizer or other nitrogen-containing material enriched in ^{15}N is applied to soil or a crop and samples are taken after various periods of time. The samples are digested using Kjeldahl methods as discussed earlier in Section 10.4. Ammonia is collected and then decomposed before injecting it into the mass spectrometer. A species enriched in ^{15}N provides a measure of the partitioning of nitrogen and its movement in the environment. Other stable isotopes can be used in a similar way to follow their movement through the environment in general and in soil in particular [22–26].

14.8   ATOMIC SPECTROSCOPY

Excitation of electrons in an atom promotes them to a higher energy level, and when they fall back to their original level, they release energy of the same wavelength as the energy absorbed. When this energy is in the visible range of the electromagnetic spectrum, it gives rise to what is termed a *line spectrum*, which is a set of discrete wavelengths, or lines of light unique to each element. Early chemists used these unique lines to identify new elements as they were discovered (see Chapter 1). Samples of a new (or thought to be new) element were put in an electrical arc and the light emitted was dispersed using prisms and recorded using photographic film. These were the original spectrometers.

Although such instruments as described earlier are available, they are not typically used in soil analysis. Today, samples are most often aspirated into a flame or torch to cause the promotion of electrons in elements, and the diagnostic wavelengths are detected and quantified by photomultipliers. Modern spectrometers are different because of the use of many different ways of heating samples and the range of wavelengths available. Today, because of increased sensitivity of instrumentation and detectors, more of the spectrum is available for this type of analysis. Thus, wavelengths from 200 to 900 nm can be used for the analysis of the elements that are present.

The basic idea of exciting electrons and measuring unique, diagnostic wavelengths of light emitted from a sample is useful for routine soil analysis. This is called the *emission mode* (EM). Potassium (one of the three most important plant nutrients), sodium (which poses problems in some arid soils), and calcium are routinely measured using the EM mode, which is also commonly used to determine these elements in food and blood. An instrument capable of determining elements in both the EM and the atomic absorption (AA) modes is shown in Figure 14.7. The composition of many samples can be determined very quickly using this instrument.

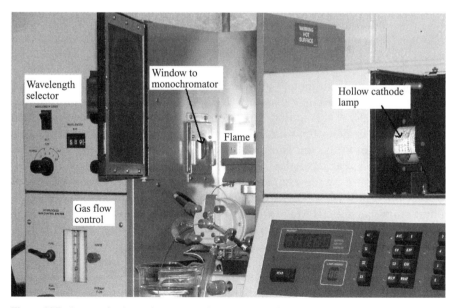

Figure 14.7. An atomic absorption spectrophotometer capable of functioning in either the EM or AA mode.

The analysis of a well-characterized sample by atomic spectroscopic analysis is straightforward. The instrument is adjusted to isolate and direct the analytical wavelength of light to the detector, typically a photomultiplier. Maximum sensitivity is achieved by introducing a known sample of the element of interest into the flame and adjusting the instrument, including the flame, to maximize the reading obtained. The sample is then introduced into the flame, and the amount of light at the analytical wavelength either emitted or absorbed is measured. A standard curve (see Section 14.9) is prepared and used to determine the amount of the element of interest in the samples.

Even when soils are from the same area, significant variations in concentration and content are found. This leads to the possibility of diagnostic wavelength overlap. Even though each element has a unique spectrum, two elements may produce wavelengths of light that are the same or very close together. This can lead to the overdetermination of the quantity of an element because the total amount of light measured is the sum of the light from both the desired element and the interfering element.

Matrix interferences can be observed in a number of different forms. Components in the extract may interfere with excitation of electrons. Absorbance of light by unexpected metals or organic compounds, generally or specifically, would also cause interference. Another source of interference is the complexation of the analyte with extract components such that the metal of interest is protected from the heat source. Other interferences such as changes in the viscosity of the extract are also possible, although they are less common. Whenever soil samples are analyzed by atomic spectroscopic methods, it is essential to make sure that no interfering species are present in the soils. If they are, then steps must be taken to correct for these interferences.

14.8.1   Excitation for Atomic Emission

For the alkaline-earth metals, as noted earlier, a simple flame of almost any type can be used to excite the metals. However, to be able to determine a wide range of metals, it is common to use either an acetylene–air or acetylene–nitrous oxide flame as the source of energy to excite the atoms. The burner is long with a slot at the top and produces a long narrow flame that is situated end-on to the optics receiving the emitted light.

Light given off by the excited electrons falling back to the ground state is passed through slits, isolated using a prism or grating, and measured using a photomultiplier tube (PMT) or other light detecting device. The wavelength of the light emitted is specific to the element, while the intensity of the light is directly related to the amount present.

Introduction of sample into the flame is accomplished by Bernoulli's principle. A capillary tube is attached to the burner head such that gases entering the burner will create a negative pressure. The sample is thus aspirated into the burner, mixed with the gases, and passed into the flame. Heating in the flame excites electrons that emit light of distinctive diagnostic wavelengths as they fall back to their original positions in the elements' orbitals. Because there

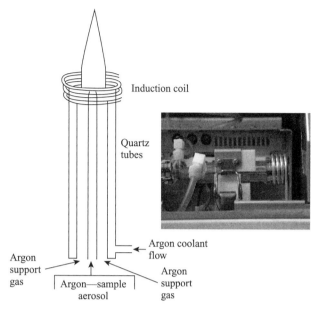

Figure 14.8. On the left is a diagram and on the right is a picture of an ICP torch.

are a number of electrons that can be excited and a number of orbitals into which they can fall, each element emits a number of different wavelengths of light. In the case of atomic emission, one of these, usually the most prominent or the strongest, is chosen as the primary analytical wavelength.

Generally, the higher the temperature of the sample, the more sensitive will be the analysis. Thus, in addition to the two types of flames discussed earlier, another excitation source, which is not a flame but a plasma from an inductively coupled plasma (ICP) torch, is used (see Figure 14.8). To create the plasma, argon gas is seeded with free electrons that interact with a high-frequency magnetic field of an induction coil, gaining energy and ionizing the argon. A rapidly reversing magnetic field causes collisions that produce more ions and intense thermal energy, resulting in a high-temperature plasma into which the sample is introduced. An ICP torch is shown in Figure 14.8.

While flames are used both for emission and absorption spectroscopy (AA), ICP is used only for emission spectroscopy. For AA the methods of heating a sample, arranged in order of increasing temperature, are acetylene–air and acetylene–nitrous oxide. The flames are used in the same instrument and can be adjusted to be oxidizing or reducing to allow for increased sensitivity for the element being analyzed. ICP is carried out using a different, separate instrument and generally has a significantly higher sensitivity than do flame instruments. There are a number of hyphenated variations on the basic EM and ICP instrumentation such as ICP-optical emission spectroscopy[3]

---

[3] ICP-OES is also sometimes called ICP-AES (atomic absorption spectroscopy).

(ICP-OES—also sometimes referred to as ICP-AES for atomic emission spectroscopy) and ICP–mass spectrometry (ICP-MS); also see Chapter 15.

AA and ICP instruments can be equipped with multiple detectors that allow for analysis of more than one element at a time [3,25–29]. Recently, an ICP instrument that used microwaves and air or nitrogen as the supporting gas has been made commercially available. Because argon gas is relatively expensive, microwave-based instruments should prove to be significantly less expensive to operate.

14.8.2   Atomic Absorption

In AA spectroscopy, light emitted by the element of interest contained in a hollow cathode lamp (HCL) is passed through the flame of an AA spectrometer (the same instrument as is used for EM except now it is configured for the AA mode). In this case, the same burner and flame as described earlier and shown in Figure 14.7 are used. The source of the light is an HCL, also shown in Figure 14.7, where the cathode is made of the element of interest and thus, when excited, emits the analytical wavelengths of light needed for the analysis of that element. In a majority of cases, a different lamp will be needed for each element for which an analysis is required. The amount of light absorbed by the element in the flame is directly proportional to the amount of that element present. AA is significantly more sensitive than flame emission for many elements [27,28].

14.9   COLOR MEASUREMENT: THE SPECTROPHOTOMETER

The method described next is used with all of the spectrophotometric methods of analysis described earlier; that is, standard solutions are prepared and analyzed, and standard curves created and statistical analyses carried out. The final step of relating the results back to the original sample and field is also essential and the same.

Color measurement is based on the Beer–Lambert law, which can be expressed as follows:

$$A = \log_{10} \frac{I_i}{I_{ex}}.$$

In this equation, A is the absorbance, also called the optical density; I_i is the intensity of the beam when no absorbing species is present; and I_{ex} is the intensity of the beam of light when some light has been absorbed. The absorbance is proportional to both the concentration of absorbing species and the path length of the light through the sample. Normally, standard size sample cells are used, and so the path length is constant for all samples.

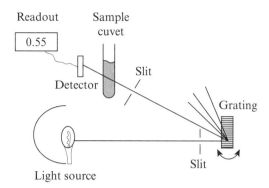

Figure 14.9. Diagram of the light path in a spectrophotometer.

For analysis involving colored solutions, the spectrophotometer needs to function only in the visible range of the spectrum. As shown in Figure 14.9, a light source is used to provide a broad range of wavelengths in the visible portion of the spectrum. The light interacts with a grating, which separates the different wavelengths of light that are present, essentially creating a rainbow. A slit is used to select the specific wavelength that is allowed to pass through the sample, which is contained in a cuvette. Lastly, a detector records the amount of light that passes through the sample. To obtain an absorbance reading, two measurements are needed: one with just the solvent in the cuvette, which determines I_i, and a second with the actual sample in place, which yields I_{ex}.

Cuvettes used with the spectrometer are not simply test tubes. They are specially made tubes or cuvettes and are often matched such that a set of tubes or cuvettes will all have the same absorbance characteristics. Cuvettes should never be used as test tubes; they must be kept clean at all times, and care must be taken not to scratch them. When using cuvettes that have not been used before, they should be tested to make sure they are all the same. This is accomplished by inserting them into the spectrophotometer and noting their absorbance. All should be the same. Keep in mind that empty cuvettes will have a higher absorbance than when filled with water. This is because light is refracted at each surface, and when filled with water or solvent, there is less refraction at the surfaces.

14.9.1 Zeroing and Blanks

When using a spectrophotometer for a colorimetric analysis, both the 0% and 100% transmittance (∞ and 0 absorbance) readings must be set. Once the instrument has warmed up, with the light beam blocked and with nothing in the sample compartment, the readout is set to 0% transmittance (∞ abs.). Again, this measurement is done to set I_i in the absorbance equation shown earlier. A blank, a solution containing all the components used in the analysis except the analyte being measured, is placed in a cuvette, placed in the sample compartment, and the instrument adjusted to 100% transmittance (0 abs.). This procedure is intended to account for all interferences that may be introduced into the measurement by components other than the analyte of interest. Once the instrument is adjusted, determination of the absorbance of standards and samples can be made.

14.9.2   Relating Component Concentration to the Extract

Performing an analysis requires the preparation of a standard curve, which is also called a calibration curve. A series of standard solutions (four to six typically) containing known amounts of the component of interest are prepared. There are two primary restrictions on these solutions: (1) they should cover the range of the expected concentrations of the component of interest, and (2) all absorbance values obtained for the extracted component must fall in between the lowest and highest absorbance values obtained for the standards. If a result is beyond these limits, its concentration cannot be determined with a high degree of confidence. In the table of data for Figure 14.10, Unk1

Concentration	optical density
10	0.15
30	0.3
60	0.65
90	0.95
Unk1	0.05
Unk2	0.97

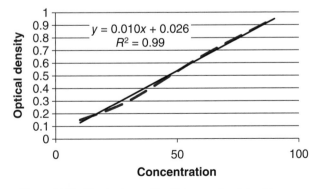

Figure 14.10. Standard or calibration curve for data at top.

(unknown 1) is 0.05, which is below the lowest standard (0.15) and so its value cannot be determined. Likewise, the Unk2 value is 0.97 is above the highest standard, and also cannot be determined with a high degree of statistical certainty.

If, as with Unk 2, the value is above the highest standard, then the sample may be diluted and measured. In this case, the concentration must be corrected for the amount of dilution. Similarly, in some cases when the concentration of an unknown is lower than the lowest standard, the solution can be concentrated and reanalyzed.

Furthermore, standards must cover a range where there is a direct or straight-line relationship between the amount of color produced and the amount of component present. It is common for data points beyond the standard curve to be part of a different straight line or simply not measurable because the solution is too light or too dark. In Figure 14.10, the distance between the points 0, 0 and 10, 0.15 may be a very different line than the one shown in the graph. We know nothing about this region because samples with concentrations and absorbance values in this range were not measured. The same can be said about the region above the points 90, 0.95. With many instruments, very low absorbance values are easily measured; however, absorbance values approaching or above 1 become increasingly less reliable the higher they are.

A blank is often used as a 0, 0 point in preparing a calibration curve as in Figure 14.11 (not shown in Figure 14.10). This is an acceptable practice as long as the calibration curve is sufficiently accurate. If it is not, or the inclusion of the 0, 0 point produces an irregular curve, its removal from the curve is warranted. In Figure 14.11, the inclusion of the 0, 0 point gives the same r^2 value (see Figure 14.11) and thus can be included in the curve. Also, note that when this is done, the Unk1 value can be determined because the number is within the range of the standards used [30].

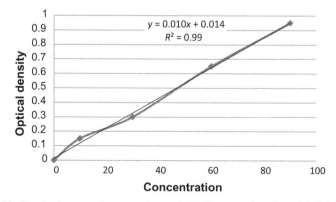

Figure 14.11. Standard curve using same data as in 14.10 except that the origin 0.0 is included.

In some cases, curved lines may be used as standard curves if they give a suitable equation for the line and an acceptable r^2 is obtained (see below). However, it is always preferable to have a straight-line relationship between concentration and absorbance.

14.10 REGRESSION ANALYSIS

The calibration curves in Figure 14.10 and Figure 14.11 were prepared using Excel (*xy* scatter chart), which has automatic features for adding a trend line (the straight line) and cell commands for generating statistical analyses for lines such as calculating the best slope and intercept and the uncertainties in these values, r^2, the standard error (SE) of the line, and the F-statistic. The r^2 value is sometimes used as an indication of how close the obtained data are to a straight line, with values closer to 1.00 being considered better than those that are further from 1.00. The F-statistic and SE of the fit, however, are better goodness-of-fit measures. Regression analysis is also sometimes referred to as *least-squares analysis* and is a standard statistical analysis [30].

Once the calibration curve has been prepared and is of sufficient accuracy, the extracted samples can be analyzed and the calibration curve used to relate the results of the analysis of the unknown to the amount of component of interest present in them [31].

These basic ideas are applicable to all the measurements made using the spectroscopic methods discussed previously.

14.11 RELATIONSHIP TO THE ORIGINAL SAMPLE

At this point, the original sample has been extracted and the amount of component of interest has been determined. But it has been determined for only a portion of the extract and a portion of the field that was sampled. It is essential to relate this amount back to the amount originally present in the sample taken in the field and ultimately to the field itself. This is then a process of working backward; that is, one must first find the total amount present in the whole extract, then determine how this is related to the amount of sample originally taken and then to the field sampled.

If a 1-g soil sample is extracted with 10 mL of extractant, then the component extracted is evenly distributed throughout the 10 mL. This means the final result will need to be multiplied by 10 because the component was diluted 1:10 (this assumes an extractant density of 1g/mL). This then is related back to the volume or mass of soil in the original sample, or it may be directly related back to the field. It may also be necessary to apply other conversion or correction factors, such as the percent water present in the original soil sample, depending on the procedure used.

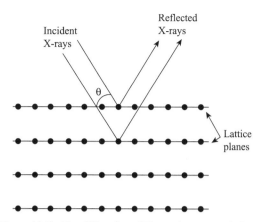

Figure 14.12. The diffraction of X-rays from crystal planes.

14.12   X-RAY DIFFRACTION

X-ray diffraction is a powerful tool used extensively to identify the crystalline clay minerals in soil. It is also used to study the characteristics of the clay minerals in terms of shrink–swell characteristics and occluded components. It is not, however, applicable to amorphous clays found in some tropical soils and common in Andisols (see Chapter 2). It is carried out by irradiating, at various incident angles, a soil clay sample. When the incident angle θ, shown in Figure 14.12, results in constructive interference of the reflected (diffracted) X-rays, the distance between the layers can be calculated using Bragg's law (see Reference 31 for an explanation of Bragg's law). In addition to the distance between lattice planes, the swelling characteristics of various clays are different when they are exposed to different solvents. This provides additional information about the type of clay present and its characteristics.

In practice, clay from a soil sample is prepared on a microscope slide, dried, and its X-ray diffraction measured; subsequently, the same clay is placed in atmospheres saturated with, for example, glycerol, with subsequent X-ray diffraction, and again the distance between layers determined. Changes in the diffraction pattern or the lack thereof will identify the type of clay present [32].

14.13   X-RAY FLUORESCENCE

Exposure of elements to a broad spectrum of X-rays results in the ejection of electrons from their inner shells. Electrons from outer shells falling into these vacancies emit radiation of specific wavelengths (see Figure 14.13). Analysis of this radiation, referred to as X-ray fluorescence (XRF), allows for the identification of the element from which the photon is emitted. Instruments for carrying out this analysis can be either laboratory sized or can be handheld

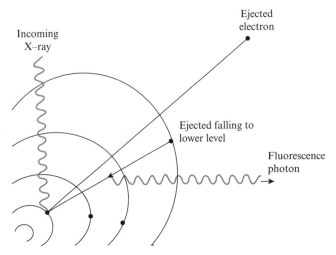

Figure 14.13. Diagram showing the source of X-ray fluorescence photons.

units that can be taken to the field. The excitation radiation for the XRF instrument must pass through a window. The window material, often a metal, will, in part, determine the range of elements that can be detected.

When used in direct soil analysis, XRF suffers from the fact that it is largely a surface phenomenon. For this reason, only surface elements will be determined. However, because it is a surface phenomenon, it has been extensively used to study sorption on the surfaces of soil components. An extensive list of investigations using XRF to investigate various sorption mechanisms is given in the text by Sparks [33].

Fluorescence determinations are best made on samples of homogeneous particle size, and this is not the normal state of soil. Grinding and carefully sieving soil before analysis can minimize problems associated with particle size heterogeneity. Another approach has been to fuse soil with borate or to dilute it with cellulose or other suitable diluents. Very thin layers of soil may also be prepared and used for quantitative analysis.

The sensitivity of XRF determinations is better for higher mass number elements and poorer for lighter elements. Often in soil analysis, lighter elements are of greater interest and this makes application of this method more difficult. Typically, determination of elemental composition in the parts per million (ppm) range is, however, achievable.

XRF characteristics that can limit its usefulness are the surface area observed and surface contamination. In XRF, the surface area measured is small, meaning that a large number of determinations must be made in order to obtain a representative sample of the elements present. In addition, transport and storage of uncovered soil samples can lead to surface contamination that will subsequently appear as part of the soil constituents.

Other limitations involve both the mass absorption coefficient of soil components and secondary and tertiary excitation. The mass absorption coefficient can be calculated and used to correct fluorescence determinations if the exact composition of the material being analyzed is known. This is not possible in soil. Secondary and tertiary excitations occur when X-rays emitted by an element other than the one of interest may cause emission or fluorescence of the element of interest. These potential sources of error are possible in any soil analysis using XRF.

Some of the above-mentioned limitations, but not all, can be overcome by making an extract of the soil and determining the elemental composition of the extract. This approach can eliminate or minimize problems associated with XRF analysis being limited to particle surface [34,35].

14.14 REMOTE SENSING

It is possible to obtain significant information about soil by remote sensing without taking samples in the field, bringing them to the laboratory, and carrying out a spectroscopic analysis in which light reflected from soil or plant surfaces is analyzed. Radiation from the UV and IR regions of the spectrum is used in this type of analysis. However, not all regions are detected or analyzed by one instrument at one time.

With this type of investigation, it is important to make detailed correlations between the observed spectra and the conditions that actually exist on the soil and plant surfaces being investigated. In addition, correlations obtained in one area must be verified if they are to be used in another area or region [36–38].

14.15 CONCLUSION

Matter interacts with all forms of electromagnetic radiation, and these interactions are used to gain information about the matter with which it interacts. Thus, X-ray, AA, and many spectroscopic methods are available for the investigation of soil. Atomic emission, adsorption, and ICP are routinely used to determine metals extracted from soil. UV and IR spectroscopies, along with mass spectrometry, are used to unequivocally identify compounds. NMR is extremely powerful in determining the structure and species of various compounds. A common method of analysis is colorimetry, in which the component of interest is reacted with a reagent to produce a colored compound. Using the visible region of the spectrum, the intensity of the color, as measured by a spectrophotometer, is directly related to the amount of component present in the sample using a calibration curve. The results are related back to the original sample. X-rays are used to determine the structure of clays and identify the elements present in soil via XRF.

PROBLEMS

14.1. Describe two ways to distinguish between an analytical signal and noise in the output of an instrument.

14.2. Describe some major limitations of XRF as a method for determining metals in soil.

14.3. What is spectral overlap and how may it affect analysis of metals extracted from soil?

14.4. List three common solvents useful in UV-Vis spectroscopy. Which of these might be more useful in the analysis of soil extracts?

14.5. Describe two ways in which spectra may be used to identify compounds.

14.6. Explain how IR and NMR spectra are different.

14.7. Soil commonly contains water. Why are water and IR spectroscopy not compatible?

14.8. Explain why water is generally not used as a solvent in NMR spectroscopy.

14.9. Mass spectroscopy is useful because it can determine what important characteristic of a compound?

14.10. Explain and give at least two reasons why using UV-Vis or mass spectrometry as a detector for chromatography might be beneficial.

14.11. A researcher randomly picks test tubes and uses them in his spectrophotometer and obtains confusing results. Explain how this can happen.

14.12. Explain what a calibration curve is and how it is used.

14.13. Two researchers prepare calibration curves. One researcher's curve has an r^2 of 0.9977, the other an r^2 of 0.9845. Which is the better calibration curve? Explain how you came to this conclusion; be specific. What statistics are better to use than r^2 when assessing the quality of a calibration curve?

REFERENCES

1. Fendorf S, Sparks DL. X-ray absorption fine structure spectroscopy. In Bartels JM (ed.), *Methods of Soil Analysis Part 3: Chemical Methods*. Madison, WI: Soil Science Society of America and Agronomy Society of America; 1996, pp. 377–416.

2. Guest CA, Schulze DG, Thompson IA, Huber DM. Correlating manganese X-ray absorption near-edge structure spectra with extractable soil manganese. *Soil Sci. Soc. Am. J.* 2002; **66**: 1172–1181.

3. Kola H, Perämäki P, Välimäki I. Correction of spectral interferences of calcium in sulfur determination by inductively coupled plasma optical emission spectroscopy using multiple linear regression. *J. Anal. At. Spectrom.* 2002; **17**: 104–108.

4. Goossens J, Moens L, Dams R. Inductively coupled plasma mass spectrometric determination of heavy metals in soil and sludge candidate reference materials. *Anal. Chim. Acta* 1995; **304**: 307–315.

5. Saxberg B, Kowalski BR. Generalized standard addition method. *Anal. Chem.* 1979; **51**: 1031–1038.

6. Gottlein A, Mataner E. Microscale heterogeneity of acidity related stress-parameters in the soil solution of a forested cambic podzol. *Plant Soil* 1997; **192**: 95–105.

7. Prabhaker M, Dubinshi MA, eds. *Ultraviolet Spectroscopy and UV Lasers*. New York: Marcel Dekker; 2002.

8. Kerven GL, Ostatek BL, Edwards DG, Asher CJ, Oweczkin J. Chromatographic techniques for the separation of Al associated organic ligands present in soil solution. *Plant Soil* 1995; **171**: 29–34.

9. Cawse PA. The determination of nitrate in soil solutions by ultraviolet spectroscopy. *Analyst* 1967; **92**: 311–315.

10. Wetters JH, Uglum KL. Direct spectrophotometric simultaneous determination of nitrite and nitrate in the ultraviolet. *Anal. Chem.* 1970; **42**: 335–340.

11. Roberts CA, Workman RJ, Jr., Reeves JB, III, eds. *Near-Infrared Spectroscopy in Agriculture*. Agronomy monograph 44. Madison, WI: American Society of Agronomy, Crop Science Society of America, and Soil Science Society of America; 2004.

12. McCarty GW, Reeves JB, III, Reeves VB, Follett RF, Kimble JM. Mid-infrared and near-infrared diffuse reflectance spectroscopy for soil carbon measurement. *Soil Sci. Soc. Am. J.* 2002; **66**: 640–646.

13. Günzler H, Gremlich H-U. *IR Spectroscopy: An Introduction*. New York: Wiley-VCH Publishing; 2002.

14. Clark RN, Swayze GA, Gallagher AJ, King TVV, Calvin WM. *The U.S. Geological Survey, Digital Spectral Library: Version 1: 0.2 to 3.0 microns*. U.S. Geological Survey Open File Report 9-592, 1340 pages. 1993.

15. Van Der Marek HW, Beutelspacher RH. *Atlas of Infrared Spectroscopy of Clay Minerals and Their Admixtures*. New York: Elsevier Science Ltd.; 1976.

16. Johnson CT, Aochi YO. Fourier transform infrared spectroscopy. In Bartels JM (ed.), *Methods of Soil analysis Part 3 Chemical Methods*. Madison, WI: Soil Science Society of America and Agronomy Society of America; 1996, pp. 269–321.

17. White JL, Roth CB. Infrared spectroscopy. In Klute A (ed.), *Methods of Soil Analysis Part 1: Physical and Mineralogical Methods*, 2nd ed. Madison, WI: American Society of Agronomy and Soil Science Society of America; 1986, pp. 291–329.

18. Chefetz B, Salloum MJ, Deshmukh AP, Hatcher PG. Structural components of humic acids as determined by chemical modifications and carbon-13 NMR pyrolysis-, and thermochemolysis-gas chromatography/mass spectrometry. *Soil Sci. Soc. Am. J.* 2002; **66**: 1159–1171.

19. Dec J, Haider K, Benesi A, Rangaswamy V, Schäffer A, Plücken U, Bollag J-M. Analysis of soil-bound residue of ^{13}C-labeled fungicide cyprodinil by NMR spectroscopy. *Environ. Sci. Technol.* 1997; **31**: 1128–1135.

20. Sutter B, Taylor RE, Hossner LR, Ming DW. Solid state ^{31}phosporous nuclear magnetic resonance of iron-, manganese-, and copper-containing synthetic hydroxyapatites. *Soil Sci. Soc. Am. J.* 2002; **66**: 455–463.

21. Turner BL, Richardson AE. Identification of scyllo-inositol phosphates in soil by solution phosphorus-31 nuclear magnetic resonance spectroscopy. *Soil Sci. Soc. Am. J.* 2004; **68**: 802–808.

22. Lee EA, Strahn AP. *Methods of analysis by the U.S. Geological Survey Organic Research Group—Determination of acetamide herbicides and their degradation products in water using online solid—phase extraction and liquid chromatography/ mass spectrometry.* 2003. Available at http://ks.water.usgs.gov/pubs/reports/of.03- 173.pdf. Accessed June 3, 2013.

23. Shen G, Lee HK. Determination of triazenes in soil by microwave-assisted extrac- tion followed by solid-phase microextraction and gas chromatography-mass spec- trometry. *J. Chromatogr. A* 2003; **985**: 167–174.

24. Wachs T, Henion J. A device for automated direct sampling and quantization from solid-phase sorbent extraction cards by electrospray tandem mass spectrometry. *Anal. Chem.* 2003; **75**: 1769–1775.

25. Encinar JR, Rodríguez-González P, Fernandez JR, Alonso JIG, Diez S, Bayona JM, Sanz-Medel A. Evaluation of accelerated solvent extraction for butyltin speciation in PACS-2 CRM using double-spike isotope dilution-GC/ICPMS. *Anal. Chem.* 2002; **74**: 5237–5242.

26. Soltanpour PN, Johnson GW, Workman SM, Jones JB, Jr., Miller RO. Inductively coupled plasma emission spectrometry and inductively coupled plasma-mass spec- troscopy. In Bartels JM (ed.), *Methods of Soil Analysis Part 3: Chemical Methods.* Madison, WI: Soil Science Society of America and Agronomy Society of America; 1996, pp. 91–139.

27. Wright RJ, Stuczynski TI. Atomic absorption and flame emission spectrometry. In Bartels JM (ed.), *Methods of Soil analysis Part 3: Chemical Methods.* Madison, WI: Soil Science Society of America and Agronomy Society of America; 1996, pp. 65–90.

28. Preetha CR, Biju VM, Rao TP. On-line solid phase extraction preconcentration of ultratrace amounts of zinc in fractionated soil samples for determination by flow injection flame AAS. *At. Spectrosc.* 2003; **24**: 118–124.

29. Willard HH, Merritt LL, Jr., Dean JA, Settle FA, Jr. *Instrumental Methods of Analy- sis*, 7th ed. Belmont, CA: Wadsworth Publishing Company; 1988, p. 159.

30. Freund JE, Simon GA. *Statistics: A First Course.* Englewood Cliffs, NJ: Prentice Hall; 1991.

31. Atkins P, de Paula J. *Physical Chemistry*, 7th ed. New York: Freeman; 2002, p. 776.

32. Whittig LD, Allardice WR. X-ray diffraction techniques. In Klute A (ed.), *Methods of Soil Analysis Part 1: Physical and Mineralogical Methods*, 2nd ed. Madison, WI: American Society of Agronomy, Soil Science Society of America, and Agronomy Society of America; 1986, pp. 331–361.

33. Sparks DL. *Environmental Chemistry of Soils*, 2nd ed. New York: Academic Press; 2003, p. 135.

34. Bernick MB, Getty D, Prince G, Sprenger M. Statistical evaluation of field- portable X-ray fluorescence soil preparation methods. *J. Hazard. Mater.* 1995; **43**: 111–116.

35. Karathanasis AD, Hajek BF. Elemental analysis by X-ray fluorescence spectros- copy. In Bartels JM (ed.), *Methods of Soil Analysis Part 3: Chemical Methods.*

Madison, WI: Soil Science Society of America and Agronomy Society of America; 1996, pp. 161–223.

36. Huete A, Tucker C. Investigation of soil influences in AVHRR red and near-infrared vegetation index imagery. *Int. J. Remote Sens.* 1991; **12**: 1223–1242.

37. Asner G, Heidebrecht K. Spectral unmixing of vegetation, soil and dry carbon cover in arid regions: comparing multispectral and hyperspectral observations. *Int. J. Remote Sens.* 2002; **23**: 3939–3958.

38. Palacios-Orueta A, Ustin S. Remote sensing of soil properties in the Santa Monica Mountains I. Spectral analysis. *Remote Sens. Environ.* 1998; **65**: 170–183.

BIBLIOGRAPHY

Environmental Protection Agency. *Test Methods SW-846 On-Line. Test Methods for Evaluating Solid Wastes Physical/Chemical Methods.* 2012. Available at http://www.epa.gov/epawaste/hazard/testmethods/sw846. Accessed June 4, 2013.

Madejová J, Komadel P. Baseline studies of the clay minerals society source clays: infrared methods. *Clays Clay Miner.* 2001; **49**: 410–432.

Willard HH, Merritt LL, Jr., Dean JA, Settle FA, Jr. *Instrumental Methods of Analysis*, 7th ed. Belmont, CA: Wadsworth Publishing Company; 1988.

CHAPTER

15

HYPHENATED METHODS IN SOIL ANALYSIS

Hyphenated methods involve both separation and identification of components in one analytical procedure and are commonly used in investigating soil chemistry. These investigations can involve one separation step and one identification step, two separation steps and one identification step, and two separation and two identification steps. Hyphenated analytical method instruments are arranged in tandem, without the analyte being isolated between the applications of the two methods. This leads to a very long list of possible combinations of instrumentation and, potentially, any separation method can be paired with any identification method. The list of hyphenated methods is long, although only a few methods are commonly used in soil analysis as can be seen in the review by D'Amore et al. [1].

Soil analysis involves not only investigation of basic soil chemistry but also determination of amendments, contaminants, pollutants, and the breakdown products of all of these in soil. There are two problems involved in these

Introduction to Soil Chemistry: Analysis and Instrumentation, Second Edition.
Alfred R. Conklin, Jr.
© 2014 John Wiley & Sons, Inc. Published 2014 by John Wiley & Sons, Inc.

investigations. First, many different compounds and sometimes isomers of compounds occur in the extracts. Second, each of the components needs to be unequivocally identified.

Chromatographic methods are used to separate the components in a mixture, but in a complex mixture, a single chromatographic method or step many not separate all components. In these cases, using simple retention time to identify the components will not suffice and the identification of components in the mixture will be incorrect. Thus, the addition of a method of identification such as mass spectrometry (MS) or Fourier transform infrared (FTIR) is essential. In some cases, it may even be necessary to confirm either an FTIR or MS identification by the same method applied in a different way. For example, FTIR may be followed by MS, or electron ionization (EI) MS followed by chemical ionization (CI) MS or by an entirely different method.

Analytical instrumental methods are commonly referred to by abbreviations using the first letters of the method's name (see Table 15.1). Thus, GC always refers to gas chromatography and MS to mass spectrometry. When these abbreviations are used, there is commonly no indication as to the type of GC (i.e., capillary, packed column, gas-solid, gas-liquid) being used. Likewise, no indication of the MS ionization method being used (i.e., EI or CI) is given.

Sometimes, the basic name abbreviation has letters added to the main abbreviation letters. These letters indicate some specific methodology being

TABLE 15.1. Instrumental Method Abbreviations and Their Explanation

Instrumental Method Abbreviation	Explanation
GC	Gas chromatography
HPLC	High-performance liquid chromatography or high-pressure liquid chromatography
LC	Liquid chromatography
TLC	Thin-layer chromatography
FTIR	Fourier transform infrared (spectroscopy)
UV-Vis	Ultraviolet-visible (spectroscopy)
NMR	Nuclear magnetic resonance (spectroscopy)
MS	Mass spectrometry
XRF	X-ray fluorescence
XRD	X-ray dispersion
AA	Atomic absorption
ICP	Inductively coupled plasma
TA	Thermal analysis
DTA	Differential thermal analysis
XAS	X-ray spectroscopy
XANES	X-ray near-edge spectroscopy
UHPLC	Ultra-high-pressure liquid chromatography
TCD	Thermal conductivity detector
HG	Hydride generator

TABLE 15.2. Common Hyphenated Method Abbreviations

Hyphenated Method	Separation Method	Modification	Identification Method
GC-MS	Gas chromatography	None	Mass spectrometry
HPLC-MS	High-performance liquid chromatography	None	Mass spectrometry
LC-ICP	Liquid chromatography	None	Inductively coupled plasma
LC-AAS	Liquid chromatography	Hydride derivatization	Atomic absorption spectroscopy
TA-MS	Thermal analysis	None	Mass spectrometry
DTA-MS	Differential thermal analysis	None	Mass spectrometry
LC-ICP-MS	Liquid chromatography	Inductively coupled plasma	Mass spectrometry
GC-IR-MS	Gas chromatography	Infrared spectroscopy	Mass spectrometry

used with the base method. This is particularly common with both atomic absorption and X-ray methods. Thus, an atomic absorption spectroscopy (AAS) method might be referred to as flame atomic absorption spectroscopy (FAAS). In X-ray work, additional letters may occur in the middle of the abbreviation. Thus, X-ray absorption spectroscopy (XAS) might be described as X-ray near-edge spectroscopy (XANES). XANES analysis could then be coupled with one of many separation methods. A compilation of the more common hyphenated method abbreviations found in this chapter is given in Table 15.2 and inside the back cover of this book. For a more complete list, see Reference 1.

Hyphenated methods can be divided into two types: those that do and those that do not destroy the sample in the process of analysis. Spectrophotometric methods, thermal conductivity, and refractive index methods of detection do not destroy the sample. Chromatographic methods using flame ionization and similar detection methods destroy the sample as it is detected. Any hyphenated method that involves MS or thermal analysis (TA) will also destroy the sample. In most cases, the identification of the components in soil is most important, so the destruction of the analyte is of less importance.

The best-known and the most commonly used hyphenated method is GC-MS; more specifically, and most commonly, capillary column GC combined with quadrupole MS. This type of instrumentation is controlled by computer and data collected and analyzed by dedicated computer programs. The mass spectra produced by the analytes can be compared to those in a library of mass spectra of known compounds using a computer search algorithm. The computer program finds known compounds that best match the spectra of the analytes of interest.

15.1 SAMPLE PREPARATION

Sample preparation for analysis by hyphenated methods requires some additional planning when compared to nonhyphenated methods. All steps, extraction, concentration, and final solvent selection must take into consideration and be compatible with all the components of the hyphenated instrumentation. For gas chromatographic methods, all the components in the mixture must be in the gaseous state. For liquid chromatography (LC) or high-performance liquid chromatography (HPLC), the samples of the analytes of interest can be solids or liquids, neutral or charged molecules, or ions, but they must be in solution. If the follow-on analysis is by MS, then each of the analytes may require a different method of introduction into the MS. Metals and metal ions may be introduced by HPLC if they are in solution but commonly are introduced via AAS or inductively coupled plasma (ICP). Other analytes may be directly introduced from HPLC to MS [2].

In addition to the above-mentioned restrictions, eluent selection for LC and HPLC is especially important. While the gas used in GC will not interfere with analysis, it is possible for eluent components used in LC or HPLC to interfere with follow-on analysis. This will be true for both MS and IR analysis. Usually, however, all samples can be accommodated if sufficient thought is exercised in selecting both the method of separation and the method of introduction into the follow-on analytical procedure.

15.2 SAMPLE DESTROYED

There are four basic hyphenated methods that result in the sample being destroyed. These are GC-MS, HPLC-MS, AAS/ICP-MS and TA/DTA-MS. All mass spectroscopic methods destroy the sample after separation; however, both AAS and ICP destroy the sample no matter what follow-on method of analysis is used. In most cases, TA and differential thermal analysis (DTA) will also destroy the sample. The follow-on methods then analyze the components that result from this decomposition. DTA may also be used to follow transitions in the sample without destroying it. Because the sample is identified, there is typically no reason to collect the analyte of interest, and so destruction is not of concern. However, if there is a limited amount of sample, care should be taken in using one of these methods.

15.2.1 GC-MS

In a GC-MS analysis, a soil extract is injected into the gas chromatograph part of the GC-MS, with an appropriate column for the desired separation.[1] The sample is separated into its component parts and as they leave the

---

[1] For an explanation of the individual methods, see the appropriate sections of the preceding chapter.

Figure 15.1. Gas chromatograph (right) attached to a mass spectrometer (left). Transfer line between GC and MS is hidden between the upper left of the GC and the upper right of the MS.

chromatographic column, they enter a heated transfer line that carries the separated components to the mass spectrometer. In the mass spectrometer, the isolated component is ionized and the mass fragments determined. A typical GC-MS setup is shown in Figure 15.1.

Using a computer and an appropriate software program, both a chromatogram and mass spectrum for each component in the mixture are produced. The chromatogram, which is a graph of signal versus time, provides the retention time of each of the components. The mass spectrum of each component is produced and gives the typical fragment pattern for the particular compound (see Figure 15.2). Examples of mass spectra signals used in identifying organic compounds and fragments are given in Table 15.3 [3].

Once retention time and fragment patterns are obtained, a database can be searched to find the identity of each component. This makes GC-MS extremely valuable because it is fast and identifies the components separated from the sample [4–6].

Most GC-MS instruments do not detect small compounds and are restricted to fragments with masses in the range of 60–600 amu.[2] However, mass spectrometers that can measure small and large masses are available, although usually not in the same instrument, and can be used as part of a GC-MS system.

Because of the limited mass range, the gas chromatograph carrier gas, which is typically either high-purity hydrogen or helium (both with amu of 2), does not interfere with the mass spectral analysis. The capillary column can be

---

[2] amu stands for atomic mass unit, which is the mass of a proton. For fragments, it represents the fragment mass.

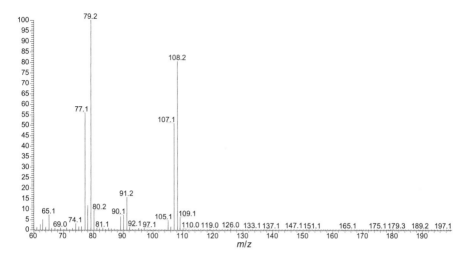

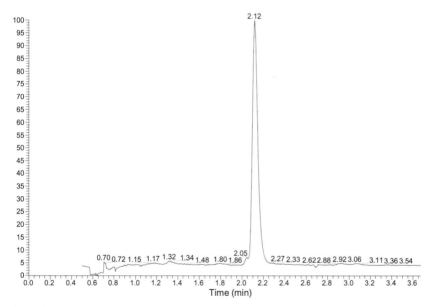

Figure 15.2. Output of GC-MS. Upper graph is the mass spectrum of the peak labeled 2.12 in the lower graph (chromatogram).

chosen so as to effect the desired separation; however, a low bleed column, the stationary phase of which does not come off the column during analysis, is essential. Stationary phase components in the carrier gas will be seen as components of the chromatogram by the mass spectrometer and thus can lead to confusion when interpreting the mass spectrum.

Eriksson et al. were able to extract, concentrate, and subsequently illustrate the utility and power of GC-MS by determining 16 hydrocarbons in old

TABLE 15.3. Examples of Infrared Absorptions and Mass Spectra Signals Used in the Identification of Organics

Type of Compound	Infrared Absorption due to Following Functional Groups	Absorption Range (cm^{-1})	Mass Spectra m/z Signal
Carbohydrates	—OH, $>$C$=$O	2500–3000, 1800–1600	56, 58, 60, 68, 72, 82, 96
Fatty acids	—COOH	2500–2700	242, 256, 284
Alkanes	—CH$_3$, —CH$_2$	3000–2700	142, 268, 282, 310
Aromatic esters	O $\parallel$ —C—O—C$\diagup$	1700 and 1100	248, 262, 276, 304, 264, 278, 292

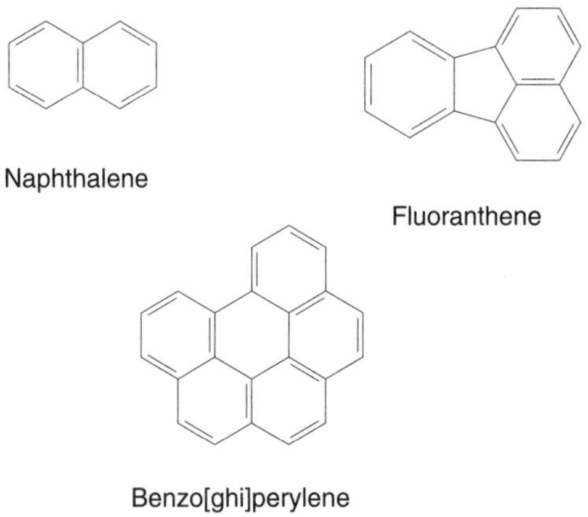

Naphthalene

Fluoranthene

Benzo[ghi]perylene

Figure 15.3. Examples of complex multicyclic compounds that can be separated and identified by GC-MS.

creosote-contaminated soil. Contaminants ranged from naphthalene to benzo[ghi]perylene with sensitivities ranging from 1.3 µg/g soil for naphthalene to 70 µg/g soil for fluoranthene and 1.6 µg/g soil for benzo[ghi]perylene. The structures of these contaminants are given in Figure 15.3 [4].

15.2.2 HPLC-MS

Samples that have low vapor pressure or extracts that are expected to contain components of low volatility can be analyzed by HPLC. The liquid sample is introduced via an injection valve and components separated by a chromatographic column. As components exit the column, they enter the MS where some of the eluent is stripped away and the remaining liquid enters the mass spectrometer. As with GC-MS, the mass spectrometer does not measure low-molecular-weight compounds, and thus, common eluents such as water, acetonitrile, and methanol do not interfere with the mass spectrum.

Unlike GC, HPLC can handle a variety of both aqueous phases (with various pHs and salt contents) and organic phases, and there are a variety of stationary phases that can be used depending on the analytes to be separated and the eluent being used. It is also possible to switch between columns and eluent composition during analysis so that a wide range of analytes can be separated and analyzed.

Both GC and HPLC columns are expensive, so it is important not to clog them during analysis. GC columns will be clogged by nonvolatile compounds. For these materials, it is important to use HPLC. Samples must also be free of suspended particles that will clog the finely packed HPLC columns. Filtering samples, especially soil extracts, before injection is essential.

Mass spectra from HPLC separations are obtained in a manner similar to those from GC-MS. Unlike GC, where both the eluent and analyte are in the gas phase, HPLC eluents are dissolved in liquids that are stripped off before mass spectroscopic analysis is carried out. Analytes must also be vaporized before analysis. For this reason, metals are usually introduced via either AA or ICP. As with GC-MS analyses, the mass spectra from each eluted compound can be compared with standards and the compound identified.

Using HPLC-MS, and including solid-phase microextraction (SPME), Möder et al. were able to extract and identify five insecticides and eight herbicides from soil. Method sensitivity ranged from 0.1–0.5 to 50ng/mL for herbicides, carbamate, and triazines, and from 0.5 to 8–9ng/mL for insecticides, carbamates, and oximes. Identification of these pesticides was accomplished using the target ion method, primarily the target ion $[M + H]^+$, but in some cases, $(M + Na)^+$ was used. This is a good example of both the versatility and the sensitivity of this hyphenated method [7].

In order to obtain better and faster separation, instruments using higher pressures than common HPLC have been developed. These are often referred to as ultra-high-pressure liquid chromatographs (UPHPLC), or some similar abbreviation. They are used in a manner similar to HPLC and can be connected to mass spectrometers.

15.2.3   AAS and ICP-MS

There are many different ways to obtain information from either AAS or ICP alone; however; when combined with MS, the mass spectrometer becomes the detector. Flame gases are taken into a mass spectrometer through a port into a low-pressure compartment and then transferred, once the pressure is low enough, to the mass spectrometer (see Figure 15.4). The interface between the AAS or ICP and the MS has been a problem in the past, but these problems have largely been overcome [8,9].

At this point, the elements in the sample are ionized and their masses determined. The mass of the most prominent isotope is determined along with

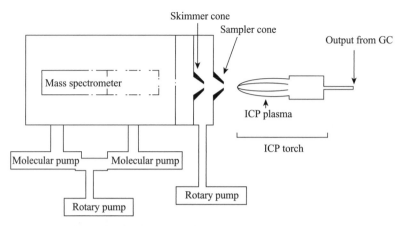

Figure 15.4. Simplified diagram of interface between GC, ICP, and MS.

the amounts of its isotopes. Isotope distribution can often tell the source of the sample and thus gives additional important information about it.

In some cases, isotope enrichment, particularly of stable and nonradioactive isotopes, is used to follow a particular element through the environment or through biological systems. The advantage of stable isotopes is that they are not radioactive and thus are easier to handle and dispose.

de Maranhão et al. investigated the use of both ICP-MS and high-resolution continuous source AAS (HR-CS AAS) for the determination of metal species in retorted shale. Ag, As, Ba, Cd, Cr, Pb, and Se were determined with ICP-MS proving to be the most sensitive method of detection. Sensitivity ranged from a high of $0.5\,\mu g/L$ for ^{52}Cr to a low of $0.004\,\mu g/L$ for ^{107}Ag [10].

15.2.4 Thermal Decomposition with Mass Spectroscopy

Heating soil to drive off, usually, organic compounds, but sometimes inorganic compounds, is termed pyrolysis. It may also be termed thermal analysis (TA) and differential thermal analysis (DTA). DTA involves measuring and comparing the changes in soil samples with the changes in some standard compound such as the common soil clay kaolinite. Temperatures used in these types of analyses are usually between 50 and 800°C. Information can be obtained by carrying out heating under either inert or oxidizing atmospheres. Under inert atmospheres, only vaporization and decomposition changes are measured. In oxidizing atmospheres, vaporization, decomposition, and oxidative changes in the sample are determined.

All these methods result in gases being given off by the samples. These gases are easy to analyze by either MS or FTIR. In either method, information about the compounds present in, primarily, the soil organic matter is obtained. In the case of FTIR, specific functionalities such as $-OH$, $-C=O$, $-NH_2$, and $-COOH$ can be obtained. If specific compounds are released, they can also be identified. In the case of MS, specific mass-to-charge (m/z) ratios can be used to infer the release of specific classes of compounds such as carbohy-drates, phenols, fatty acids, and alkenes.

Using pyrolysis combined with MS, which is called pyrolysis field ionization mass spectrometry (Py-FIMS), Gillespie et al. were able to differentiate organic matter occurring in different topographic locations in a landscape in a St. Denis site in Saskatoon, Saskatchewan. While this approach gives an insight into the composition of soil organic matter in different toposites, it provides little information about the reactions leading to the differences observed [3,11].

15.2.5   MS-MS

In some cases, more than one mass spectrometer may be used to carry out the analysis of a sample. Thus, the components from the first MS analysis may be passed to another MS for further analysis. In this type of analysis, a fragment from the first MS analysis is further broken down and the resulting new frag-ments are analyzed. This allows for analysis of complex samples. Because there are several types of sample ionization (e.g., EI, CI, and electrospray) and dif-ferent types of mass spectrometers (e.g., quadrupole, time of flight [TOF], and magnetic sector) there are several different ways an MS-MS analysis can be carried out.

It is possible to use the same mass spectrometer (i.e., a quadrupole) with different methods of ionization. In the simplest terms, the first fragmentation can be accomplished by the impact of electrons (EI), while the second can be accomplished by impact with molecular ions known as chemical ionization (CI). This later ionization often preserves more of the original molecule and yields larger fragments. In any case, the two ionization methods give different fragment patterns, which, in turn, can give additional information about the material undergoing analysis [12]. Analysis using the same ionization could be coupled to two different MS techniques, for example, a quadrupole coupled to a TOF.

In a typical analysis, one approach would be to carry out the analysis by first using CI and quadrupole MS. The fragments from this first MS would then be directed to an EI and a TOF mass spectrometer. Different fragments will be observed and this will yield additional information about the sample. In many cases, the MS-MS analysis is applied to samples eluting from either LC, HPLC, or GC chromatographic separation techniques. For additional informa-tion on this topic, see "Triple Hyphenated Methods."

15.3   NONDESTRUCTIVE METHODS

Nondestructive hyphenated methods of analysis require follow-on spectro-photometric methods. The most common spectrophotometric analysis is FTIR, and it is most commonly associated with GC. It is, however, possible to combine it with other separation methods under specific conditions.

15.3.1   GC-FTIR

As each component exits the chromatographic column, it is channeled into an infrared (IR) gas cell and the component's IR spectrum obtained. A thermal conductivity detector (TCD) (see Chapter 13) can be used to determine when a component is emerging from the column. The TCD detector does not destroy the sample and none of the common gases used in GC have IR spectra, and thus do not interfere with the spectrum of eluting components. Half-peak height is a common time to obtain the spectrum of that component and the setup for detection and obtaining the spectrum can thus be automated [6,13].

It is also possible to carry out a GC-GC-FTIR analysis (see "Triple Hyphenated Methods"). FTIR is very rapid, which means that several scans of the eluting material can usually be obtained. These are automatically combined by the spectrometer to give the final spectrum. The more scans that can be obtained, the more noise can be reduced, producing a cleaner spectrum.

15.3.2   HPLC-FTIR

Some HPLC eluents can interfere with obtaining a usable FTIR spectrum. Water, alcohol, and acid, which are common eluents or components of HPLC mobile phases, have strong absorption, particularly in the $4000-3000\,cm^{-1}$ region. Other common solvents also have strong absorption strongly in the $3000-2700\,cm^{-1}$ and in the $1700\,cm^{-1}$ regions of the IR. However, some solvents, particularly halogenated compounds, have minimal IR spectra and may thus be used allowing useful IR spectra to be obtained.

Another approach is to separate the solvent from the sample before a spectrum is obtained. Because HPLC is often used with a combination of volatile eluents and organic compounds that are not volatile, the solvent can be removed and the isolated component analyzed by FTIR [13]. FTIR is not particularly useful for the identification of inorganic components, particularly ions.

15.4   TRIPLE HYPHENATED METHODS

Triply hyphenated methods are not common. However, they do exist and have been used in certain applications to elicit information about soil chemistry. LC linked to ICP spectroscopy linked to MS is one example. The eluent from a liquid chromatograph is easily directed into an ICP torch and the gases from the ICP are then directed into an MS. Because the components from the LC are converted to gases in the ICP torch, they are thus easily analyzed by the MS [14].

Another approach that has been reported is GC linked to IR spectroscopy linked to MS. The gas phase components from a GC analysis can easily be directed into an IR gas cell for analysis. Note that the carrier gases used in GC will not generally interfere with or be detected by IR spectroscopy and thus are ideal for this analysis; the output from the IR cell can easily be directed to the MS for further analysis.

15.4.1 LC-ICP-MS

In LC-ICP-MS, samples are separated on a chromatographic column, which may be a simple silica or alumina column with a relatively simple eluent. As the components elute from the column, they enter the ICP and the identity of the elements present and their concentration are determined based on the wavelengths of light (identity) and intensity of light (quantification) they emit. The exhaust from the ICP then enters the mass spectrometer, where the metals and their isotopic composition are determined based on their characteristic m/z ratios. The metals are thus identified and verified by two methods, ICP and MS [15].

This is a powerful method for the determination of the species of elements present in a specific environment. For example, it has been used to determine the arsenic species present in particular environmental conditions [16].

15.4.2 GC-IR-MS

In some cases, confirming identification of components obtained from soil, such as pesticides, is essential. Thus, the uncertainty in some analyses needs to be addressed. This can be accomplished by identifying the components using two entirely different methods such as IR spectroscopy and MS. Although GC-IR-MS methods can positively identify separated components, the IR component of the system is not nearly as sensitive as are the GC and MS components. This detracts from the usefulness of this method. However, in cases where the level of analyte is not limiting, which frequently occurs in soil extracts, this can be an excellent method to use. Also, with modern concentration techniques, it is neither difficult nor time-consuming to concentrate analytes to a level that is identifiable by IR spectroscopy [17,18].

15.4.3 GC-GC-MS

Two-dimensional GC can be used to separate complex mixtures of polyaromatic compounds, and MS used to subsequently identify the compounds. In this method, the original sample is injected into a gas chromatograph with one type of column. As the components exit the first GC, they are fed into a second GC, with a different column, for further separation and finally into a mass spectrometer. In this way, compounds that coeluted from the first column are separated on the second. Focant et al. [19] were able to separate polychlorinated dibenzo-*p*-dioxin (PCDD), polychlorinated dibenzofuran (PCDF), and coplanar polychlorinated biphenyl (cPCB) using this type of analytical procedure, including isotope dilution TOF-MS. These compounds are frequently found as contaminants in soils surrounding industrial settings; thus, the ability to separate and identify them is extremely important [6,12,19].

15.4.4   GC-GC-FTIR

GC-GC-FTIR is similar to the one described previously except that the separated components are fed into an FTIR cell for identification. To effect a more complete separation of components, two different GC columns are used; for instance, a 5% phenyl/phenyl-methyl silicone column might be followed by a 50% phenyl/phenyl-methyl silicone column [12].

15.4.5   (Cp and Py)-GC-MS

Pyrolysis analysis followed by mass spectrometry is carried out by heating a soil organic matter sample to a specified temperature (the Curie point [Cp]) and analyzing the compounds pyrolyzed (Py) at this temperature using GC and MS. The gases given off by pyrolysis are passed into a gas chromatograph for separation and then into a mass spectrometer for identification. Using this method, Schulten and Schnitzer were able to identify a number of alkanes and aromatic compounds released by humic acids from two different soils [20].

15.5   CONCLUSIONS

Hyphenated methods of instrumental analysis are extremely powerful for separating and identifying compounds. Inorganic and organic constituents, both natural and contaminant, can be separated and identified in one continuous instrumental analysis. Volatile organic constituents, both natural and synthetic, are commonly determined by GC-MS. Nonvolatile organic compounds are determined using HPLC-MS. Inorganic constituents are most commonly isolated and identified using HPLC-AAS, HPLC-ICP, or HPLC-MS. The various hyphenated methods are essential in determining the identity of the species present in a mixture, particularly the ionic species in the soil solution. Triple hyphenated methods add even more separation and identification to the basic double hyphenated methodologies. These triple hyphenated methods are not, however, as well developed or as commonly used as are the double hyphenated methods.

PROBLEMS

15.1. Describe a typical GC-MS instrument and explain why these two methods go together well.

15.2. Explain why a GC-IR-MS instrumental setup works well in terms of identifying analytes.

15.3. Considering mass alone, why will the gases used in GC typically not interfere with either follow-on MS or IR analysis.

15.4. MS detects on the basis of mass-to-charge ratio. Describe a scenario where the mass-to-charge ratio of a component of an LC eluent interferes with the analysis of an analyte.

15.5. Explain why the eluents used in HPLC might interfere with FTIR analysis.

15.6. A sample containing both metallic ions and metals is to be analyzed by LC. Explain why the output from the LC cannot be introduced directly into an MS.

15.7. Describe the advantages and disadvantages of triple hyphenated analytical methods.

15.8. Describe the drawback with IR identification of chromatographic eluents.

15.9. Under what conditions would the eluent from LC or HPLC not be compatible with MS analysis?

15.10. Describe a situation where a nondestructive determination of an analyte might be important.

REFERENCES

1. D'Amore JJ, Al-Abed SR, Scheckel KG, Ryan JA. Methods for speciation of metals in soils: a review. *J. Environ. Qual.* 2005; **34**: 1707–1745.

2. Alder L, Greulich K, Kempe G, Vieth B. Residue analysis of 500 high priority pesticides: better by GC–MS or LC–MS/MS? *Mass Spectrom. Rev.* 2006; **25**: 838–865.

3. Leinweber P, Schulten HR. Differential thermal analysis, thermogravimetry and in-source pyrolysis-mass spectrometry studies on the formation of soil organic matter. *Thermochim. Acta* 1992; **200**: 151–167.

4. Eriksson M, Fäldt J, Dalhammar G, Borg-Karlson A-K. Determination of hydrocarbons in old creosote contaminated soil using headspace solid phase microextraction and GC–MS. *Chemosphere* 2001; **44**: 71641–71648.

5. Qin BH, Yu BB, Zhang Y, Lin XC. Residual analysis of organochlorine pesticides in soil by gas chromatograph–electron capture detector (gc–ecd) and gas chromatograph–negative chemical ionization mass spectrometry (GC–NCI–MS). *Environ. Forensics* 2009; **10**: 331–335.

6. Bhatt BD, Prasad JV, Kalpana G, Ali S. Separation and characterization of isomers of para-nonylphenols by capillary GC/GC-MS/GC-FTIR techniques. *J. Chromatogr. Sci.* 1992; **30**: 203–210.

7. Möder M, Popp P, Eisert R, Pawliszyn J. Determination of polar pesticides in soil by solid phase microextraction coupled to high-performance liquid chromatography-electrospray/mass spectrometry. *Fresenius. J. Anal. Chem.* 1999; **363**: 680–685.

8. Newman A. Elements of ICPMS. *Anal. Chem.* 1996; **68**: A46–A51.

9. Olesik JW. Fundamental research in ICP-OES and ICPMS. *Anal. Chem.* 1996; **68**: 469A–474A.

10. de Maranhão TA, Silva JSA, Bascuñan VLAF, Oliveira FJS, Curtius AJ. Analysis of acetic acid extraction solutions by inductively coupled plasma spectrometry for the classification of solid waste. *Microchem. J.* 2011; **98**: 32–38.

11. Gillespie AW, Walley FL, Farrell RE, Leinweber P, Eckhardt K-U, Regier TZ, Blyth RIR. XANEX and pyrolysis-FIMS evidence of organic matter composition in a hummocky landscape. *Soil Sci. Soc. Am. J.* 2011; **75**: 1741–1755.

12. Hashimoto S, Takazawa Y, Fushimi A, Tanabe K, Shibata Y, Leda T, Ochiai N, Kanda H, Ohura T, Tao Q, Reichenbach S. Global and selective detection of organo-halogens in environmental samples by comprehensive two-dimensional gas chromatography-tandem mass spectrometry and high-resolution time-of-flight mass spectrometry. *J. Chromatogr. A* 2011; **1218**: 3799–3810.

13. Norton KL, Lange AJ, Griffiths PR. A unified approach to the chromatography-FTIR interface: GC-FTIR, SFC-FTIR and HPLC-FTIR with subnanogram detection limits. *HRC-J. High Resolut. Chromatogr.* 1991; **14**: 225–229.

14. Bayon M, Camblor MG, Alonso JI, Sanz-Medel A. An alternative GC-ICP-MS interface design for trace element speciation. *J. Anal. At. Spectrom.* 1999; **14**: 1317–1322.

15. Carignan J, Hild P, Mevel G, Morel J, Yeghicheyan D. Routine analysis of trace elements in geological samples using flow injection and low pressure on-line liquid chromatography coupled to ICP-MS: a study of geochemical reference materials BR, DR-N, UB-N, AN-G, and GH. *Geostandards Newsletter* 2001; **25**: 187–198.

16. Jain CK, Ali I. Arsenic: occurrence, toxcity and speciation techniques. *Water. Res.* 2000; **34**: 4304–4312.

17. Gurka DF, Titus R. Rapid nontarget screening of environmental extracts by directly linked gas chromatography/Fourier transform infrared/mass spectrometry. *Anal. Chem.* 1986; **58**: 2189–2194.

18. Wilkins CL. Recent technological advances improve prospects for the widespread acceptance of GC/IR/MS instrumentation. *Anal. Chem.* 1987; **59**: 571A–581A.

19. Focant JF, Reiner E, MacPerson K, Kolic T, Sjödin A, Patterson DG, Reese SL, Dorman FL, Cochran J Measurement of PCDDs, PCDFs, and non-ortho-PCBs by comprehensive two-dimensional gas chromatography-isotope dilution time-of-flight mass spectrometry (GC × CG-IDTOFMS). *Talanta* **63**, 1231–1240, 2004.

20. Schulten H-F, Schnitzer M. Structural studies on soil humic acids by Curie-point pyrolysis-gas chromatography/mass spectroscopy. *Soil Sci.* 1992; **153**: 205–224.

BIBLIOGRAPHY

Alder L, Kerstin G, Kempe G, Vieth B. Residue analysis of 500 high priority pesticides: better by GC–MS or LC–MS/MS? *Mass Spectrom. Rev.* 2006; **25**: 838–865.

B'Hymer C, Brisbin J, Sutton K, Caruso JA. New approaches for elemental speciation using plasma mass spectrometry. *Am. Lab.* 2000; **32**: 17–39.

Donard OFX, Martin FM. Hyphenated techniques applied to environmental speciation studies. *TRAC Trends Analyt. Chem.* 1992; **11**: 17–26.

Lespes G, Gigault J. Hyphenated analytical techniques for multidimensional characterisation of submicron particles: a review. *Anal. Chim. Acta* 2011; **692**: 26–41.

Shaliker A, ed. *Hyphenated and Alternative Methods of Detection in Chromatography.* Boca Raton, FL: CRC Press; 2011.

Wilkins CL. Directly-Linked Gas Chromatography–Infrared–Mass Spectrometry (GC/IR/MS). Published Online: 15 AUG 2006.

Worthy W. Scope of ICP/MS expands to many fields. *Chem. Eng. News* 1988; **66**: 33–34.

INDEX

Multiple
  analyses, 110
  detectors, 292
Munsell color book, 36–37
Muscovite, 116

^{15}N, 288
Naked proton, 179
Naphthalene, 310
Natural
  attenuation, 185
  levels, 138
  lighting, 200
  organic components, 262
  organic matter, 233–234
  state, 138
Nausea, 116
N-containing pesticides, 249
Near-infrared, 280–281
  spectroscopy, 161
Needle hub, 262
Negative
  charge, 104
  potential, 187
  pressure, 290
  terminal, 268
Nematodes, 70
Nernst equation, 185
Neutral
  ammonium acetate, 215
  cation exchange solutions, 215
  molecules, 102, 199
Neutralization, 197
Neutralized, 223
Neutron
  activation, 161, 163
  activation analysis, 163
  scattering, 109, 111–112
New Zealand, 40
Nickel, 103, 120, 210
Ninhydrin, 266
Nitrate, 10–13, 52, 62–63, 95, 103–104,
  123–126, 149, 164, 171, 184, 203–205,
  279
  anions, 104
Nitric acid, 3, 196, 212, 218, 221
Nitrite, 52, 62–63, 95, 103–104, 123–126,
  171, 203–204, 279
Nitrobacter, 63

Nitro compound, 196, 204
Nitrogen, 3, 9, 12–13, 15, 19, 52, 54, 57, 72,
  75, 78, 81, 89, 95, 99, 124–125, 162,
  170, 195, 197, 204, 245, 258, 278, 286,
  292
  compounds, 207, 288
  cycle, 13, 78, 126, 164, 288
  determination, 170
  fertilizers, 164
  fixation, 124, 126
  gas, 95, 104, 117
  oxides, 95, 117, 164
  oxyanions, 125
  in soil, 171
  species, 119, 126
Nitrogenase, 124
Nitrogen-containing
  compounds, 195
  lipids, 126
Nitrosomonas, 63
Nitrous oxide, 55, 245
NMR, 3, 13, 287, 299
  analysis, 171
  experiments, 286
  sample preparation, 287
  sample tube, 287
  solids, 286
  spectroscopy, 286
Noble gas, 54
Noble metal, 99
Nodules, 75
Noise, 276
Nonbonding electron pairs, 278
Nondestructive methods, 313
Nonexchangeable cation, 216
Nonexchangeable ions, 210
Nonhyphenated methods, 307
Nonionic species, 123
Nonmetal cations, 102
Nonmetals, 115, 210
Nonpoint
  sampling, 146
  source, 137, 144145, 147, 155
Nonradiative interactions, 164
Nonradioactive isotopes 312
Nonsoil, 221
North America, 40
Nuclear magnetic resonance
  spectroscopy, 275. *See also* NMR

Swelling, 52
  clay, 52
Synchrotron
  radiation, 14, 161
  radiation spectroscopy, 161
Synge, R. L. M., 1, 12
Syringe, 163, 257
  filters, 154
  needle, 257, 262
  plunger, 257

TA, 307
TA/DTA-MS, 307
Target ion, 311
Tars, 233
Tartrate, 55
Teflon-backed septa, 257
Teflon vessels, 241
Temperature, 147, 149
  correction, 182
  probe, 182
Temperature-sensitive, 182
Tensiometers, 109, 111
Tentative identified, 262
Tertiary
  excitation, 299
  protons, 162
Test methods for evaluating solid waste, 149
Test tubes, 293
Tetrachloroethylene, 283
Tetrahedral
  sheets, 51
  silica, 51
Tetramethylsilane, 287
Textural
  name, 53
  triangle, 53
Texture, 6, 20, 35, 94
Thalidomide, 116
Thawing, 35, 147
Thermal
  analysis, 306, 312
  conductivity, 306
  conductivity detector, 169, 259, 314
  decomposition, 312
Thermocouple psychrometer, 189
Thermodynamically stable, 61

Thermodynamic equilibrium constant, 216
Thermogravimetric analysis, 170
Thermophilic, 75
Thimble, 238
Thin layer chromatography, 12, 169, 265, 267, 270. *See also* TLC
Thin-layer plates, 267
Thin section, 168
Thompson, 2, 11
  thermocouple, 189
Thomson, J. J., 9
Time-domain reflectometry, 109, 111, 188
Time of flight, 313, 315
Time-of-flight mass spectrometer, 287
Time-resolved X-ray diffraction, 275
Titanium, 100
  oxides, 100
Titrant, 193, 195, 197, 205
Titrating soil, 167, 200
Titration, 88, 168, 193, 197, 200, 203–204
  curve, 193–194, 197, 199
  end point, 193
Titrimetric determination, 207
TLC, 255, 269, 271
  detection, 266
  mobile phases, 266
  sample application, 266
  stationary phases, 266
  visualization reagent, 266
Toluene, 286
Topography, 27, 144, 146, 313
  features, 137
Torch, 289
Total
  acidity, 106
  nitrogen, 204
  recoverable petroleum hydrocarbons, 282
Toxic, 118, 125, 155, 168, 220, 233
  elements, 154, 210
  levels, 210
Toxicity, 126, 247, 284
Trace ions, 227
Trans, 116
Transects, 143
  line, 137
  samples, 138
  sampling, 142, 145–146, 155

CHEMICAL ANALYSIS

A SERIES OF MONOGRAPHS ON ANALYTICAL CHEMISTRY
AND ITS APPLICATIONS

Series Editor
MARK F. VITHA